Springer Series on

Atoms+Plasmas 21

Editor: I. I. Sobel'man

Springer
Berlin
Heidelberg
New York
Barcelona
Budapest
Hong Kong
London
Milan
Paris
Santa Clara
Singapore
Tokyo

Springer Series on

Atoms+Plasmas

1 **Polarized Electrons** 2nd Edition
By J. Kessler

2 **Multiphoton Processes**
Editors: P. Lambropoulos and S. J. Smith

3 **Atomic Many-Body Theory**
2nd Edition
By I. Lindgren and J. Morrison

4 **Elementary Processes in Hydrogen-Helium Plasmas**
Cross Sections and Reaction Rate Coefficients
By R. K. Janev, W. D. Langer, K. Evans Jr., and D. E. Post Jr.

5 **Pulsed Electrical Discharge in Vacuum**
By G. A. Mesyats and D. I. Proskurovsky

6 **Atomic and Molecular Spectroscopy**
2nd Edition
Basic Aspects and Practical Applications
By S. Svanberg

7 **Interference of Atomic States**
By E. B. Alexandrov, M. P. Chaika and G. I. Khvostenko

8 **Plasma Physics** 2nd Edition
Basic Theory with Fusion Applications
By K. Nishikawa and M. Wakatani

9 **Plasma Spectroscopy**
The Influence of Microwave and Laser Fields
By E. Oks

10 **Film Deposition by Plasma Techniques**
By M. Konuma

11 **Resonance Phenomena in Electron-Atom Collisions**
By V. I. Lengyel, V. T. Navrotsky and E. P. Sabad

12 **Atomic Spectra and Radiative Transitions** 2nd Edition
By I. I. Sobel'man

13 **Multiphoton Processes in Atoms**
By N. B. Delone and V. P. Krainov

14 **Atoms in Plasmas**
By V. S. Lisitsa

15 **Excitation of Atoms and Broadening of Spectral Lines**
2nd Edition
By I. I. Sobel'man, L. Vainshtein, and E. Yukov

16 **Reference Data on Multicharged Ions**
By V. G. Pal'chikov and V. P. Shevelko

17 **Lectures on Non-linear Plasm Kinetics**
By V. N. Tsytovich

18 **Atoms and Their Spectroscopic Properties**
By V. P. Shevelko

19 **X-Ray Radiation of Highly Charged Ions**
By H. F. Beyer, H.-J. Kluge, and V. P. Shevelko

20 **Electron Emission in Heavy Ion-Atom Collision**
By N. Stolterfoht, R. D. DuBois, and R. D. Rivarola

21 **Molecules and Their Spectroscopic Properties**
By S. V. Khristenko, A. I. Maslov, and V. P. Shevelko

S. V. Khristenko A. I. Maslov
V. P. Shevelko

Molecules and Their Spectroscopic Properties

With 66 Figures and 78 Tables

Springer

Dr. Sergei V. Khristenko

Russian Academy of Sciences,
Institute of Chemical Physics,
Ul. Kosygina 4,
117977 Moscow, Russia

Dr. Alexander I. Maslov
Dr. Viatcheslav P. Shevelko

Lebedev Physics Institute,
Russian Academy of Sciences,
Leninsky Prospect 53,
117924 Moscow, Russia

ISSN 0177-6495
ISBN-13:978-3-642-71948-6 e-ISBN-13:978-3-642-71946-2
DOI: 10.1007/978-3-642-71946-2

Library of Congress Cataloging-in-Publication Data. Khristenko, S. V. (Sergei V.). 1943– Molecules and their spectroscopic properties / S. V. Kristenko, A. I. Maslov, V. P. Shevelko. p. cm. – (Springer series on atoms + plasmas, ISSN 0177-6495; 21) Includes bibliographical references and index. ISBN-13:978-3-642-71948-6 1. Molecular spectra. 2. Molecular structure. I. Maslov, A. I. (Alexander I.), 1941– . II. Shevel'ko, V. P. (Viacheslav Petrovich) III. Title. IV. Series. QC454.M6K47 1998 539'.6–dc21 97-46883

Softcover reprint of the hardcover 1st edition 1998

Typesetting: Scientific Publishing Services (P)LTD, India

SPIN 10059360 54/3144 - 5 4 3 2 1 0 - Printed on acid-free paper

Preface

Interactions of molecules with photons and atomic particles (electrons, atoms, ions, and molecules) play a key role in elementary processes occurring in laboratory and astrophysical plasma sources, such as low-temperature discharges, gas lasers, chemical reactors, tokamaks, astrophysical objects, the Earth's higher atmosphere, and many others. Detailed information on molecular radiative constants and collisional characteristics is required for investigations performed in chemical physics, low-temperature plasmas, molecular spectroscopy and astrophysics.

The structure of molecules is more complicated than the structure of atoms and atomic ions, due to a larger number of degrees of freedom in motion of molecular composite particles. In addition, the understanding of collisions of molecules with electrons and other particles is a difficult task because one faces the problem of scattering on a non-spherical multicentral molecular potential with additional channels for dissociation and ionization of the target molecules into different fragments which may also be molecules. Therefore, the range of collisional processes involving molecules is extremely large and covers such aspects as elastic and momentum-transfer scattering, rotational, vibrational and electronic excitation as well as dissociation and dissociative attachment reactions.

This monograph aims to help in solving such problems. It contains the necessary information on molecular structure, i.e., equilibrium internuclear distances, interaction constants, energy levels and ionization potentials, as well as on molecular spectra including selection rules for radiation, fundamental frequencies, dipole and multipole moments, electric polarizabilities and, finally, the collisional characteristics like excitation, ionization, dissociation and photoionization cross sections.

Obviously, it is nearly impossible to give complete information on molecules and their structure. However, we make an effort to present a short 'guide' to molecules for researchers. Besides purely tabulated and graphical material, we present in brief the physical background of the processes we present together with the basic theoretical approaches used in calculations of the molecular wave functions and different characteristics. Because of the limited volume of the book, only the constants of relatively simple molecules are considered which are required as a starting point for further investigations.

The monograph is composed in the form of a reference data book on molecules similar to our previous monographs *Reference Data on Multicharged*

Ions by V.G. Pal'chikov and V.P. Shevelko (Springer, Berlin, Heidelberg 1995) and *Atoms and Their Spectroscopic Properties* by V.P. Shevelko (Springer, Berlin, Heidelberg 1997). It is intended for research students and specialists dealing with chemical physics, astrophysics, controlled plasma fusion, spectroscopy and related fields of atomic and molecular physics.

The authors acknowledge very much the help of Ludmila Khristenko and Natalia Kozulina in preparing the manuscript.

Moscow
December 1997

S.V. Khristenko
A.I. Maslov
V.P. Shevelko

Contents

Glossary of Terms

The fundamental constants used in atomic and molecular physics are listed in the table on the hardcover. Other fundamental physical constants are given in the report of the CODATA Task Group on Fundamental Constants, CODATA Bulletin No. 63, E.R. Cohen, B.N. Taylor: Rev. Mod. Phys. **59**, 1121 (1987). A list of symbols used in the book is also given in a Table.

List of Symbols

a, b, c	Inertial axes of molecule
A,B,C	Rotational constants
C_σ, C_{12}	Lennard-Jones interaction constants
C_{n},	$n = 6$, 8 and 10 van-der-Waals interaction constants
d_k	Degeneracy of the k-th vibration
D_{e}	Dissociative energy
f	Oscillator strength
I_a, I_b, I_c	Inertia molecular moments
J	Rotational quantum number
$K = \Lambda + 1$	Projection of the vibronic angular momentum onto the axis of symmetry
L	Electron orbital momentum
M_L	Projection of L onto the axis of symmetry
m	Electron mass
q	Franck-Condon factor
Q	Nuclear quadrupole moment
r_e	Equilibrium internuclear distance
R	Rotational momentum
S	Electron spin
v	Vibrational quantum number
α_{e}	Rotational-vibrational quantum number
σ	Bonding orbital; effective cross section
θ	Valence angle of molecule
Λ	Sum of M_L projections
ω	Frequency (wavenumber)

1 Molecular Structure

The types of chemical bonds and the classification of molecular terms are given in this introductory chapter. The coupling schemes for the addition of angular momenta in molecules, and quantum-chemistry methods for the calculation of wave functions and energy levels are considered as well.

1.1 Classification of Chemical Bonds and Molecular Terms

The chemical bond of atoms in a molecule is called *ordinary*, *double* or *triple* depending on the number of valence electrons involved [1.1]. For example, in the ethane molecule C_2H_6, the carbon atoms are connected by an ordinary C–C bond, in the ethylene molecule C_2H_4 by a double bond C=C and in the acetylene molecule C_2H_2 by a triple C≡C bond, respectively. Therefore, the ordinary, double and triple bonds are created by two, four and six valence electrons. As a rule, the inner-shell electrons do not participate in chemical bonding. This classification of chemical bonds is, to some extent, a matter of convention, not always being applicable. For instance, six equal bonds between carbon atoms in cyclic hydrocarbons of the benzene type (C_6H_6) are generated by 18 valence electrons ($4 \times 6 - 6 = 18$). Following this classification we would have six "one-and-a-half" bonds ($18/(6 \times 2) = 1.5$).

In some cases, chemical bonds are created by an odd number of electrons. For example, the bond in the H_2^+ ion is created by one electron, while the three H–H bonds in the H_3^+ ion are created by two electrons, and so forth. The cases also exist when not all outer electrons participate in the chemical bond. For example, in the NH_3 molecule two of the five outer electrons of the nitrogen atom remain outside the generated bonds. Such electrons are called *lone pairs*. A bond between atoms A and B is called *ionic* if in its formation one or more electrons are transferred from A to B (and vice versa), such as, for example, in the NaCl molecule. An A–B bond is called *covalent* if the valence electrons are uniformly distributed between A and B.

According to the symmetry of equilibrium configuration of nuclei all molecules can be divided into three main groups [1.2, 3]: (i) molecules having no axes of symmetry of the third or higher order, (ii) molecules having one axis of the third or higher order and (iii) molecules which have more than one axis of the third or higher order. These molecules are known, respectively, as *asymmetric top*, *symmetric top* and *spherical top*. A linear molecule is a special case of a symmetric top.

The molecular energy levels are determined by solving the Schrödinger equation using quantum-chemistry methods [1.4, 5]. Numerical solutions of the Schrödinger equation have been found only for the simplest molecules such as H_2^+ and H_2. For all other molecules, the Schrödinger equation is usually split into separate equations for electronic and nuclear subsystems on the basis of the Born-Oppenheimer theorem [1.5].

The solution of the Schrödinger equation for the electrons gives the electronic energy levels, called *potential surfaces* [1.6–8] which depend on the nuclear coordinates. The Schrödinger equation for nuclei gives vibrational-rotational energy levels for different electronic states. This equation can be divided into two parts: vibrational and rotational ones. The electronic energy of a molecule is $(\mathrm{M}/\mathrm{m})^{1/2}$ times larger than the vibrational one which, in turn, is $(\mathrm{M}/\mathrm{m})^{1/2}$ times larger than the rotational one. Here, m is the electron mass and M is the mass of the nucleus. These relations make it possible to divide the movement of a molecule into electronic, vibrational and rotational components.

Complete electronic-vibrational-rotational (*rovibronic*) energy levels of a molecule are classified according to irreducible representations (types of symmetry) of its symmetry group [1.9]. The separation of the complete movement into different types allows one to introduce the approximate quantum numbers for the classification of the molecule's energy levels [1.2]. In the majority of cases, these numbers correspond to the squares and z-projections of the corresponding angular momenta. Angular momenta and their quantum numbers used in the spectroscopy of diatomic molecules are given in Table 1.1.

For linear polyatomic molecules, symmetric and spherical top molecules, aside from the angular momentum (Table 1.1), the vibrational angular momenta l_i for each degenerate vibration mode i and the total vibrational angular momenta $l = \Sigma_i l_i$ are also used [1.2]. For symmetric top molecules, the quantum number K of vibrational angular momenta projection on the assigned axis of symmetry is the most important: $K = 0$ for non-degenerate vibrational states of linear molecules and $K = 1$ for degenerate vibrational states of linear molecules.

The quantity K has no meaning for the asymmetric top molecules; in this case, the rotational levels are designated by the number $J_{K_a K_c}$, where K_a and K_c are the projection quantum numbers for the limiting cases of oblate (a) and prolate (c) tops. For spherical tops the K number also has no meaning. In this case, the symmetry types for a given J are used.

Different electronic levels in a linear molecule with Σ, Π, Δ, Φ terms have electron orbital momenta $\Lambda = 0, 1, 2, 3, \ldots$, respectively. There is a one-to-one correspondence between the symmetry types and Λ values. Therefore,

Table 1.1. Angular momenta and their quantum numbers for classification of diatomic molecule states

Angular momentum	Operator	Quantum number: Total momentum	Projection of momentum on the molecular axis
Electron orbital momentum	$\boldsymbol{L}$	L	Λ
Electron spin	$\boldsymbol{S}$	S	Σ
Rotational momentum	$\boldsymbol{R}$	R	K
Total momentum without electron and nuclear spin	$\boldsymbol{N} = \boldsymbol{R}+\boldsymbol{L}$	N	$K+\Lambda$
Total momentum without nuclear spin	$\boldsymbol{J} = \boldsymbol{R}+\boldsymbol{L}+\boldsymbol{S}$	J	$\Omega = K+\Lambda+\Sigma$
Nuclear spin	$\boldsymbol{I}$	I	
Total momentum	$\boldsymbol{F} = \boldsymbol{J}+\boldsymbol{I}$	F	

irreducible representations $D_{\infty h}$ and $C_{\infty v}$ are also designated as Σ, Π, Δ, Φ... [1.2, 9]. Multiplicity of the level, defined by $2S+1$ values, is written as left upper subscript of Λ. For example, $^3\Sigma$ defines the level with $\Lambda = 0$ and $S = 1$, and $^2\Pi$ the level with $\Lambda = 1$ and $S = 1/2$, respectively. This symbol is added with values of J, N and F for every rotational sublevel and, if necessary, the number of the vibrational level v is added.

For nonlinear molecules, the number Λ has no meaning; the type of symmetry is used instead and the remaining designations are retained. In the simplest approximation, the normal vibration of a molecule is described by the harmonic oscillator with the energy

$$E_k = \omega_k\left(v_k + \tfrac{1}{2}\right) , \qquad (1.1.1)$$

where ω_k is the wave number and v_k is the vibrational quantum number. The state of a molecule in which some vibrations are excited is designated by a set of v_k numbers. For example, the state (1, 2, 1) of the H_2O molecule corresponds to the numbers $v_1 = 1$, $v_2 = 2$ and $v_3 = 1$ (sometimes this state is denoted as $\omega_1 + 2\omega_2 + \omega_3$). If the degenerate vibrations are excited, the quantum numbers v_k are provided by the upper index l_k as well, which gives a quantum number of vibrational angular momentum and is equal to $\pm v_k, \pm(v_k - 2), \ldots$. For example, the quantum numbers $v_1 = 2$, $v_2 = 3$, $l_2 = \pm 1$ and $v_3 = 1$ correspond to the state $(2, 3^{\pm 1}, 1)$.

1.1.1 Electronic-Energy Levels

Molecular-energy levels are calculated by sophisticated quantum-chemistry methods, but the number of levels of different types of symmetry and their relative position can be found using theoretical models and assumptions on the symmetry [1.9]. If a molecule is considered as a united atom or a united molecule with the same number of electrons, all possible electronic levels of different point-symmetry groups of the molecule can be determined by considering a splitting of the electronic-energy level of the united atom (or united molecule) in an electric field caused by the molecule, or simply, by a correlation between levels in the united atom or molecule, which can be determined from the characters of the point-symmetry group. For example, the united atom of the CH_4 molecule is the Ne atom, and its first three levels ${}^1S^{\circ}$, ${}^3P$ and ${}^1P$ correlate, respectively, with ${}^1A_1$, ${}^3F_2$, ${}^1F_2$ levels of the CH_4 molecule. In other words, the ground level of CH_4 is the ${}^1A_1$ level, the first excited state ${}^3P$ of Ne gives the ${}^3F_2$ level of the CH_4 molecule, etc.

Obviously, only levels of the same multiplicity of the united atom and molecule can correlate with each other. This holds if the spin-orbit splitting is small. In the case of large splitting, it is necessary to consider the correlation between the total wave functions including the spin component [1.10].

The electronic levels of a molecule can also be determined from the levels of separated atoms or group of atoms using the vector model. Then the quantum number Λ is the sum of projections of M_{L_i} for all atoms

$$\Lambda = \left| \sum_i M_{L_i} \right| , \tag{1.1.2}$$

and the total electron spin of the molecule is obtained as a vector sum of spins of all atoms

$$\boldsymbol{S} = \sum_i \boldsymbol{s}_i . \tag{1.1.3}$$

For example, the HCN molecule is created by H, C and N atoms in their ground states ${}^2S_g$, ${}^3P_g$, ${}^4S_u$, corresponding to $\Lambda = 1$ or 0 and $S = 3$, 2, 2, 1, 1, 0. Here, g corresponds to the even (gerade) and u to the odd (ungerade) states. As a result, one has the HCN levels: ${}^1\Sigma$, ${}^1\Pi$, two levels of each type ${}^3\Sigma$, ${}^3\Pi$, ${}^5\Sigma$, ${}^5\Pi$, levels ${}^7\Sigma$, ${}^7\Pi$. The levels of the HCN molecule can also be obtained from the levels of H and CN.

United and separated models allow one to obtain the number of electronic levels of different type, but no reliable information on their energies. More detailed information on disposition of levels can be obtained only by direct quantum-chemical calculations.

1.1.2 Vibrational-Energy Levels

In the harmonic-oscillator approximation, the energy of vibrational levels of a molecule is determined by a sum of the type (1.1.1) over all normal vibrations excited into a given electronic state

$$E_{\mathrm{v}}^{(0)} = \sum_k \omega_k \left(v_k + \frac{d_k}{2} \right) , \tag{1.1.4}$$

where d_k is the degeneracy of the k-th vibration. For the asymmetric top molecule all harmonic vibrations are non-degenerate and, as a rule, the *anharmonicity* effects only shift them down. The contribution of the cubic and quartic terms of an expansion of the potential surface in second-order perturbation theory gives the following expression for the energy:

$$E_{\mathrm{v}}^{(2)} = \sum_{jk} x_{jk}\left(v_j + \tfrac{1}{2}\right)\left(v_k + \tfrac{1}{2}\right) \ , \tag{1.1.5}$$

where x_{jk} are anharmonicity constants, usually ranging in the limits $1\,\mathrm{cm}^{-1} < x_{jk} < 100\,\mathrm{cm}^{-1}$. In the harmonic-oscillator approximation (1.1.4), the vibrational-level energies are independent on the vibrational angular momenta l_k. Therefore, the states of symmetric and spherical tops in which the degenerated vibrations are excited, can also be degenerate. This degeneration is partly removed by the anharmonicity effects and causes the so-called *anharmonic splitting*. For molecules of this type, the values of vibrational energy levels can be obtained by using a more general equation than (1.1.5):

$$E_{\mathrm{vl}}^{(2)} = \sum_{jk} x_{jk}\left(v_j + \frac{d_j}{2}\right)\left(v_k + \frac{d_k}{2}\right) + \sum_{tt'} g_{tt'} l_t l_{t'} \ . \tag{1.1.6}$$

Here, the anharmonic splitting is represented by the second sum, where $g_{tt'}$ is an anharmonicity coefficient. For example, in molecules of C_{3v} symmetry (e.g., NH_3) the level with $v_k = 2$, $l_t = 0, \pm 2$ (other $v_k = 0$) is splitted into $l_t = 0$ and $l_t = \pm 2$ components with the $4g_{tt'}$ interval between them, and the level with $v_k = 1$, $l_t = \pm 1$ and $v_{t'} = 1$, $l_{t'} = \pm 1$ is splitted into sublevels with $l_t + l_{t'} = 0$ and $l_t = \pm 2$ with a $2g_{tt'}$ interval between them. Equations (1.1.5 and 6) are valid when the so-called anharmonic resonances are absent. If the fundamental harmonic frequency ω_k is in random resonance (equal or close to) with some of the composite harmonic frequencies $\omega_i + \omega_j$ and if the anharmonic coefficients $K_{ijk} \neq 0$, where K_{ijk} describes the interaction of i-th, j-th and k-th normal modes, then the anharmonicity causes a strong mixing of the states and an anomalous shift of the levels, the so-called *accidental resonance* (or *Fermi resonance*) [1.2].

For example, Fermi resonance in the CO_2 molecule creates two bands, ω_1 and $2\omega_2$, in the Raman spectrum with almost equal intensities (off resonance, the $2\omega_2$ band would be much weaker). In general, anharmonic resonances may be observed also for $\omega_i \pm \omega_j \approx \omega_k \pm \omega_l$ if the corresponding anharmonicity coefficients are non-zero. Anharmonic resonances occur only between the vibrational states of the same symmetry.

The vibrational structure of the singlet electronic states of the molecules is described by (1.1.4–6) in which one should take into account the dependence of the vibrational frequencies and the anharmonicity constants of electronic states. These formulae also describe either the levels of non-degenerate vibrations in the degenerate electronic states, or the levels of degenerate vibrations in non-degenerate electronic states. The new effects can appear in degenerate electronic states while exciting degenerate vibrations, principally due to interaction of vibrational angular momenta of the degenerate vibrations with the electronic orbital angular momentum [1.11].

Electron-vibrational (vibronic) energy levels of the symmetric top or linear molecule can be classified by the values of quantum numbers $K = \Lambda + l$ of the projection of the vibronic angular momentum onto the axis of molecular symmetry. Electron–vibrational interaction removes the degeneration on Λ and l, the vibronic energy levels being splitted [1.11].

For symmetric and spherical top molecules, the linear terms of the expansion of the electronic Hamiltonian operator on the degenerate vibration coordinates have non-zero values [1.2]. In this case, the vibronic level splitting is referred to the so-called *Jahn-Teller linear effect*. When the values of the Jahn-Teller parameter D are small, the energy of the splitted sublevels is described by

$$G(v_2,\ l_2) = \omega_2(v_2 + 1) \mp 2D\omega_2(l_2 \pm 1) \tag{1.1.7}$$

For linear molecules the linear terms in question are zero and the splitting is described by the square terms of the expansion (*Renner effect*). In general, the Jahn-Teller and Renner effects should be accounted for together with the anharmonicity effect and the spin–orbital interaction.

1.1.3 Rotational Energy Levels

Rotational energy levels of a diatomic molecule in the $^1\Sigma$ state are qualitatively described in terms of the rigid top model:

$$E_r = BJ(J + 1) \ , \tag{1.1.8}$$

where J is the *rotational quantum number*, $B = \hbar/4\pi Ic$ (cm^{-1}) is the rotational constant and I is the *moment* of *inertia* of a molecule [1.3]. Equation (1.1.8) is also valid for rigid linear molecules and for rigid spherical top $^1\Sigma$ states when each J level of the spherical top molecule is $(2J + 1)$-time degenerate on J projection on one of the axes of the molecule (for a linear molecule this projection is zero). For rigid symmetric tops, two of three principal momenta of inertia are equal to each other, the rotational energy is

$$E_r = B_xJ(J + 1) + (B_z - B_x)K^2 \ , \tag{1.1.9}$$

where z is the selected axis of symmetry of a top and x is orthogonal to z axis [1.12, 13]. The axes of inertia of a molecule are usually denoted by the letters a, b, c for $I_a \leq I_b \leq I_c$, and rotational constants by $A \geq B \geq C$. Here, I are momenta of inertia of a molecule. Depending on conformity between axes x, y and z and a, b, c, symmetric tops are divided in two classes: *prolate*, for which energy is equal to

$$E_r = BJ(J + 1) + (A - B)K_a^2 \tag{1.1.10}$$

and *oblate* when

$$E_r = BJ(J+1) + (C-B)K_c^2 \ . \qquad (1.1.11)$$

In (1.1.10), the quantization axis of the rotational angular momentum is the selected axis a $(I_B = I_C)$. In (1.1.11), the axis c $(I_A = I_B)$ is a selected one.

The model of the rigid top is a rough approximation for the description of the real molecule. In reality, the rotating molecule is deformed, and this rotational distortion gives some essential contribution to its energy [1.12, 13]. In the case of a diatomic molecule, the ground (quartic) centrifugal correction to the energy (1.1.9) is given by

$$\Delta E_J = -D_v J^2 (J+1)^2 \ , \qquad (1.1.12)$$

where $D_v = 4B^3/\omega^2$. If $B = 1\,\mathrm{cm}^{-1}$ and $\omega = 1000\,\mathrm{cm}^{-1}$, then $D_v = 4 \times 10^{-6}\,\mathrm{cm}^{-1}$ $= 120\,\mathrm{kHz}$. For $J = 10$, the energy correction is 1.2 GHz. In the spherical top case, the quartic rotational correction consists of two parts

$$\Delta E_J = -D_v J^2 (J+1)^2 - D_t f(J,K) \ , \qquad (1.1.13)$$

where the first term is isotropic and is independent of the projection of J, and the second one is anisotropic, which causes a splitting of the level with a given J into sublevels with different K. For example, $D_t = 132\,\mathrm{kHz}$ for methane molecule and the level with $J = 2$ is splitted into components with a 60 D_t interval between them. The function $f(J,K)$ is proportional to J^4 and is found numerically. The splitting rapidly increases with increasing J. For a given J, the minimal value is $-4J^2(J+1)^2$ and its maximum is $+8J^2(J+1)^2$.

In the case of symmetric tops, the rotational correction also consists of two parts, the first of which is defined by

$$-D_v J^2 (J+1)^2 - D_{JK} J(J+1) K^2 - D_K K^4 \ . \qquad (1.1.14)$$

This sum leads to the level shift, the second part depends on the molecule symmetry and can remove the level degeneracy on the sign of K.

For asymmetric tops, the rotational correction to the energy can be obtained only numerically [1.2, 12]. The constants of a quartic rotational distortion depend on quadratic terms in the expansion of the potential surface and can be used to determine the harmonic force constants of the molecule.

Rotational levels of excited non-degenerate vibrational states are distinguished from rotational levels of the ground state only in the sense that the values of rotational and centrifugal constants differ in 0.1–1% from their values for the ground state, although in the presence of accidental resonances the rotational structure of the excited state can be strongly deformed [1.12]. Qualitatively, the difference of the rotational structure of the degenerate vibrational states from the rotational structure of non-degenerate states is defined, first of all, by the presence of the vibrational angular momenta l_t in the degenerate states. For the symmetric top, the interaction of l_t with the rotational angular momentum $\boldsymbol{J}$ (the so-called *Coriolis interaction*) gives a contribution to the energy, which, in the first approximation, is given by

$$\Delta E = -2B_z K \sum_t S_t^z l_t \tag{1.1.15}$$

Here S_t^z are the *Coriolis interaction constants* and B_z is the structure parameter of the molecule. The Coriolis interaction gives a contribution to the anharmonic level splitting with different values of $|l_t|$. As a result of this interaction, the levels (k, l), $(-k, -l)$, $(-k, l)$ and $(k, -l)$ form two doubly-degenerated levels: $\{(k, l), (-k, -l)\}$ and $\{(k, -l), (-k, l)\}$. This degeneration is removed by effects of vibrational–rotational interaction of higher orders [1.12]. In the particular case of molecules of C_{3v} symmetry, when $K = l_t == \pm 1$, the sublevel splitting is of the order of $q_t J(J+1)$. This effect is called *l-doubling*. The constant q_t of *l*-doubling depends on the cubic coefficient of anharmonicity. Such doubling takes place also for linear molecules, but, in this case, q_t depends only on the harmonic part of the potential surface.

1.2 Coupling Schemes

In interaction of the rotational angular momentum with electronic spin, orbital and vibronic angular momenta play the main role in the formation of rotational structure of degenerate electronic and vibronic states [1.12–14]. In the general case of polyatomic molecules, accounting for these interactions is sufficiently complex. In the case of diatomic molecules, it is important to distinguish the limiting cases of coupling between angular momenta, the so-called *Hund cases* [1.15]. The Hund case (a) corresponds to a strong spin-orbital interaction, which takes place in heavy molecules. The coupling between momenta $\boldsymbol{L}$ and $\boldsymbol{S}$ is considered first, and then the coupling of the momentum $\boldsymbol{J} = \boldsymbol{L} + \boldsymbol{S}$ and the rotational momentum $\boldsymbol{R}$. In the Hund case (b), the spin–orbital interaction is weak, which is typical for light molecules and for all Σ states. Here, the coupling between $\boldsymbol{R}$ and $\boldsymbol{L}$ is considered first and then the summation of the momentum $\boldsymbol{N} = \boldsymbol{R} + \boldsymbol{L}$ with $\boldsymbol{S}$ follows to obtain the total momentum $\boldsymbol{J}$. Other Hund cases also exist (Table 1.2).

Interaction of the rotational angular momentum with electronic momenta removes the degeneration on L and S that is shown in the last column of

Table 1.2. The Hund cases (a, b, c and d) for rotational energies of the rigid top molecules

Hund case	Rotational energy	Quantum numbers of electronic level	Degeneracy number of electronic level
(a)	$BJ(J+1)$	Λ, S, Σ	2 or 1
(b)	$BN(N+1)$	Λ, S, Σ	2 (2S + 1) or (2S + 1)
(c)	$BJ(J+1)$	Ω	2 or 1
(d)	$BR(R+1)$	L, Λ, S, Σ	(2L + 1) (2S + 1)

Table 1.2. In the case of $^1\Pi$ states, the Hamiltonian terms proportional to L^2J^2 cause the Λ-doubling effect with a splitting $\Delta\omega = q_e J(J+1)$, where $q_e \approx 8\pi^2 B_e^2/\omega_e$, and ω_e is the difference between a given state and the nearest $^1\Sigma$ state [1.13].

1.3 Quantum Chemistry Methods

Quantum chemistry is the field of theoretical chemistry studying the structure and chemical transformations of atoms, molecules and other polyatomic systems on the basis of quantum mechanics [1.4, 5]. The main equation used in quantum chemistry is the non-relativistic Schrödinger equation:

$$\hat{\mathscr{H}}\Psi = E\Psi \ , \tag{1.3.1}$$

where Ψ is the wave function of a system depending on the space and spin coordinates of all particles of the system, and E is the total internal energy of the system. The Hamiltonian $\hat{\mathscr{H}}$ includes the operators of kinetic energy of all particles in the molecule and also all Coulomb interactions between the charged particles involved.

The wave equation for two or more electrons can be solved only using some approximation. A common approach is to consider a molecular wave function for each electron as a linear combination of atomic orbitals (LCAO approximation):

$$\Phi = c_1\psi_1 + c_2\psi_2 + \cdots + c_n\psi_n \ , \tag{1.3.2}$$

where $\psi_1, \ldots, \psi_n$ is a basis of atomic orbitals, and the constants $c_1 \ldots c_n$ are chosen to minimize the total electronic energy of a molecule.

Figure 1.1 depicts a variation of the potential energy V as a function of interatomic distance for a pair of atoms forming a molecule. The curve is quite well represented by the Morse potential:

$$V(r) = D_e\{1 - \exp[-a(r - r_e)]\}^2 - D_e \ . \tag{1.3.3}$$

Here, D_e is the dissociation energy of a molecule, r_e is the equilibrium interatomic distance, and a is a constant. At infinite separation, the potential energy is zero. Let us consider two atoms, e.g., hydrogen atoms, described by the functions $\phi_A(1)$ with electron 1 of atom A, and $\psi_B(2)$ with electron 2 of atom B. As the atoms are put together, one can assume that they lose much of their individual electronic character and that a molecular bond is formed due to the overlapping of atomic orbitals since the two atoms are close enough to each other. The potential energy $V(r)$ of the pair of atoms decreases as the molecule is formed, with a minimum at the equilibrium interatomic distance r_e. At distances $r < r_e$, the potential rapidly increases because of internuclear and electron–electron repulsions (Fig. 1.1).

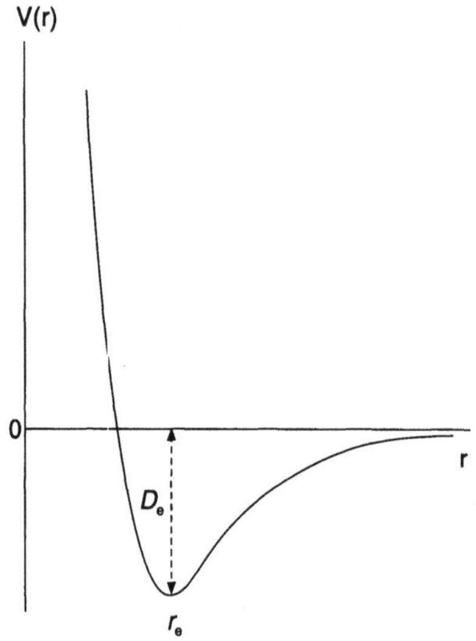

Fig. 1.1. Dependence of the potential energy $V(r)$ on the interatomic distance r. At the equilibrium distance r_e, the Coulomb forces of attraction balance the forces of repulsion. The energy D_e corresponds to the dissociation energy of the paired atoms

If a molecule is formed, electron 1 may be found closer to nucleus B, and vice versa. Because of the indistinguishability of electrons, the total wave function can be written in the form:

$$\Phi = c_1\psi_A(1)\psi_B(2) + c_2\psi_A(2)\psi_B(1) \tag{1.3.4}$$

where, again, c_1 and c_2 are chosen to produce a minimum-energy conformation. In the case of the hydrogen molecule, from symmetry properties and because the probabilities are proportional to $|\psi|^2$, one has $c_1^2 = c_2^2$, and

$$\Phi_\pm = \psi_A(1)\psi_B(2) \pm \psi_A(2)\psi_B(1) \; . \tag{1.3.5}$$

This type of wave functions have been used in [1.16] to obtain the *valence-bond wave functions* and energies E_+ and E_- (Fig. 1.2). The wave functions Φ_+ corresponding to the spin-paired electrons leads to the bond formation, whereas Φ_- (parallel spins) does not. The calculated value of the energy E_+ was found to be −303 kJ/mol (experimental value is −458 kJ/mol) and the equilibrium distance r_e is 0.087 nm (experimental value 0.074 nm), respectively. More accurate approaches lead to excellent agreement with experimental data [1.2].

Let us consider a simple extension of (1.3.5). It is possible to include the ionic contributions to bonding in molecular hydrogen when both electrons move around a single hydrogen atoms, A or B. One can obtain the following possible structures

$$H_A^1H_B^2, \; H_A^2H_B^1, \; H_A^{1,2}H_B, \; H_AH_B^{1,2}$$

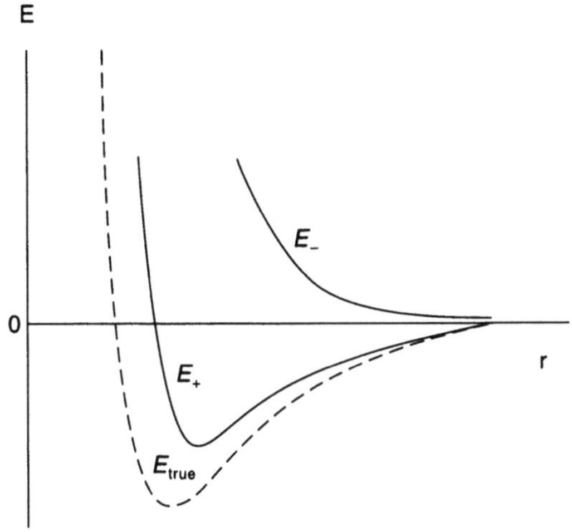

Fig. 1.2. Valence-bond energies for the hydrogen molecule: E_+, E_- and E_{true} are shown as a function of the internuclear distance r

and a molecular wave function of the form:

$$\Phi_+ = \psi_A(1)\psi_B(2) + \psi_A(2)\psi_B(1) + \lambda[\psi_A(1)\psi_A(2) + \psi_B(1)\psi_B(2)] \ . \quad (1.3.6)$$

The right-hand side of (1.3.6) can be considered as a sum $\psi_{cov} + \lambda\psi_{ionic}$, where the parameter λ is found from the minimum-energy condition. The structures considered above for the hydrogen molecule correspond to the same energy, and are known as *canonical structures* [1.17].

An alternative approach is termed the *molecular-orbital method* [1.18], which offers more general applicability for solving chemical problems. It forms a more logical extension of the atomic-orbital treatment of atoms than the valence-bond method. It takes the H_2^+ ion as the basic structure rather than the bond in molecular hydrogen, used by the valence-bond method. The probability of the electron is then given by

$$\Phi = \psi(1s_A) + \psi(1s_B) \ . \quad (1.3.7)$$

When the electron is close to the nucleus A, the amplitude of $\psi(1s_B)$ is small and $\Phi \approx \psi(1s_A)$, and vice versa. The wave function (1.3.7) is similar to the atomic orbital, but covers the whole molecular volume; it is the so-called *molecular orbital* [1.18]. The probability of finding the electron at any point is given by the expression

$$|\Phi|^2 d\tau = \left[|\psi|^2(1s_A) + |\psi|^2(1s_B) + 2\psi(1s_A)\psi(1s_B)\right] d\tau \ , \quad (1.3.8)$$

where $d\tau$ is the volume element.

In the region of nucleus A, this function resembles the atomic orbital for hydrogen, and similarly nearby the nucleus B. In the intermediate region there

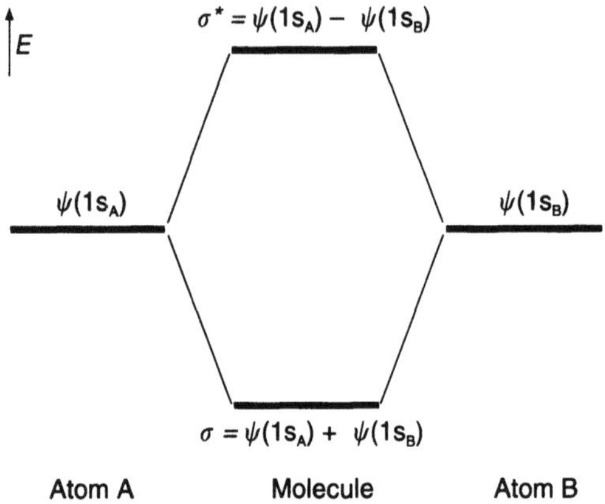

Fig. 1.3. Molecular-orbital energy-level diagram

is an enhancement, or accumulation, of electron density given by the factor $2\psi(1s_A)\psi(1s_B)d\tau$, expected for the hydrogen atoms alone. The reason for this enhancement is the constructive interference of the electron wave functions in the intermediate region between the nuclei: each contribution has the same phase so that the total amplitude is the sum of its components.

The molecular orbital $\psi(1s_A) + \psi(1s_B)$ responsible for the bond is called a σ-bonding orbital [1.18]; it has a cylindrical symmetry relative to the internuclear axis. Another molecular orbital, the next-highest energy solution of the wave equation, is the combination $\psi(1s_A) - \psi(1s_B)$. In this case, the atomic orbitals interfere destructively in the intermediate region and the resultant molecular orbital is called *antibonding* ($\sigma*$). These two molecular orbitals are illustrated in Fig. 1.3, exhibiting a molecular-orbital energy-level diagram.

Molecular hydrogen can be considered in terms of the H_2^+ ion and the *aufbau principle* [1.17]. Two electrons should be fed into the molecular orbital of the lowest energy. The configuration now is $(1s\sigma)^2$ with the two electrons having paired spins. The instability of the helium molecule (He_2), can be explained as follows: The first two electrons form the $(1s\sigma)^2$ orbital; the next electron must enter the first antibonding orbital, so that the bond is weakened. The fourth electron leads to the $(1s\sigma*)^2$ orbital, and the effects of the $(1s\sigma)^2$ and $(1s\sigma*)^2$ orbitals are cancelled. (In fact, the antibonding effect is a little stronger). The configuration of the Li_2 molecule is $(1s\sigma)^2(1s\sigma*)^2(2s\sigma)^2$ and a bond is formed. Homonuclear diatomic molecules (X_2) can easily be considered in this way.

When the second-row elements are introduced into a molecule, p orbitals may be involved in bonding. Two $2p$ orbitals may overlap along the direction of the internuclear axis, as it takes place, for example, in the fluorine molecule. The bonding molecular orbital is denoted by $2p\sigma$ and the corresponding antibonding molecular orbital is $2p\sigma*$, both having cylindrical symmetry

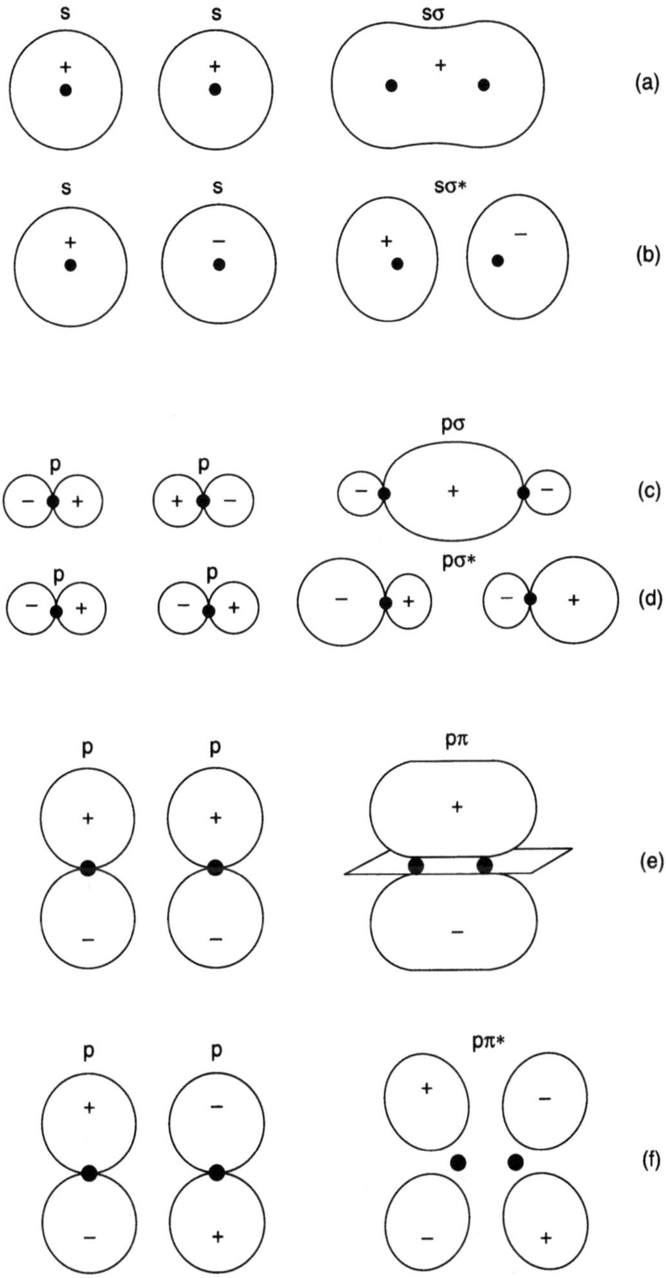

Fig. 1.4a–f. Simple atomic and molecular orbitals (homonuclear species): (**a**) $s+s \rightarrow s\sigma$, (**b**) $s-s \rightarrow s\sigma^*$, (**c**) $p+p \rightarrow p\sigma$, (**d**) $p-p \rightarrow p\sigma^*$, (**e**) $p+p \rightarrow p\pi$, (**f**) $p-p \rightarrow p\pi^*$. In general, Ψ, though not $|\Psi|^2$, being an amplitude, has regions (lobes) of positive and negative values

around the internuclear axis. When p orbitals overlap in directions other than along the internuclear axis, the molecular orbitals so formed have reflection symmetry across a nodal plane; they are called π molecular orbitals. Figure 1.4 depicts some examples of simple molecular orbitals.

Similar considerations can be used [1.8] for heteronuclear diatomic molecules (XY), for example hydrogen fluoride. The $\psi(1s)$ orbital of the hydrogen atom overlaps with the $\psi(2p)$ one of fluorine giving rise to the $p\sigma$ molecular orbitals of cylindrical symmetry.

In general, one can use one-electron MO functions obtained in the Hartree-Fock approximation. However, this problem can be solved only for simple diatomic molecules. For other systems, the approximation proposed in [1.19] is used, in which molecular orbitals are expanded on the Slater- or Gauss-type functions χ_μ

$$\varphi_i = \sum_{\mu=1}^{N} c_{\mu i}\chi_\mu, \quad i = 1, 2, \ldots, n \ , \tag{1.3.9}$$

and instead of the functions φ_i the coefficients $c_{\mu i}$ are optimized.

A qualitative representation of the MO structure (structure of nodes, nodal structure) constitutes the basis for many approaches in molecular physics. The most known qualitative theory of molecule formation is the *Walsh* theory [1.10] in the MO approximation based on the relation:

$$\frac{\partial E}{\partial \theta} \approx \frac{\partial}{\partial \theta}\left(\sum_i^{\text{val}} \varepsilon_i\right) \ , \tag{1.3.10}$$

where θ is the valence angle of the molecule, the sum is over all occupied valence MOs, and ε_i are their energies. To predict the molecular form one has to know how the energy of MO depends on the internal coordinates: if the sum of the energies of MO is lowered when the molecule is bending, then the stable molecular configuration is bending. The angle of bending, θ, may be calculated too.

The run of chemical reactions may be qualitatively explained on the basis of the MO approach using the *Woodward-Hoffman* or *Pirson* rules [1.21]. According to the *Woodward-Hoffman* rule, one has to construct the correlation diagrams for chemical reactions. As a first step, one has to choose chemical bonds and orbitals which are changed most in the course of the reaction. Then, one should set the way in which reagents are brought together in view of the expected structure of the transient states, which classify the MO orbitals of reagents and products concerning the symmetry of the chosen structure of the transient state. Then one should establish the correspondence between the MO of reagents and products in accordance with the chosen classification of MO by symmetry. It is necessary to take into account the axis and the symmetry planes with the components which pass through the breaking and creating bonds. If in the course of the chemical reaction all the filled MO of reagents convert to the filled MO of products in the ground state (the orbital symmetry is conserved), then the reaction is called *allowed.* If crossing of the filled and vacant MOs take place, such a reaction is termed *forbidden.*

Complex systems are usually described by the semiempirical methods when the majority of molecular integrals is replaced by experimental data (e.g., ionization potential and electron affinity) [1.18, 22].

In many cases, the Hartree-Fock approximation or its modifications give large uncertainties [1.22]. For example, it yields a negative value for the bonding energy for the F_2 molecule, incorrect symmetry for the ground electronic state of the C_2 molecule, and so on. For the removal of these disadvantages of the method, one should account for the so-called *residual* or *correlation interaction* between electrons using the configuration interaction or perturbation approaches.

In the configuration–interaction method, the wave functions have the form of linear combinations of Slater determinants ψ_K corresponding to different ways for filling molecular orbitals:

$$\Psi = \sum_{k=1}^{M} A_k \psi_k \ , \tag{1.3.11}$$

where M is the number of configurations possible. The wave functions are obtained by solving the electronic part of the molecular Schrödinger equation using the variational method.

1.4 Equilibrium Form of the Molecules

The form of a molecule is determined by the space disposition of atoms in the molecule which, in turn, is specified by bond lengths and angles between bonds [1.1, 4, 23]. There is a correlation between multiplicity and bond length: the bond between definite atoms is shortened with the increase of its multiplicity. For example, typical lengths of C—C, C=C, and C≡C bonds are 1.50, 1.35 and 1.20 Å, respectively. The length of the bond depends on its chemical environment in the molecules. For example, the bond length C—H in the CH_3 group is approximately 1.10 Å [1.23], but in the groups $=CH_2$ and ≡CH is 1.08 and 1.05 Å, respectively. There are similar relationships for valence angles as well. For example, usually, the angle between the CH bonds in CH_3 groups in different molecules is about 109°. The carbon atom has four bonds, and the angles between these bonds are close to tetrahedral 109°30′.

Some preliminary view of the molecular form created by a given set of atoms can be obtained on the basis of the symmetry of the molecule and the concept of *Molecular Orbitals* (MO) [1.18]. Depending on the sign of the energy introduced by electrons in the given MO to the total molecular energy, MO are called *bonding* or *antibonding* [1.2]. The bonding MOs take part in the formation of strong chemical bonds while antibonding orbitals do not. The number of bonding and antibonding MOs depends only on symmetry disposition of the atoms in the molecule. Therefore, the definition of a stable form of a molecule reduces to the problem of finding such a disposition of atoms which correspond to the maximum number of bonding MO. In fact, different MO give different contributions and, therefore, this method is not always applicable, but in the majority of cases it correctly predicts the geometrical molecular structures. For example, it allows one to establish that the H_2O molecule is nonlinear, but the CO_2 molecule is linear. This method is especially effective for excited electronic molecular states.

1.5 Potential-Energy Curves

Except for the case of a strong vibronic interaction, the Schrödinger equation for electronic motion can be solved for a set of fixed nuclear positions. The electronic wave function contains the molecule-fixed nuclear coordinates as parameters instead of variables because the influence of the momentum of the nuclear motion acting on the electronic wave function is neglected. The electronic energy in a certain electronic state represented as a function of the nuclear coordinates corresponds to the potential-energy function for the nuclear motion (i.e., the intramolecular vibration) [1.6–8].

In the Born-Oppenheimer approximation, the potential function for the nuclear motion is assumed to be independent of the nuclear masses and, therefore, the potential function should not change in the case of the isotopic substitution [1.2]. Detailed spectroscopic studies of diatomic potential functions have confirmed that the isotopic differences in physical properties are really small [1.24].

Modern experimental methods, particularly laser spectroscopy [1.25], are able to accurately determine the potential functions for many electronic states including weakly-bound repulsive states for stable or short-lived diatomic molecules and ions [1.24, 26]. For many excited electronic states, the potential energy curves have complicated crossings and different problems of molecular dynamics, such as dissociation, predissociation and bound-free transition, arise [1.7, 27].

The potential-energy functions can be obtained using accurate ab-initio MO calculations for the electronic energies with fixed nuclear positions. These calculations have been performed for many electronic states of different diatomic species, and the potential-energy functions have been obtained with accuracies comparable with experimental ones [1.7, 28].

Many molecules of high chemical importance have at least one large-amplitude intramolecular motion such as internal rotation, inversion, ring puckering, and pseudorotation, while other molecules have weak chemical bonds which result in large-amplitude vibrations. For example, carbon suboxide, O=C=C=C=O, is a typical quasilinear molecule, that has a very anharmonic, low-frequency vibrational mode (ν_7) related to the C=C=C bending; the average amplitude of displacement is estimated to be more than 20° [1.30]. Complexes, caused by the hydrogen bonds or Van-der-Waals forces are also very non-rigid [1.31]. The Van-der-Waals constants C_n and the molecular terms of molecular hydrogen and its ions are given in Chap. 7.

1.6 Corrections to the Electronic Structure

There are two main reasons to correct the non-relativistic consideration of molecular structure, which was the topic of the preceding part in this chapter. The first one is related to the some characteristics of nuclei which are not considered here. They give rise to a *hyperfine structure* of molecular levels. The second reason is due to the presence of relativistic and quantum-electrodynamical effects in the exact treatment of molecular systems, which give the *fine structure* and also the Lamb shift of molecular levels.

1.6.1 Hyperfine Interaction and Isotope Effect

Each energy level of a molecule has a nuclear HyperFine Structure (HFS) which is caused by the presence of the nuclear electric and magnetic momenta [1.3]. In the electronic $^1\Sigma$ state the nuclear HFS is caused mainly by

(i) electrostatic interaction of the nuclear electric quadrupole moment with the electric field of the molecule (quadrupole interaction or quadrupole coupling)
(ii) interaction of the nuclear magnetic dipole moment with the magnetic field, which is produced by the rotation of the molecule (nuclear spin–rotation interaction) and
(iii) interaction of magnetic moments of different nuclei with each other (nuclear spin–spin interaction).

Usually, the quadrupole interaction gives the main contribution to HFS; but this interaction takes place only for nuclei with spin $I > 1/2$ [1.32]. In the simplest case of a quadrupole interaction of a single quadrupole nucleus with a corresponding gradient q of the electronic density at the nucleus, its energy is given by [1.32]

$$W_Q = -e^2qQ\left[\tfrac{3}{4}C(C+1) - I(I+1)J(J+1)\right] \times \left[2I(2I-1)\,(2J-1)\,J\right]^{-1} . \tag{1.6.1}$$

Here, e is the electron charge, q is the electric field gradient, Q is the nuclear electric quadrupole moment, J is the rotational quantum number, and I is the nuclear spin. The constant C is given by

$$C = F(F+1) - I(I+1) - J(J+1) \ , \tag{1.6.2}$$

where F is the quantum number of the total angular momentum $\boldsymbol{F} = \boldsymbol{J} + \boldsymbol{I}$, with $F = J+I, J+I-1, J+I-2, \ldots, J-I$. In this case, the momentum J loses its meaning of a good quantum number and HFS levels are numbered according to F. For example, the rotational level with $J = 1$ of a nucleus having spin $I = 5/2$ splits into three sublevels $F = 5/2$, $7/2$ and $3/2$ with energies $W_Q = +4e^2qQ/25$, $-e^2qQ/20$, and $-7e^2qQ/50$, respectively. The quadrupole coupling interaction e^2qQ depends on the type of nucleus and on the molecular environment and can change in a wide range. A quadrupole HFS is usually observed in high-resolved spectra [1.33]. Spin–rotational and spin–spin interactions give a small contribution to HFS and take place for all nuclei with a spin $I \geq 1/2$. Splitting of rotational levels due to this interactions is usually not higher than 100 kHz and may be observed by using unique experiments. Experimental data for the relative constants of quadrupole coupling and spin–rotational interaction comprise valuable information on electronic molecule structure. Constants of the spin–spin interaction depend only on geometrical parameters of molecules.

In the degenerate electronic states, the interaction of electronic spin with nuclear spins play an important role, the energy is $g_e\mu_B/g_I\mu_N$ times higher than that of purely nuclear spin–spin interactions. Here, g_e and g_I are electronic and nuclear g-factors, μ_B is the Bohr magneton and μ_N is the nuclear magneton.

Electron-nuclear spin–spin interactions are of two types [1.32, 33]

(i) classical dipole–dipole interaction; the energy, in the general case, being defined by the tensor of second rank with 9 components
(ii) isotropic contact Fermi interaction which is due to the availability of an electronic spin density at the place of the nucleus, which has no classical analog. In contrast to anisotropic spin–spin interaction, the contact interaction takes place only for the states with $\Lambda = 0$, which is similar to the s states in atoms. This is due to that only atomic s orbitals have non-zero spin density at the place of the atom. Constants of both types of interactions depend on the electronic density of the molecule and give valuable information about electronic wave functions.

1.6.2 Relativistic Effects

The electronic-energy levels of the molecule are splitted due to the relativistic spin–orbital interaction into so-called multiplets [1.34]. In the case of normal coupling, the splitting is given by

$$\Delta E = A \cdot \Lambda \cdot \Sigma \ . \tag{1.6.3}$$

The constant A of the spin–orbital interaction increases rapidly with the increase of the nuclear charges of the atoms constituting a molecule. The quantum numbers Λ and Σ lose their meaning as good quantum numbers. The splitted sublevels are characterized by the quantum number $\Omega = \Lambda + \Sigma$ of the projection of the total electronic angular momentum on the molecular axis. For example, the level $^3\Pi$ of a linear molecule is splitted due to the spin–orbital interaction into sublevels Σ^+, Σ^-, Π and Δ, corresponding to $\Omega = 0, 0, 1$ and 2, respectively. The value of the constant A varies from a few cm^{-1} for light molecules up to hundreds of cm^{-1} for heavy molecules.

2 Spectroscopic and Geometrical Constants of Molecules

The spectroscopic and geometrical constants of diatomic molecules and the most important parameters for simple polyatomic molecules (the fundamental vibrational frequencies, rotational and geometrical constants) are presented in this chapter.

2.1 Constants of Diatomic Molecules

The vibrational-rotational energy of diatomic molecules is given by

$$T = G(v) + F_v(J) \ ,$$

where

$G(v) = \omega_e(v + 1/2) - \omega_e x_e(v + 1/2)^2 + \cdots$ – vibrational energy,
$F_v(J) = B_v J(J + 1) - D_v J^2(J + 1)^2 + \cdots$ – rotational energy,
v – vibrational quantum number,
ω_e – vibrational frequency,
$\omega_e x_e$ – anharmonic constant,
J – rotational quantum number,
$B_v = B_e - \alpha_e(v + 1/2) + \cdots$ and B_e – rotational constants,
$D_v = D_e - \beta_e(v + 1/2) + \cdots$ and D_e – centrifugal distortion constants,
α_e, β_e – rotation-vibration interaction constants.

The subscript e indicates that the value corresponds to the minimum of the potential curve. The vibrational energy $G_0(v)$, counted from the lowest vibrational level, is given by

$$G_0(v) = \omega_0 v - \omega_0 x_0 v^2 + \cdots \ ,$$

where

$$\omega_0 = \omega_e - \omega_e x_e + \cdots \ ,$$

$$\omega_0 x_0 = \omega_e x_e - \cdots \ .$$

In this case, the appropriate values are designated by subscript 0. Neglecting rotation, the transition energy between successive vibrational levels (equal to the separation of these levels) is given by

$$\Delta G_{v+1/2} = G(v+1) - G(v) = G_0(v+1) - G_0(v)$$

$$\cong \omega_e - 2\omega_e x_e - 2\omega_e x_e v \cong \omega_0 - \omega_0 x_0 - 2\omega_0 x_0 v \ .$$

Thus, for the $v = 0$ level it follows that the fundamental frequency in the infrared spectrum of a diatomic molecule is given by

$$\Delta G_{1/2} \cong \omega_e - 2\omega_e x_e \cong \omega_0 - \omega_0 x_0 \ .$$

Table 2.1 presents the data on vibrational and rotational constants for the ground states of diatomic molecules and radicals [2.1–11]. The following quantities are given in Table 2.1:

- molecular formula with the mass numbers of constituent atoms,
- reduced mass μ in atomic mass units [amu],
- the ground state electronic term of the molecule,
- vibrational frequency ω_e [cm^{-1}],
- anharmonic constant $\omega_e x_e$ [cm^{-1}],
- rotational constant B_e [cm^{-1}],
- rotation-vibration interaction constant α_e [cm^{-1}],
- centrifugal distortion constant D_e [cm^{-1}],
- equilibrium internuclear distance r_e [Å], defined by the relation

$$B_e = \hbar/4\pi c \mu r_e^2 \ ,$$

is calculated from

$$r_e[\text{Å}] = 4.106(\mu[\text{amu}] B_e[\text{cm}^{-1}])^{-1/2} \ .$$

The force constants $k_e = \mu\omega_e^2$ and the inertial momenta I_e (not listed in Table 2.1) can be found from relations:

$$k_e[\text{dyn/cm}] = 5.89.10^{-2}\mu[\text{amu}] \cdot \omega_e^2[\text{cm}^{-1}] \ ,$$

$$I_e[\text{g cm}^2] = 2.80 \cdot 10^{-39}/B_e[\text{cm}^{-1}]$$

The molecules in the Table 2.1 are given by their chemical formulae in alphabetical order. The error of the numerical values listed is estimated to be less than one or two in the last digits of the given value. The asterisk (*) denotes that the appropriate value is related to the lowest vibrational level or to the lowest vibrational level observed. Vibrational frequencies with an asterisk (*) mean $\Delta G_{1/2}$ values.

The dissociation energies D and ionization potentials IP of diatomic molecules are given in Chap. 3.

Table 2.1. Constants of diatomic molecules

Molecule	μ [amu]	Ground state	ω_e [cm^{-1}]	$\omega_e x_e$ [cm^{-1}]	B_e [cm^{-1}]	α_e [cm^{-1}]	D_e [cm^{-1}]	r_e [Å]
$^{107.109}Ag_2$	53.948	$^1\Sigma_g^+$	192.4	0.643	0.496	0.19	–	2.68
$^{109}Ag^{81}Br$	46.424	$^1\Sigma^+$	247.7	0.679	0.063	0.00023	$1.7 \cdot 10^{-8}$	2.39
$^{107}Ag^{35}Cl$	26.350	$^1\Sigma^+$	343.49	1.17	0.123	0.0006	$6.3 \cdot 10^{-8}$	2.28
$^{107}Ag^{19}F$	16.132	$^1\Sigma^+$	513.45	2.59	0.266	0.00192	$2.84 \cdot 10^{-7}$	1.98
$^{107}Ag^{1}H$	0.9984	$^1\Sigma^+$	1759.9	34.06	6.449	0.201	$3.44 \cdot 10^{-4}$*	1.62
$^{107}Ag^{2}H$	1.977	$^1\Sigma^+$	1250.7	17.17	3.257	0.0722	$0.86 \cdot 10^{-4}$*	1.62
$^{107}Ag^{127}I$	58.02	$^1\Sigma^+$	206.52	0.445	0.045	0.0001	$0.85 \cdot 10^{-8}$	2.54
$^{107}Ag^{16}O$	13.91	$^2\Pi_{1/2}$	490.2	3.06	0.302	0.0025	$4.5 \cdot 10^{-7}$	2.00
$^{27}Al_2$	13.49	$^3\Sigma_g^-$	350.01	2.022	0.205	0.0012	$3.0 \cdot 10^{-7}$	2.47
$^{27}Al^{79}Br$	20.11	$^1\Sigma^+$	378.0	1.28	0.159	0.0009	$1.13 \cdot 10^{-7}$	2.29
$^{27}Al^{35}Cl$	15.23	$^1\Sigma^+$	481.30	1.95	0.244	0.0016	$2.50 \cdot 10^{-7}$	2.13
$^{27}Al^{19}F$	11.15	$^1\Sigma^+$	802.2	4.77	0.552	0.0050	$1.05 \cdot 10^{-6}$	1.65
$^{27}Al^{1}H$	0.972	$^1\Sigma^+$	1682.6	29.09	6.391	0.1858	$3.56 \cdot 10^{-4}$	1.65
$^{27}Al^{2}H$	1.874	$^1\Sigma^+$	1211.9	15.14	3.319	0.0697	$0.97 \cdot 10^{-4}$*	1.65
$^{27}Al^{127}I$	22.25	$^1\Sigma^+$	316.1	1.0	0.1177	0.0006	–	2.54
$^{27}Al^{16}O$	10.04	$^2\Sigma^+$	979.2	6.97	0.6413	0.0058	$1.08 \cdot 10^{-6}$	1.62
$^{27}Al^{32}S$	14.63	$^2\Sigma^+$	617.1	3.33	0.2799	0.0018	$2.20 \cdot 10^{-7}$	2.03
$^{40}Ar_2$	19.98	$^1\Sigma_g^+$	25.74*	2.6	0.0597	0.0037	$1.13 \cdot 10^{-6}$*	3.76
$^{75}As_2$	37.46	$^1\Sigma_g^+$	429.6	1.117	0.1018	0.0003	–	2.10
$^{75}As^{19}F$	15.16	$^3\Sigma_{0^-}^-$	685.8	3.12	0.3648	0.0024	$4.50 \cdot 10^{-7}$	1.74
$^{75}As^{14}N$	11.80	$^1\Sigma^+$	1068.5	5.41	0.5455	0.0034	$0.53 \cdot 10^{-6}$	1.62
$^{75}As^{16}O$	13.18	$^2\Pi_{1/2}$	967.08	4.85	0.4848	0.0033	$4.90 \cdot 10^{-7}$	1.62
$^{75}As^{31}P$	21.91	$^1\Sigma^+$	604.02	1.98	0.1925	0.0008	$7.80 \cdot 10^{-8}$	2.00
$^{75}As^{32}S$	22.41	$^2\Pi_{1/2}$	567.9	1.97	0.1848	0.0008	$7.80 \cdot 10^{-8}$	2.02
$^{197}Au_2$	98.48	$^1\Sigma_g^+$	190.9	0.42	0.0280	0.0001	$0.25 \cdot 10^{-8}$*	2.47
$^{197}Au^{27}Al$	23.73	$^1\Sigma^+(0^+)$	333.0	1.16	0.1299	0.0007	$0.71 \cdot 10^{-7}$	2.34
$^{197}Au^{9}Be$	8.62	$^2\Sigma^+$	607.68	3.53	0.4607	0.0040	$1.04 \cdot 10^{-6}$*	2.06
$^{197}Au^{1}H$	1.00	$^1\Sigma^+$	2305.0	43.1	7.2401	0.2136	$2.79 \cdot 10^{-4}$	1.52
$^{197}Au^{2}H$	1.99	$^1\Sigma^+$	1635.0	21.7	3.6415	0.0761	$0.71 \cdot 10^{-4}$	1.52
$^{197}Au^{24}Mg$	21.38	$^2\Sigma^+$	307.9	1.1	0.1321	0.0007	$1.02 \cdot 10^{-7}$*	2.44
$^{11}B_2$	5.50	$^3\Sigma_g^-$	1051.3	9.35	1.212	0.014	–	1.59
$^{138}Ba^{19}F$	16.70	$^2\Sigma^+$	468.9	1.79	0.216*	0.0012	$1.75 \cdot 10^{-7}$*	2.16
$^{138}Ba^{1}H$	1.00	$^2\Sigma^+$	1168.3	14.5	3.3828	0.0660	$1.13 \cdot 10^{-4}$*	2.23
$^{138}Ba^{2}H$	1.98	$^2\Sigma^+$	829.77	7.32	1.7071	0.0236	$2.88 \cdot 10^{-5}$*	2.23
$^{138}Ba^{16}O$	14.33	$^1\Sigma^+$	669.76	2.02	0.3126	0.0014	$2.72 \cdot 10^{-7}$	1.94
$^{138}Ba^{32}S$	25.95	$^1\Sigma^+$	379.42	0.88	0.1033	0.0003	$3.06 \cdot 10^{-8}$	2.51
$^{11}B^{79}Br$	9.66	$^1\Sigma^+$	684.31	3.52	0.4894	0.0035	$1.00 \cdot 10^{-6}$	1.89
$^{11}B^{35}Cl$	8.37	$^1\Sigma^+$	839.12	5.11	0.6838	0.0065	$1.72 \cdot 10^{-6}$	1.72
$^{9}Be^{79}Br$	8.09	$^2\Sigma^+$	715.0	3.8	0.546*	–	$1.31 \cdot 10^{-6}$*	1.95
$^{9}Be^{35}Cl$	7.17	$^2\Sigma^+$	846.7	4.8	0.7285	0.0069	$2.50 \cdot 10^{-6}$	1.80
$^{9}Be^{19}F$	6.11	$^2\Sigma^+$	1247.4*	9.12	1.4889	0.0176	$8.28 \cdot 10^{-6}$	1.36
$^{9}Be^{1}H$	0.91	$^2\Sigma^+$	2060.8	36.31	10.316	0.3030	$1.02 \cdot 10^{-3}$	1.34
$^{9}Be^{2}H$	1.65	$^2\Sigma^+$	1530.3	20.71	5.6872	0.1225	$3.14 \cdot 10^{-4}$	1.34
$^{9}Be^{3}H$	2.26	$^2\Sigma^+$	1305	15	4.142	0.064	$1.67 \cdot 10^{-4}$*	1.34
$^{9}Be^{127}I$	8.41	$^2\Sigma^+$	611.7	1.6	0.422*	–	$0.82 \cdot 10^{-6}$*	2.18
$^{9}Be^{16}O$	5.76	$^1\Sigma^+$	1487.3	11.83	1.6510	0.0190	$0.82 \cdot 10^{-5}$	1.33
$^{9}Be^{32}S$	7.03	$^1\Sigma^+$	997.94	6.137	0.7906	0.0066	$2.00 \cdot 10^{-6}$	1.74
$^{11}B^{19}F$	6.97	$^1\Sigma^+$	1402.1	11.8	1.507*	0.0198	$0.76 \cdot 10^{-5}$*	1.26
$^{11}B^{1}H$	0.92	$^1\Sigma^+$	2366.9	49.39	12.021	0.412	$1.24 \cdot 10^{-3}$	1.23

Table 2.1. *Continued*

Molecule	μ [amu]	Ground state	ω_e [cm^{-1}]	$\omega_e x_e$ [cm^{-1}]	B_e [cm^{-1}]	α_e [cm^{-1}]	D_e [cm^{-1}]	r_e [Å]
$^{11}B^2H$	1.70	$^1\Sigma^+$	1703.2*	28	6.54	0.17	$0.4 \cdot 10^{-3}$	1.23
$^{209}Bi_2$	104.49	$^1\Sigma_g^+$	172.71	0.341	0.023*	0.0001	$0.15 \cdot 10^{-8*}$	2.66
$^{209}Bi^{79}Br$	57.29	0^+	209.5	0.45	0.0432	0.0001	$0.73 \cdot 10^{-8}$	2.61
$^{209}Bi^{35}Cl$	29.96	0^+	308.4	0.96	0.0921	–	$0.31 \cdot 10^{-7}$	2.47
$^{209}Bi^{19}F$	17.42	0^+	510.7	2.05	0.230	0.0015	–	2.05
$^{209}Bi^1H$	1.003	$^3\Sigma^-(0^+)$	1635.7*	31.6	5.137	0.148	$1.83 \cdot 10^{-4*}$	1.80
$^{209}Bi^2H$	1.995	$^3\Sigma^-(0^+)$	1173.3*	16.1	2.592	0.054	$0.51 \cdot 10^{-4*}$	1.80
$^{209}Bi^{127}I$	78.96	0^+	163.8	0.28	0.0272	0.0001	$0.03 \cdot 10^{-7}$	2.80
$^{209}Bi^{16}O$	14.86	$^2\Pi_{1/2}$	692.4	4.34	0.3034	0.0022	$2.21 \cdot 10^{-7*}$	1.93
$^{209}Bi^{32}S$	27.73	$^2\Pi_{1/2}$	408.71	1.46	0.113*	0.0005	$3.34 \cdot 10^{-8*}$	2.32
$^{11}B^{14}N$	6.16	$^3\Pi$	1514.6	12.3	1.666	0.025	$0.81 \cdot 10^{-5}$	1.28
$^{11}B^{16}O$	6.52	$^2\Sigma^+$	1885.7	11.81	1.782	0.0166	$6.32 \cdot 10^{-6}$	1.20
$^{79}Br_2$	39.46	$^1\Sigma_g^+$	325.32	1.077	0.0821	0.0003	$0.21 \cdot 10^{-7}$	2.28
$^{79}Br^{35}Cl$	24.23	$^1\Sigma^+$	444.3	1.843	0.1525	0.0008	$0.72 \cdot 10^{-7}$	2.14
$^{79}Br^{19}F$	15.31	$^1\Sigma^+$	670.7	4.054	0.3558	0.0026	$0.04 \cdot 10^{-5}$	1.76
$^{79}Br^{16}O$	13.30	$^2\Pi_{3/2}$	778	6.8	0.4296	0.0036	$0.05 \cdot 10^{-5}$	1.72
$^{11}B^{32}S$	8.19	$^2\Sigma^+$	1180,2	6.3	0.795	0.006	$0.14 \cdot 10^{-5*}$	1.61
$^{12}C_2$	6.00	$^1\Sigma_g^+$	1854.7	13.3	1.82	0.018	$0.69 \cdot 10^{-5}$	1.24
$^{40}Ca_2$	19.98	$^1\Sigma_g^+$	64.9	1.06	0.046	0.0007	$0.09 \cdot 10^{-6}$	4.28
$^{40}Ca^{35}Cl$	18.65	$^2\Sigma^+$	367.5*	1.3	0.152	0.0008	$0.01 \cdot 10^{-5}$	2.44
$^{40}Ca^{19}F$	12.88	$^2\Sigma^+$	581*	2.7	0.34	0.003	$0.45 \cdot 10^{-6}$	1.97
$^{40}Ca^1H$	0.983	$^2\Sigma^+$	1298.3	19.1	4.277	0.097	$1.84 \cdot 10^{-4}$	2.00
$^{40}Ca^2H$	1.917	$^2\Sigma^+$	910*	–	2.177*	0.04	$0.48 \cdot 10^{-4}$	2.00
$^{40}Ca^{16}O$	11.42	$^1\Sigma^+$	732	4.8	0.444	0.0034	$0.06 \cdot 10^{-5}$	1.82
$^{40}Ca^{32}S$	17.76	$^1\Sigma^+$	462.2	1.8	0.1767	0.0008	$0.01 \cdot 10^{-5}$	2.32
$^{12}C^{35}Cl$	8.93	$^2\Pi_{1/2}$	866.7*	6.2	0.694	0.0067	$0.19 \cdot 10^{-5*}$	1.64
$^{114}Cd^1H$	0.999	$^2\Sigma^+$	1337*	–	5.32*	–	$0.31 \cdot 10^{-3*}$	1.78
$^{114}Cd^2H$	1.979	$^2\Sigma^+$	–	–	2.70*	–	$0.08 \cdot 10^{-3*}$	1.77
$^{12}C^{19}F$	7.35	$^2\Pi_r$	1308	11.1	1.417	0.018	$0.06 \cdot 10^{-4}$	1.27
$^{12}C^1H$	0.930	$^2\Pi_r$	2858	63	14.46	0.53	$0.14 \cdot 10^{-2}$	1.12
$^{12}C^2H$	1.725	$^2\Pi_r$	2100	34.0	7.81	0.21	$0.42 \cdot 10^{-3}$	1.12
$^{35}Cl_2$	17.48	$^1\Sigma_g^+$	559	2.7	0.244	0.0015	$0.19 \cdot 10^{-6}$	1.99
$^{35}Cl^{19}F$	12.31	$^1\Sigma^+$	786.2	6.2	0.5165	0.0044	$0.08 \cdot 10^{-5}$	1.63
$^{35}Cl^{16}O$	10.97	$^2\Pi_i$	854	5.5	0.6234	0.006	$0.13 \cdot 10^{-5*}$	1.57
$^{12}C^{14}N$	6.46	$^2\Sigma^+$	2068.6	13.09	1.900	0.0174	$0.64 \cdot 10^{-5}$	1.17
$^{12}C^{16}O$	6.86	$^1\Sigma^+$	2169.8	13.288	1.9313	0.0175	$6.12 \cdot 10^{-6}$	1.13
$^{59}Co^1H$	0.991	$^3\Phi_4$	1890	–	7.15*	–	$0.40 \cdot 10^{-3}$	1.54
$^{59}Co^2H$	1.948	$^3\Phi_4$	1373.2	17.6	3.757	0.075	$1.12 \cdot 10^{-4*}$	1.52
$^{12}C^{31}P$	8.65	$^2\Sigma^+$	1239.7	6.9	0.799	0.0060	$0.13 \cdot 10^{-5}$	1.56
$^{52}Cr^1H$	0.989	$^6\Sigma^+$	1581*	32	6.22	0.18	$0.35 \cdot 10^{-3*}$	1.66
$^{52}Cr^{16}O$	12.23	$^5\Pi$	898	6.7	0.53	0.005	–	1.62
$^{12}C^{32}S$	8.73	$^1\Sigma^+$	1285.1	6.5	0.8200	0.0059	$0.14 \cdot 10^{-5}$	1.53
$^{133}Cs_2$	66.45	$^1\Sigma_g^+$	42.02	0.082	0.013	0.00003	$0.46 \cdot 10^{-8}$	4.47
$^{133}Cs^{79}Br$	49.52	$^1\Sigma^+$	149.7	0.37	0.0361	0.0001	$8.38 \cdot 10^{-9}$	3.07
$^{133}Cs^{35}Cl$	27.68	$^1\Sigma^+$	214.2	0.73	0.0721	0.0003	$3.27 \cdot 10^{-8}$	2.91
$^{133}Cs^{19}F$	16.62	$^1\Sigma^+$	352.6	1.6	0.1844	0.0012	$2.02 \cdot 10^{-7}$	2.35
$^{133}Cs^1H$	1.000	$^1\Sigma^+$	891	13	2.71	0.058	$1.13 \cdot 10^{-4*}$	2.49
$^{133}Cs^2H$	1.984	$^1\Sigma^+$	619*	–	1.35*	–	$0.20 \cdot 10^{-4*}$	2.50
$^{133}Cs^{127}I$	64.92	$^1\Sigma^+$	119.18	0.250	0.0236	0.00007	$3.71 \cdot 10^{-9}$	3.32

Table 2.1. *Continued*

Molecule	μ [amu]	Ground state	ω_e [cm^{-1}]	$\omega_e x_e$ [cm^{-1}]	B_e [cm^{-1}]	α_e [cm^{-1}]	D_e [cm^{-1}]	r_e [Å]
$^{63}Cu^2$	31.46	$^1\Sigma_g^+$	264.6*	1.02	0.1087	0.0006	$0.72 \cdot 10^{-7}$	2.22
$^{63}Cu^{79}Br$	35.01	$^1\Sigma^+$	315	1.0	0.1019	0.0004	$4.27 \cdot 10^{-8}$	2.17
$^{65}Cu^{35}Cl$	22.73	$^1\Sigma^+$	415.3	1.6	0.1763	0.001	$1.27 \cdot 10^{-7}$	2.05
$^{63}Cu^{19}F$	14.59	$^1\Sigma^+$	623	4.0	0.3794	0.0032	$0.56 \cdot 10^{-6}$	1.74
$^{63}Cu^1H$	0.992	$^1\Sigma^+$	1941.3	37.5	7.944	0.256	$0.52 \cdot 10^{-3}$	1.46
$^{63}Cu^2H$	1.952	$^1\Sigma^+$	1384.1	19.0	4.038	0.092	$1.36 \cdot 10^{-4}$*	1.46
$^{63}Cu^{127}I$	42.07	$^1\Sigma^+$	264	0.6	0.0733	0.0003	$2.24 \cdot 10^{-8}$	2.34
$^{63}Cu^{16}O$	12.75	$^2\Pi_{3/2}$	640.2	4.4	0.4445	0.0046	$0.08 \cdot 10^{-5}$	1.72
$^{63}Cu^{32}S$	21.20	$^2\Pi_i$	415	2	0.189*	–	$0.02 \cdot 10^{-5}$*	2.05
$^{63}Cu^{80}Se$	35.21	$^2\Pi_i$	302	1.0	0.108*	–	$0.06 \cdot 10^{-6}$*	2.11
$^{63}Cu^{130}Te$	42.39	$^2\Pi$	252.7	0.70	0.072	0.0002	$0.26 \cdot 10^{-7}$	2.35
$^{19}F_2$	9.499	$^1\Sigma_g^+$	916.6	11.24	0.8902	0.0138	$0.03 \cdot 10^{-4}$	1.41
$^{69}Ga^{81}Br$	37.22	$^1\Sigma^+$	263	0.8	0.0818	0.0003	$0.03 \cdot 10^{-6}$	2.35
$^{69}Ga^{35}Cl$	23.20	$^1\Sigma^+$	365	1.2	0.1499	0.0008	$0.10 \cdot 10^{-6}$	2.20
$^{69}Ga^{19}F$	14.89	$^1\Sigma^+$	622	3.2	0.3595	0.0029	$0.05 \cdot 10^{-5}$	1.77
$^{69}Ga^1H$	0.993	$^1\Sigma^+$	1604.5	28.8	6.14	0.18	$0.34 \cdot 10^{-3}$	1.66
$^{69}Ga^2H$	1.957	$^1\Sigma^+$	–	–	3.08*	0.06	$0.08 \cdot 10^{-3}$*	1.66
$^{69}Ga^{127}I$	44.67	$^1\Sigma^+$	217	0.5	0.0569	0.0002	$0.16 \cdot 10^{-7}$	2.57
$^{74}Ge^{19}F$	15.11	$^2\Pi_{1/2}$	665.7	3.15	0.3658	0.0027	$0.45 \cdot 10^{-6}$	1.74
$^{72}Ge^1H$	0.994	$^2\Pi_r$	1833.8*	37	6.73	0.19	$0.33 \cdot 10^{-3}$*	1.59
$^{72}Ga^2H$	1.959	$^2\Pi_r$	1320.1*	19	3.41	0.07	$0.83 \cdot 10^{-4}$*	1.59
$^{74}Ge^{16}O$	13.15	$^1\Sigma^+$	986	4.3	0.4857	0.0031	$0.47 \cdot 10^{-6}$	1.62
$^{74}Ge^{32}S$	22.32	$^1\Sigma^+$	576	1.8	0.1866	0.0007	$7.88 \cdot 10^{-8}$	2.01
$^{74}Ge^{80}Se$	38.40	$^1\Sigma^+$	409	1.4	0.0963	0.0003	$2.21 \cdot 10^{-8}$	2.13
$^{74}Ge^{130}Te$	47.11	$^1\Sigma^+$	324	0.8	0.0653	0.0002	$0.01 \cdot 10^{-6}$	2.34
$^1H_2$	0.504	$^1\Sigma_g^+$	4401.2	121.3	60.85	3.06	$0.47 \cdot 10^{-1}$	0.74
$^1H^2H$	0.672	$^1\Sigma_g^+$	3813	91.6	45.66	1.99	$0.26 \cdot 10^{-1}$	0.74
$^2H_2$	1.007	$^1\Sigma_g^+$	3115.5	61.8	30.44	1.079	$1.14 \cdot 10^{-2}$	0.74
$^3H_2$	1.508	$^1\Sigma_g^+$	2546	41.2	20.34	0.589	–	0.74
$^1H^{81}Br$	0.995	$^1\Sigma^+$	2649.0	45.22	8.465	0.2333	$3.46 \cdot 10^{-4}$	1.41
$^2H^{81}Br$	1.965	$^1\Sigma^+$	1884.8	22.7	4.246*	0.08	$0.88 \cdot 10^{-4}$*	1.41
$^1H^{35}Cl$	0.980	$^1\Sigma^+$	2990.9	52.82	10.593	0.3072	$5.32 \cdot 10^{-4}$	1.27
$^2H^{35}Cl$	1.904	$^1\Sigma^+$	2145.2	27.18	5.4488	0.1133	$0.14 \cdot 10^{-3}$	1.27
$^1H^{19}F$	0.957	$^1\Sigma^+$	4138.3	89.9	20.956	0.80	$2.15 \cdot 10^{-3}$	0.92
$^2H^{19}F$	1.821	$^1\Sigma^+$	2998.2	45.8	11.010	0.302	$0.59 \cdot 10^{-3}$	0.92
$^{180}Hf^{16}O$	14.69	$^1\Sigma^+$	974.1	3.23	0.3865	0.0017	$2.44 \cdot 10^{-7}$	1.72
$^1H^{127}I$	0.9999	$^1\Sigma^+$	2309.0	39.64	6.426*	0.169	$2.07 \cdot 10^{-4}$*	1.61
$^2H^{127}I$	1.983	$^1\Sigma^+$	1639.6	19.9	3.253*	0.0608	$5.26 \cdot 10^{-5}$*	1.61
$^{165}Ho^{19}F$	17.04	–	615.3	2.6	0.2630	0.0014	$0.18 \cdot 10^{-6}$*	1.94
$^{127}I_2$	63.45	$^1\Sigma_g^+$	214.5	0.61	0.0374	0.0001	$0.04 \cdot 10^{-7}$	2.67
$^{127}I^{79}Br$	48.66	$^1\Sigma^+$	268.6	0.81	0.0568	0.0002	$0.10 \cdot 10^{-7}$	2.47
$^{127}I^{35}Cl$	27.41	$^1\Sigma^+$	384.3	1.50	0.1142	0.0005	$0.40 \cdot 10^{-7}$	2.32
$^{127}I^{19}F$	16.52	$^1\Sigma^+$	610.2	3.12	0.2797	0.0019	$0.24 \cdot 10^{-6}$	1.91
$^{115}In^{81}Br$	47.48	$^1\Sigma^+$	221	0.6	0.0549	0.0002	$0.14 \cdot 10^{-7}$	2.54
$^{115}In^{35}Cl$	26.81	$^1\Sigma^+$	317	1.0	0.1091	0.0005	$0.05 \cdot 10^{-6}$	2.40
$^{115}In^{19}F$	16.30	$^1\Sigma^+$	535	2.6	0.2623	0.0019	$0.25 \cdot 10^{-6}$	1.98
$^{115}In^1H$	0.999	$^1\Sigma^+$	1476	25.6	4.99	0.14	$0.22 \cdot 10^{-3}$	1.84
$^{115}In^2H$	1.979	$^1\Sigma^+$	1048	12	2.52	0.05	$0.58 \cdot 10^{-4}$*	1.84
$^{115}In^{127}I$	60.30	$^1\Sigma^+$	177	0.4	0.0369	0.0001	$0.08 \cdot 10^{-7}$	2.75

Table 2.1. *Continued*

Molecule	μ [amu]	Ground state	ω_e [cm^{-1}]	$\omega_e x_e$ [cm^{-1}]	B_e [cm^{-1}]	α_e [cm^{-1}]	D_e [cm^{-1}]	r_e [Å]
$^{127}I^{16}O$	14.20	$^2\Pi_{3/2}$	681	4	0.3403	0.0027	$0.04 \cdot 10^{-5}$	1.87
$^{193}Ir^{12}C$	11.30	$^2\Delta_{5/2}$	1060	4	0.527	0.003	$0.05 \cdot 10^{-5}$	1.68
$^{39}K_2$	19.48	$^1\Sigma_g^+$	92.02	0.283	0.0567	0.0002	$0.86 \cdot 10^{-7}$	3.91
$^{39}K^{79}Br$	26.08	$^1\Sigma^+$	213	0.76	0.0812	0.0004	$4.46 \cdot 10^{-8}$	2.82
$^{39}K^{35}Cl$	18.43	$^1\Sigma^+$	281	1.17	0.1286	0.00008	$1.09 \cdot 10^{-7}$	2.67
$^{39}K^{19}F$	12.77	$^1\Sigma^+$	428	2.4	0.2799	0.0023	$4.83 \cdot 10^{-7}$	2.17
$^{39}K^1H$	0.982	$^1\Sigma^+$	983	14	3.41	0.08	$0.15 \cdot 10^{-3}$	2.24
$^{39}K^{127}I$	29.81	$^1\Sigma^+$	186.5	0.57	0.0609	0.0003	$2.59 \cdot 10^{-8}$	3.05
$^{84}Kr_2$	41.96	$^1\Sigma_g^+$	24	1.3	–	–	–	4
$^{139}La^{16}O$	14.34	$^2\Sigma^+$	812*	2.2	0.353	0.001	$0.02 \cdot 10^{-5}$	1.82
$^7Li_2$	3.51	$^1\Sigma_g^+$	351.4	2.61	0.6726	0.007	$0.99 \cdot 10^{-5}$	2.67
$^7Li^{79}Br$	6.44	$^1\Sigma^+$	563	3.5	0.5554	0.0056	$2.16 \cdot 10^{-6}$	2.17
$^7Li^{35}Cl$	5.84	$^1\Sigma^+$	643.3	4.5	0.7065	0.0080	$3.41 \cdot 10^{-6}$	2.02
$^7Li^{19}F$	5.12	$^1\Sigma^+$	910.3	7.93	1.3452	0.0203	$1.18 \cdot 10^{-5}$	1.56
$^7Li^1H$	0.881	$^1\Sigma^+$	1405.6	23.2	7.513	0.213	$8.62 \cdot 10^{-4}$	1.60
$^7Li^2H$	1.565	$^1\Sigma^+$	1054.8	12.9	4.24	0.10	$2.76 \cdot 10^{-4}$	1.59
$^7Li^{127}I$	6.65	$^1\Sigma^+$	498	3	0.4432	0.0041	$1.45 \cdot 10^{-6}$	2.39
$^7Li^{23}Na$	5.38	$^1\Sigma^+$	257	1.61	0.40	0.004	–	2.8
$^{175}Lu^{19}F$	17.14	$^1\Sigma^+$	611.8	2.5	0.2676	0.0016	$0.20 \cdot 10^{-6}$	1.92
$^7Lu^1H$	1.002	$^1\Sigma^+$	1559	22	4.602	0.10	$0.17 \cdot 10^{-3}$	1.91
$^{175}Lu^2H$	1.991	$^1\Sigma^+$	1075	10	2.32	0.036	$0.43 \cdot 10^{-4}$	1.91
$^{175}Lu^{16}O$	14.66	$^2\Sigma^+$	842	3	0.358*	0.002	$0.26 \cdot 10^{-6*}$	1.79
$^{24}Mg_2$	11.99	$^1\Sigma_g^+$	51.1	1.64	0.0929	0.0038	$0.12 \cdot 10^{-5}$	3.89
$^{24}Mg^{35}Cl$	14.23	$^2\Sigma^+$	462.1*	2.1	0.2450	0.002	$0.02 \cdot 10^{-5}$	2.20
$^{24}Mg^{19}F$	10.60	$^2\Sigma^+$	711.7*	4.9	0.5192	0.0047	$1.08 \cdot 10^{-6*}$	1.75
$^{24}Mg^1H$	0.967	$^2\Sigma^+$	1495.2	31.89	5.826	0.186	$0.34 \cdot 10^{-3}$	1.73
$^{24}Mg^2H$	1.858	$^2\Sigma^+$	1078	16	3.031	0.0629	$0.09 \cdot 10^{-3}$	1.73
$^{24}Mg^{16}O$	9.60	$^1\Sigma^+$	785	5	0.574	0.005	$0.12 \cdot 10^{-5}$	1.75
$^{24}Mg^{32}S$	13.70	$^1\Sigma^+$	528.7	2.70	0.2680	0.0018	$0.28 \cdot 10^{-6}$	2.14
$^{55}Mn^1H$	0.990	$^7\Sigma^+$	1548	29	5.684	0.157	$3.04 \cdot 10^{-4*}$	1.73
$^{55}Mn^2H$	1.943	$^7\Sigma^+$	1102	14	2.896	0.05	$0.80 \cdot 10^{-4*}$	1.73
$^{14}N_2$	7.00	$^1\Sigma_g^+$	2358.6	14.32	1.998	0.0173	$0.58 \cdot 10^{-5*}$	1.10
$^{23}Na_2$	11.49	$^1\Sigma_g^+$	159.12	0.725	0.1547	0.0009	$0.58 \cdot 10^{-6}$	3.08
$^{23}Na^{79}Br$	17.80	$^1\Sigma^+$	302	1.5	0.1512	0.0009	$1.55 \cdot 10^{-7}$	2.50
$^{23}Na^{35}Cl$	13.87	$^1\Sigma^+$	366	1.8	0.2181	0.0016	$3.12 \cdot 10^{-7}$	2.36
$^{23}Na^{19}F$	10.40	$^1\Sigma^+$	536	3.4	0.4369	0.0046	$0.12 \cdot 10^{-5}$	1.93
$^{23}Na^1H$	0.9655	$^1\Sigma^+$	1172	19.7	4.901	0.135	$0.33 \cdot 10^{-3}$	1.89
$^{23}Na^2H$	1.852	$^1\Sigma^+$	826*	–	2.558	0.052	$0.92 \cdot 10^{-4}$	1.89
$^{23}Na^{127}I$	19.46	$^1\Sigma^+$	258	1.0	0.1178	0.0006	$0.97 \cdot 10^{-7}$	2.71
$^{23}Na^{39}K$	14.46	$^1\Sigma^+$	124.13	0.51	0.090	0.0005	$0.08 \cdot 10^{-5}$	3.6
$^{93}Nb^{16}O$	13.65	$^4\Sigma^-$	989	3.8	0.432	0.002	$0.22 \cdot 10^{-6}$	1.69
$^{14}N^{79}Br$	11.89	$^3\Sigma^-(0^+)$	691.8	4.72	0.44	0.004	–	1.8
$^{14}N^{35}Cl$	9.999	$^3\Sigma^-$	827	5.1	0.647*	–	$0.18 \cdot 10^{-5*}$	1.61
$^{14}N^{19}F$	8.06	$^3\Sigma^-$	1141.4	9.0	1.206	0.0149	$0.54 \cdot 10^{-5}$	1.32
$^{14}N^1H$	0.940	$^3\Sigma^-$	3282	78	16.699	0.649	$1.71 \cdot 10^{-3*}$	1.04
$^{14}N^2H$	1.761	$^3\Sigma^-$	2398	42	8.791*	0.253	$4.90 \cdot 10^{-4*}$	1.04
$^{58}Ni^1H$	0.991	$^2\Delta_{5/2}$	1926*	38	7.70*	0.23	$0.48 \cdot 10^{-3*}$	1.48
$^{58}Ni^2H$	1.946	$^2\Delta_{5/2}$	1390*	19	3.99*	0.09	$0.13 \cdot 10^{-3*}$	1.46
$^{14}N^{16}O$	7.47	$^2\Pi_r$	1904.2	14.08	1.672*	0.017	$0.05 \cdot 10^{-5*}$	1.15

Table 2.1. *Continued*

Molecule	μ [amu]	Ground state	ω_e [cm^{-1}]	$\omega_e x_e$ [cm^{-1}]	B_e [cm^{-1}]	α_e [cm^{-1}]	D_e [cm^{-1}]	r_e [Å]
$^{14}N^{32}S$	9.74	$^2\Pi_r$	1218	7.3	0.770*	0.006	$0.01 \cdot 10^{-4}$	1.49
$^{14}N^{80}Se$	11.92	$^2\Pi_{1/2}$	956.8	5.6	0.518	0.004	$0.06 \cdot 10^{-5}$	1.65
$^{16}O_2$	7.997	$^3\Sigma_g^-$	1580.2	12.0	1.438*	0.016	$4.84 \cdot 10^{-6*}$	1.21
$^{16}O^1H$	0.948	$^2\Pi_i$	3737.8	84.88	18.91	0.724	$1.94 \cdot 10^{-3}$	0.97
$^{16}O^2H$	1.789	$^2\Pi_i$	2720.2	44.06	10.02	0.28	$5.37 \cdot 10^{-4*}$	0.97
$^{31}P_2$	15.49	$^1\Sigma_g^+$	780.8	2.8	0.3036	0.0015	$0.19 \cdot 10^{-6}$	1.89
$^{208}Pb^{19}F$	17.41	$^2\Pi_{1/2}$	502.7*	2.3	0.2288	0.0015	$0.18 \cdot 10^{-6}$	2.06
$^{208}Pb^1H$	1.003	$^2\Pi_{1/2}$	1564	29.8	4.97	0.14	$0.20 \cdot 10^{-3*}$	1.84
$^{208}Pb^{16}O$	14.85	$^1\Sigma^+$	721	3.5	0.3073	0.0019	$0.02 \cdot 10^{-5}$	1.92
$^{208}Pb^{32}S$	27.71	$^1\Sigma^+$	429.4	1.3	0.1163	0.0004	$3.42 \cdot 10^{-8}$	2.29
$^{208}Pb^{80}Se$	57.73	$^1\Sigma^+$	277	0.5	0.0506	0.0001	$0.07 \cdot 10^{-7}$	2.40
$^{208}Pb^{130}Te$	79.96	$^1\Sigma^+$	211.9	0.4	0.0313	0.0001	$0.03 \cdot 10^{-7}$	2.59
$^{108}Pd^2H$	1.977	$^2\Sigma^+$	1446.0	19.6	3.649	0.081	$0.93 \cdot 10^{-4}$	1.53
$^{31}P^{19}F$	11.78	$^3\Sigma^-$	846.8	4.49	0.566	0.0046	–	1.59
$^{31}P^1H$	0.976	$^3\Sigma^-$	2365	44.5	8.537	0.251	$0.44 \cdot 10^{-3}$	1.42
$^{31}P^2H$	1.891	$^3\Sigma^-$	1699	23	4.408	0.093	$0.12 \cdot 10^{-3}$	1.42
$^{31}P^{14}N$	9.64	$^1\Sigma^+$	1337.2	6.98	0.7865	0.0055	$1.09 \cdot 10^{-6}$	1.49
$^{31}P^{16}O$	10.55	$^2\Pi_r$	1233.3	6.6	0.734	0.006	$0.13 \cdot 10^{-5}$	1.48
$^{31}P^{32}S$	15.73	$^2\Pi_r$	739	3.0	0.296	–	$0.02 \cdot 10^{-5*}$	1.90
$^{195}Pt^{12}C$	11.30	$^1\Sigma^+$	1051.1	4.9	0.5304	0.0033	$0.55 \cdot 10^{-6}$	1.68
$^{195}Pt^1H$	1.003	$^2\Delta_{5/2}$	2294.7*	46	7.196	0.200	$0.26 \cdot 10^{-3*}$	1.53
$^{195}Pt^2H$	1.994	$^2\Delta_{5/2}$	1644*	23	3.64	0.07	$0.07 \cdot 10^{-3}$	1.52
$^{195}Pt^{16}O$	14.78	$^1\Sigma$	851.1	5.0	0.3822	0.0028	$0.03 \cdot 10^{-5}$	1.73
$^{85}Rb^{79}Br$	40.90	$^1\Sigma^+$	169.5	0.46	0.0475	0.0002	$1.50 \cdot 10^{-8}$	2.94
$^{85}Rb^{35}Cl$	24.77	$^1\Sigma^+$	228	0.9	0.0876	0.0005	$0.49 \cdot 10^{-7}$	2.79
$^{85}Rb^{19}F$	15.52	$^1\Sigma^+$	376	1.9	0.2107	0.0015	$2.68 \cdot 10^{-7}$	2.27
$^{85}Rb^1H$	0.996	$^1\Sigma^+$	936.9	14.2	3.02	0.07	$0.12 \cdot 10^{-3}$	2.37
$^{85}Rb^{127}I$	50.87	$^1\Sigma^+$	138.5	0.34	0.0328	0.0001	$7.38 \cdot 10^{-9}$	3.18
$^{103}Rh^{12}C$	10.75	$^2\Sigma^+$	1049.9	4.94	0.603	0.0040	$0.08 \cdot 10^{-5}$	1.61
$^{32}S_2$	15.99	$^3\Sigma_g^-$	725.6	2.84	0.295	0.0016	$0.19 \cdot 10^{-6}$	1.89
$^{121}Sb^{19}F$	16.42	$^3\Sigma^-$	605*	2.6	0.279	0.002	$0.23 \cdot 10^{-6}$	1.92
$^{121}Sb^{16}O$	14.13	$^2\Pi_r$	816	4.2	0.358	0.002	$0.03 \cdot 10^{-5}$	1.83
$^{121}Sb^{31}P$	24.66	$^1\Sigma^+$	500.1	1.63	0.141	0.0005	$0.04 \cdot 10^{-6}$	2.20
$^{45}Sc^{35}Cl$	19.67	$^1\Sigma^+$	447.4	1.8	0.172	0.001	$0.11 \cdot 10^{-6}$	2.23
$^{45}Sc^{19}F$	13.35	$^1\Sigma^+$	735.6	3.8	0.395	0.0027	–	1.79
$^{45}Sc^{16}O$	11.80	$^2\Sigma^+$	965	4.2	0.5134*	0.003	$0.58 \cdot 10^{-6*}$	1.67
$^{80}Se_2$	39.96	$^3\Sigma_g^-(0_g^+)$	385.30	0.964	0.0899	0.0003	$0.24 \cdot 10^{-7}$	2.17
$^{80}Se^{16}O$	13.33	$^3\Sigma^-(0^+)$	914.7	4.5	0.466	0.0032	$0.05 \cdot 10^{-5}$	1.65
$^{78}Se^{32}S$	22.67	$^3\Sigma^-(0^+)$	555.6	1.85	0.179	0.0008	$0.06 \cdot 10^{-6}$	2.04
$^{32}S^1H$	0.977	$^2\Pi_i$	2711	60	9.46*	0.27	$0.48 \cdot 10^{-3}$	1.34
$^{32}S^2H$	1.895	$^2\Pi_i$	1885	31	4.90*	0.10	$0.13 \cdot 10^{-3}$	1.34
$^{28}Si_2$	13.99	$^3\Sigma_g^-$	511.0	2.0	0.239	0.001	$0.02 \cdot 10^{-5}$	2.25
$^{28}Si^{35}Cl$	15.54	$^2\Pi_r$	535.6	2.2	0.256	0.002	$0.25 \cdot 10^{-6}$	2.06
$^{28}Si^{19}F$	11.31	$^2\Pi_r$	857.2	4.74	0.581	0.0049	$0.11 \cdot 10^{-5}$	1.60
$^{28}Si^1H$	0.973	$^2\Pi_r$	2041.8	35.5	7.500	0.219	$0.40 \cdot 10^{-3*}$	1.52
$^{28}Si^2H$	1.879	$^2\Pi_r$	1469.3	18.2	3.884	0.078	$1.05 \cdot 10^{-4*}$	1.52
$^{28}Si^{14}N$	9.33	$^2\Sigma^+$	1151	6.5	0.731	0.0056	$0.12 \cdot 10^{-5}$	1.57
$^{28}Si^{16}O$	10.18	$^1\Sigma^+$	1241.6	5.97	0.7268	0.0050	$0.98 \cdot 10^{-6}$	1.51
$^{28}Si^{32}S$	14.92	$^1\Sigma^+$	749.6	2.58	0.3035	0.0015	$0.20 \cdot 10^{-6}$	1.93

Table 2.1. *Continued*

Molecule	μ [amu]	Ground state	ω_e [cm^{-1}]	$\omega_e x_e$ [cm^{-1}]	B_e [cm^{-1}]	α_e [cm^{-1}]	D_e [cm^{-1}]	r_e [Å]
$^{28}Si^{80}Se$	20.72	$^1\Sigma^+$	580	1.8	0.1920	0.0008	$0.84 \cdot 10^{-7}$	2.06
$^{120}Sn^{35}Cl$	27.07	$^2\Pi_{1/2}$	351	1.1	0.112	0.0004	–	2.36
$^{118}Sn^{19}F$	16.36	$^2\Pi_{1/2}$	577*	2.7	0.273	0.0014	$0.03 \cdot 10^{-5*}$	1.94
$^{120}Sn^{1}H$	0.999	$^2\Pi_r$	–	–	5.315*	–	$2.08 \cdot 10^{-4*}$	1.78
$^{120}Sn^{2}H$	1.981	$^2\Pi_r$	1188*	–	2.695*	0.05	$0.53 \cdot 10^{-4*}$	1.78
$^{120}Sn^{16}O$	14.11	$^1\Sigma^+$	815*	3.7	0.3557	0.0021	$0.27 \cdot 10^{-6}$	1.83
$^{120}Sn^{32}S$	25.24	$^1\Sigma^+$	487.3	1.36	0.1369	0.0005	$0.42 \cdot 10^{-7}$	2.21
$^{120}Sn^{80}Se$	47.95	$^1\Sigma^+$	331	0.74	0.0650	0.0002	$0.11 \cdot 10^{-7}$	2.33
$^{32}S^{16}O$	10.66	$^3\Sigma^-$	1149	5.6	0.7208	0.0057	$1.13 \cdot 10^{-6*}$	1.48
$^{88}Sr^{19}F$	15.62	$^2\Sigma^+$	502.4	2.2	0.2505	0.0015	$0.25 \cdot 10^{-6}$	2.08
$^{88}Sr^{1}H$	0.996	$^2\Sigma^+$	1206	17	3.675	0.081	$0.14 \cdot 10^{-3}$	2.15
$^{88}Sr^{2}H$	1.969	$^2\Sigma^+$	841*	8.6	1.861	0.029	$0.35 \cdot 10^{-4}$	2.14
$^{88}Sr^{16}O$	13.53	$^1\Sigma^+$	653	4.0	0.3380	0.0022	$0.36 \cdot 10^{-6}$	1.92
$^{88}Sr^{32}S$	23.44	$^1\Sigma^+$	388.4	1.3	0.1207	0.0004	$0.48 \cdot 10^{-7}$	2.44
$^{181}Ta^{16}O$	14.70	$^2\Delta_{3/2}$	1028.7	3.5	0.4028	0.0018	$2.45 \cdot 10^{-7*}$	1.69
$^{130}Te_2$	64.95	$^3\Sigma_g^-(0_g^+)$	247.1	0.515	0.0397	0.0001	$0.04 \cdot 10^{-7*}$	2.56
$^{130}Te^{16}O$	14.24	0^+	797.1	4.0	0.355	0.0024	$0.27 \cdot 10^{-6}$	1.82
$^{130}Te^{32}S$	25.66	0^+	471.2	1.6	0.1322	0.0005	$0.04 \cdot 10^{-6}$	2.23
$^{232}Th^{16}O$	14.96	$^1\Sigma^+$	895.8	2.4	0.3326	0.0013	$1.83 \cdot 10^{-7*}$	1.84
$^{48}Ti^{16}O$	11.99	$^3\Delta_r$	1009.0	4.50	0.5354	0.0030	$0.60 \cdot 10^{-6}$	1.62
$^{48}Ti^{32}S$	19.18	$^3\Delta_1$	558.4*	2.0	0.2018	0.0009	$0.10 \cdot 10^{-6*}$	2.08
$^{205}Tl^{81}Br$	58.01	$^1\Sigma^+$	192.1	0.4	0.0424	0.0001	$0.83 \cdot 10^{-8}$	2.62
$^{205}Tl^{35}Cl$	29.87	$^1\Sigma^+$	284	0.8	0.0914	0.0004	$0.38 \cdot 10^{-7}$	2.48
$^{205}Tl^{19}F$	17.39	$^1\Sigma^+$	477.3	2.3	0.2232	0.0015	$1.95 \cdot 10^{-7}$	2.08
$^{205}Tl^{1}H$	1.003	$^1\Sigma^+$	1391	23	4.81	0.15	$0.25 \cdot 10^{-3*}$	1.87
$^{205}Tl^{2}H$	1.995	$^1\Sigma^+$	988	12.0	2.42	0.06	$0.06 \cdot 10^{-3}$	1.87
$^{51}V^{16}O$	12.17	$^4\Sigma^-$	1011	4.9	0.5483	0.0035	$0.06 \cdot 10^{-5}$	1.59
$^{174}Yb^{19}F$	17.13	$^2\Sigma^+$	501.9*	2.2	0.2414*	0.0015	$0.22 \cdot 10^{-6*}$	2.02
$^{174}Yb^{1}H$	1.002	$^2\Sigma^+$	1249.5	21.1	3.993	0.096	$1.62 \cdot 10^{-4}$	2.05
$^{174}Yb^{2}H$	1.991	$^2\Sigma^+$	887	10.6	2.0116	0.0342	$4.16 \cdot 10^{-5*}$	2.05
$^{89}Y^{19}F$	15.65	$^1\Sigma^+$	631.3*	2.5	0.2904	0.0016	$0.24 \cdot 10^{-6*}$	1.93
$^{89}Y^{16}O$	13.56	$^2\Sigma^+$	861	2.9	0.388*	0.0018	$0.32 \cdot 10^{-6*}$	1.79
$^{64}Zn^{1}H$	0.992	$^2\Sigma^+$	1608	55.1	6.679	0.250	$0.47 \cdot 10^{-3*}$	1.59
$^{64}Zn^{2}H$	1.953	$^2\Sigma^+$	1072*	28	3.35*	–	$0.12 \cdot 10^{-3*}$	1.61
$^{90}Zr^{16}O$	13.58	$^1\Sigma^+$	970*	4.9	0.4226*	0.0023	$0.32.10^{-6*}$	1.71

2.2 Fundamental Vibrational Frequencies of Polyatomic Molecules

Tables 2.2–4 list the fundamental vibrational frequencies (or more precisely, wavenumbers) of triatomic, four- and five-atomic molecules. The point symmetry groups to which the molecules belong are indicated in the second column. The numbering of the vibrations is in accordance with the international agreement [2.12]. The molecules in Tables 2.2–4 are arranged in alphabetical order of the atoms. Atom D is counted as H in order to place

Table 2.2. Fundamental vibrational frequencies of triatomic molecules

Molecule	Point group	ν_1 [cm^{-1}]	ν_2 [cm^{-1}]	ν_3 [cm^{-1}]
XY_2 Molecules				
BO_2	$D_{\infty h}$	1056	447	1278
BS_2	$D_{\infty h}$	510	120	1015
C_3	$D_{\infty h}$	1224	63	2040
CBr_2	C_{2v}	595	196	641
CCl_2	C_{2v}	721	333	748
CH_2	C_{2v}		963	
CD_2	C_{2v}		752	
CF_2	C_{2v}	1225	667	1114
ClF_2	C_{2v}	500	242	576
Cl_2O	C_{2v}	639	296	686
ClO_2	C_{2v}	945	445	1111
CNC	$D_{\infty h}$		321	1453
CO_2	$D_{\infty h}$	1333	667	2349
CS_2	$D_{\infty h}$	658	397	1535
F_2O	C_{2v}	928	461	831
GeH_2	C_{2v}	1887	920	1864
$GeCl_2$	C_{2v}	399	159	374
H_2O	C_{2v}	3657	1595	3756
D_2O	C_{2v}	2671	1178	2788
H_2S	C_{2v}	2615	1183	2626
D_2S	C_{2v}	1896	855	1999
H_2Se	C_{2v}	2345	1034	2358
D_2Se	C_{2v}	1630	745	1696
KrF_2	$D_{\infty h}$	449	233	590
NCN	$D_{\infty h}$	1197	423	1476
NF_2	C_{2v}	1075	573	942
NH_2	C_{2v}	3219	1497	3301
NO_2	C_{2v}	1318	750	1618
O_3	C_{2v}	1103	701	1042
PbF_2	C_{2v}	531	165	507
$PbCl_2$	C_{2v}	314	99	299
SCl_2	C_{2v}	525	208	535
SF_2	C_{2v}	838	357	813
$SiBr_2$	C_{2v}	403		400
$SiCl_2$	C_{2v}	515	248	505
SiF_2	C_{2v}	855	345	870
SiH_2	C_{2v}	2032	990	2022
SiD_2	C_{2v}	1472	729	1468
$SnBr_2$	C_{2v}	244	80	231
$SnCl_2$	C_{2v}	352	120	334
SnF_2	C_{2v}	593	197	571
SO_2	C_{2v}	1151	518	1362
$XeCl_2$	$D_{\infty h}$	316		481
XeF_2	$D_{\infty h}$	515	213	555

Table 2.2. *Continued*

Molecule	Point group	ν_1 [cm^{-1}]	ν_2 [cm^{-1}]	ν_3 [cm^{-1}]
XYZ Molecules				
BrBO	$C_{\infty v}$	535	374	1937
BrCN	$C_{\infty v}$	575	342	2198
BrNO	C_s	542	266	1799
BrOO	C_s			1487
CCN	$C_{\infty v}$	1060	230	1917
CCO	$C_{\infty v}$	1063	379	1967
ClBO	$C_{\infty v}$	676	404	1958
ClCN	$C_{\infty v}$	744	378	2216
ClNO	C_s	596	332	1800
ClOO	C_s	407	373	1443
FBO	$C_{\infty v}$		500	2075
FCN	$C_{\infty v}$	1077	451	2323
FNO	C_s	766	520	1844
FOO	C_s	579	376	1490
HBO	$C_{\infty v}$		754	1817
HCC	$C_{\infty v}$	3612		1848
HCCl	C_s		1201	815
HCF	C_s		1407	1181
HCN	C_s	3311	712	2097
DCN	C_s	2630	569	1925
HCO	C_s	2485	1081	1868
HNB	$C_{\infty v}$	3675		2035
HNC	$C_{\infty v}$	3653		2032
HNF	C_s		1419	1000
HNO	C_s	2684	1501	1565
HNSi	$C_{\infty v}$	3583	523	1198
HOCl	C_s	3609	1242	725
HOF	C_s	3537	886	1393
HOO	C_s	3436	1392	1098
HPO	C_s	2095	983	1179
HSO	C_s		1063	1009
HSiBr	C_s	1548	774	408
HSiCl	C_s		808	522
HSiF	C_s	1913	860	834
ICN	$C_{\infty v}$	486	305	2188
NCO	$C_{\infty v}$	1270	535	1921
NNO	$C_{\infty v}$	2224	589	1285
NSCl	C_s	1325	273	414
NSF	C_s	1372	366	640
OCS	$C_{\infty v}$	2062	520	859

Table 2.3. Fundamental vibrational frequencies of four-atomic molecules

Molecule	Point group	ν_1 [cm^{-1}]	ν_2 [cm^{-1}]	ν_3 [cm^{-1}]	ν_4 [cm^{-1}]	ν_5 [cm^{-1}]	ν_6 [cm^{-1}]
Symmetric XY_3 Molecules							
$AlCl_3$	C_{3v}	375	183	595	150		
$AsCl_3$	C_{3v}	410	193	370	159		
AsF_3	C_{3v}	741	337	702	262		
AsH_3	C_{3v}	2116	906	2123	1003		
AsD_3	C_{3v}	1523	660	1529	714		
AsI_3	C_{3v}	219	94	224	71		
BBr_3	D_{3h}	279	372	802	151		
BCl_3	D_{3h}	471	460	956	243		
BF_3	D_{3h}	888	691	1449	480		
BH_3	D_{3h}		1125	2808	1640		
$BiCl_3$	C_{3v}	288	130	242	96		
CH_3	D_{3h}		606	3161	1396		
CD_3	D_{3h}		453	2369	1029		
CF_3	C_{3v}	1090	701	1260	510		
NF_3	C_{3v}	1032	647	907	492		
NH_3	C_{3v}	3337	950	3444	1627		
ND_3	C_{3v}	2420	748	2564	1191		
PBr_3	C_{3v}	380	162	400	116		
PCl_3	C_{3v}	504	252	482	198		
PF_3	C_{3v}	892	487	860	344		
PH_3	C_{3v}	2323	992	2328	1118		
PD_3	C_{3v}	1694	730	1700	806		
PI_3	C_{3v}	303	111	325	79		
$SbCl_3$	C_{3v}	360	165	320	134		
SbH_3	C_{3v}	1891	782	1894	831		
SbD_3	C_{3v}	1359	561	1362	592		
SiF_3	C_{3v}	830	427	937	290		
SO_3	D_{3h}	1065	498	1391	530		
Linear $XYYX$ Molecules							
C_2H_2	$D_{\infty h}$	3374	1974	3289	612	730	
C_2D_2	$D_{\infty h}$	2701	1762	2439	505	537	
C_2I_2	$D_{\infty h}$	2113	191	718	307	115	
C_2N_2	$D_{\infty h}$	2330	851	2158	507	233	
Planar X_2YZ Molecules							
Cl_2CO	C_{2v}	567	1827	285	849	440	580
Cl_2CS	C_{2v}	505	1137	220	473	816	220
F_2CO	C_{2v}	965	1928	584	1249	626	774
H_2CO	C_{2v}	2783	1746	1500	2843	1249	1167
D_2CO	C_{2v}	2056	1700	1106	2160	990	938
O_2NCl	C_{2v}	1286	793	370	1685	408	652
O_2NF	C_{2v}	1310	822	568	1792	560	742

Table 2.4. Fundamental vibrational frequencies of five-atomic molecules

Molecule	Point group	ν_1 [cm^{-1}]	ν_2 [cm^{-1}]	ν_3 [cm^{-1}]	ν_4 [cm^{-1}]	ν_5 [cm^{-1}]	ν_6 [cm^{-1}]
CH_3D	C_{3v}	2944	2200	1300	3017	1471	1155
CHD_3	C_{3v}	2993	2142	1003	2263	1291	1036
CH_3Br	C_{3v}	2973	1306	611	3057	1443	955
CH_3Cl	C_{3v}	2968	1356	731	3040	1452	1017
CH_3F	C_{3v}	2965	1459	1049	3007	1468	1182
CH_3I	C_{3v}	2971	1251	533	3060	1438	883
$CHBr_3$	C_{3v}	3050	543	223	1149	669	155
$CHCl_3$	C_{3v}	3033	674	365	1220	774	259
CHF_3	C_{3v}	2991	1141	700	1378	1158	508
CHI_3	C_{3v}	3050	440	153	1067	580	110
CF_3Br	C_{3v}	1084	761	351	1209	548	303
CF_3Cl	C_{3v}	1108	782	475	1216	562	347
CF_3I	C_{3v}	1075	742	286	1187	536	258
$CFBr_3$	C_{3v}	1080	398	218	750	306	150
$CFCl_3$	C_{3v}	1080	536	350	846	395	243
SiH_3Br	C_{3v}	2200	930	431	2209	946	633
SiH_3Cl	C_{3v}	2201	949	551	2195	954	664
SiH_3F	C_{3v}	2206	991	875	2209	961	729
SiH_3I	C_{3v}	2192	903	362	2205	941	592
GeH_3Br	C_{3v}	2116	833	305	2127	871	578
GeH_3Cl	C_{3v}	2121	848	423	2129	874	602
GeH_3F	C_{3v}	2121	859	689	2132	874	643
GeH_3I	C_{3v}	2112	812	248	2121	854	558
Tetrahedral XY_4 Molecules							
CBr_4	T_d	267	122	672	182		
CCl_4	T_d	459	217	776	314		
CF_4	T_d	909	435	1281	632		
CH_4	T_d	2917	1534	3019	1306		
CD_4	T_d	2109	1092	2259	996		
CI_4	T_d	178	90	555	125		
$GeCl_4$	T_d	396	134	453	172		
GeH_4	T_d	2106	931	2114	819		
GeD_4	T_d	1504	665	1522	596		
$HfCl_4$	T_d	382	102	390	112		
OsO_4	T_d	965	333	960	329		
RuO_4	T_d	885	322	921	336		
$SiBr_4$	T_d	247	85	494	134		
$SiCl_4$	T_d	424	150	621	221		
SiF_4	T_d	800	268	1032	389		
SiH_4	T_d	2187	975	2191	914		
SiD_4	T_d	1558	700	1597	681		
$SnCl_4$	T_d	366	104	403	134		
$TiCl_4$	T_d	389	114	498	136		
$ZrCl_4$	T_d	377	98	418	113		

the deuterated molecules after the corresponding non-deuterated ones. Elements symbols without mass number refer to the most abundant isotope. The numerical values in the tables contain uncertainties in the last digits. The data presented are taken from [2.4, 13–15].

2.3 Rotational Constants and Geometrical Parameters of Polyatomic Molecules

The rotational energy of polyatomic molecules is given by

$$F_v(J) = B_v J(J+1) - D_v J^2(J+1)^2 + \cdots$$

for polyatomic linear and spherical top molecules,

$$F_v(J,K) = B_v J(J+1) - [A_v(C_v) - B_v]K^2 + \cdots$$

for polyatomic symmetric top molecules and

$$F_v(J) = 1/2(B_v + C_v)J(J+1) + [A_v - 1/2(B_v + C_v)]W_J + \cdots$$

for polyatomic asymmetric top molecules. Here, J, K are the rotational quantum numbers; $A_v = \hbar/(4\pi c I_A^v)$, $B_v = \hbar/(4\pi c I_B^v)$ and $C_v = \hbar/(4\pi c I_C^v)$ are the rotational constants, which refer to the vibrational level v; I_A^v, I_B^v and I_C^v are the principal momenta of inertia of the molecules; the values of the momenta are determined by geometrical parameters of the molecules in the vibrational state v. The three principal momenta of inertia are defined in such a way that

$$I_A^v \leqslant I_B^v \leqslant I_C^v$$

and consequently

$$A_v \geqslant B_v \geqslant C_v \; .$$

A linear molecule has two equal rotational constants: $B_v = C_v$. The I_A values are considered to be zero. A spherical top molecule has three equal rotational constants. They are usually denoted by B. The symmetric top molecule has by definition two (of three) equal rotational constants: $A_v > B_v = C_v$ for a prolate symmetric top and $A_v = B_v > C_v$ for an oblate symmetric top molecule. The rotational energy of a symmetric top molecule is determined by two independent constants, but only one rotational constant B can be obtained from the pure rotational spectra because of selection rules. An asymmetric top molecules has, in general, three independent rotational constants, but only two constants are independent for the planar molecules. The significance of geometrical parameters of molecules and the experimental methods for their determination are described in [2.16].

The rotational constants A_0, B_0, C_0 $[\mathrm{cm}]^{-1}$ for selected triatomic, four-, five- and six-atomix molecules are given in Tables 2.5–9. The constants refer to the lowest vibrational level. The geometrical parameters of these molecules in the lowest vibrational level: internuclear distances r_0 [Å] and bond angles (in degrees) also given in Tables 2.5–9. The order of the molecules arrangement and the accuracy of the numerical values listed are the same as in Tables 2.2–4. The data presented in Tables 2.5–9 are taken from [2.4, 13, 14, 16–23].

Table 2.5. Geometrical parameters and rotational constants of triatomic linear molecules

Molecule	Point group	B_0 [cm^{-1}]	Geometrical parameters
BO_2	$D_{\infty h}$	0.328	r_0 (BO) = 1.265
Br^{79} CN	$C_{\infty v}$	0.1374	r_0(CBr) = 1.790; r_0(CN) = 1.159
Br^{81} CN	$C_{\infty v}$	0.1367	r_0(CBr) = 1.790; r_0(CN) = 1.159
C_3	$D_{\infty h}$	0.424	r_0(CC) = 1.277;
Cl^{35} CN	$C_{\infty v}$	0.1992	r_0(CCl) = 1.631; r_0(CN) = 1.159
Cl^{37} CN	$C_{\infty v}$	0.1950	r_0(CCl) = 1.631; r_0(CN) = 1.159
CNC	$D_{\infty h}$	0.454	r_0(NC) = 1.245
CO_2	$D_{\infty h}$	0.390	r_0(CO) = 1.162
CS_2	$D_{\infty h}$	0.109	r_0(CS) = 1.554
FCN^{14}	$C_{\infty v}$	0.3520	r_0(CF) = 1.262; r_0(CN) = 1.159
FCN^{15}	$C_{\infty v}$	0.3398	r_0(CF) = 1.262; r_0(CN) = 1.159
HCN	$C_{\infty v}$	1.4782	r_0(CH) = 1.064; r_0(CN) = 1.156
DCN	$C_{\infty v}$	1.2078	r_0(CD) = 1.064; r_0(CN) = 1.156
HCP	$C_{\infty v}$	0.6663	r_0(CH) = 1.067; r_0(CP) = 1.542
DCP	$C_{\infty v}$	0.5665	r_0(CD) = 1.067; r_0(CP) = 1.542
ICN	$C_{\infty v}$	0.1076	r_0(CI) = 1.995; r_0(CN) = 1.159
KOH	$C_{\infty h}$	0.2738	r_0(OH) = 0.910; r_0(KO) = 2.212
KrF_2	$D_{\infty h}$	0.1263	r_0(KrF) = 1.89
N_2^{14} O	$C_{\infty v}$	0.4190	r_0(NN) = 1.126; r_0(NO) = 1.191
N_2^{15} O	$C_{\infty v}$	0.4049	r_0(NN) = 1.126; r_0(NO) = 1.191
NCN	$D_{\infty h}$	0.3968	r_0(CN) = 1.232
O^{16} C^{12} S^{32}	$C_{\infty v}$	0.2029	r_0(CO) = 1.154; r_0(CS) = 1.563
O^{16} C^{12} S^{36}	$C_{\infty v}$	0.1935	r_0(CO) = 1.154; r_0(CS) = 1.563
O^{16} C^{14} S^{32}	$C_{\infty v}$	0.2016	r_0(CO) = 1.154; r_0(CS) = 1.563
O^{18} C^{12} S^{32}	$C_{\infty v}$	0.1903	r_0(CO) = 1.154; r_0(CS) = 1.563
$OCSe^{74}$	$C_{\infty v}$	0.1366	r_0(CO) = 1.159; r_0(CSe) = 1.709
$OCSe^{79}$	$C_{\infty v}$	0.1344	r_0(CO) = 1.159; r_0(CSe) = 1.709
$OCSe^{82}$	$C_{\infty v}$	0.1332	r_0(CO) = 1.159; r_0(CSe) = 1.709
Te^{122} CS	$C_{\infty v}$	0.0528	r_0(TeC) = 1.904; r_0(CS) = 1.557
Te^{128} CS	$C_{\infty v}$	0.0522	r_0(TeC) = 1.904; r_0(CS) = 1.557
Te^{130} CS	$C_{\infty v}$	0.0520	r_0(TeC) = 1.904; r_0(CS) = 1.557
XeF_2	$D_{\infty h}$	0.1135	r_0(XeF) = 1.977

Table 2.6. Geometrical parameters and rotational constants of triatomic asymmetric top molecules

Molecule	Point group	A_0 [cm^{-1}]	B_0 [cm^{-1}]	C_0 [cm^{-1}]	Geometrical parameters
BH_2	C_{2v}	41.64	7.24	6.00	r_0(BH) = 1.18; ∠HBH = 131
CF_2	C_{2v}	2.947	0.417	0.365	r_0(CF) = 1.30; ∠FCF = 104.9
ClO_2	C_{2v}	1.737	0.332	0.278	r_0(ClO) = 1.48; ∠OClO = 117.5
F_2O	C_{2v}	1.961	0.363	0.306	r_0(OF) = 1.41; ∠FOF= 103.1
HCCl	C_s	15.75	0.605	0.588	r_0(CH) = 1.12; r_0(CCl) = 1.69; ∠HCCl = 103
HCF	C_s	15.5	1.163	1.106	r_0(CH) = 1.12; r_0(CF) = 1.31; ∠HCF = 101.8
HCO	C_s	24.29	1.494	1.399	r_0(CH) = 1.12; r_0(CO) = 1.18; ∠HCO = 124
DCO	C_s	13.64	1.281	1.171	r_0(CD) = 1.12; r_0(CO) = 1.18; ∠DCO = 124
HNO	C_s	18.479	1.412	1.307	r_0(NH) = 1.06; r_0(NO) = 1.21; ∠HNO = 108.6
DNO	C_s	10.522	1.292	1.146	r_0(ND) = 1.06; r_0(NO) = 1.21; ∠DNO = 108.6
HOCl	C_s	20.46	0.506	0.492	r_0(OH) = 0.97; r_0(ClO) = 1.69; ∠HOCl = 104.8
HOF	C_s	19.68	0.893	0.851	r_0(OH) = 0.96; r_0(FO) = 1.44; ∠HOF = 97.2
HOO	C_s	20.358	1.118	1.057	r_0(OH) = 0.98; r_0(OO) = 1.34; ∠HOO = 104.1
HPO	C_s	8.855	0.702	0.649	r_0(PH) = 1.43; r_0(PO) = 1.51; ∠HPO = 104.7
H_2O	C_{2v}	27.878	14.509	9.287	r_0(OH) = 0.96; ∠HOH = 105
D_2O	C_{2v}	15.385	7.272	4.846	r_0(OD) = 0.96; ∠DOD = 105
H_2S	C_{2v}	10.360	9.016	4.732	r_0(HS) = 1.34; ∠HSH = 92.1
H_2Se^{80}	C_{2v}	8.170	7.727	3.901	r_e(HSe) = 1.46; $∠_e$HSeH = 90.9
H_2Te	C_{2v}	6.248	6.097	3.036	r_0(HTe) = 1.65; ∠HTeH = 90.2
NF_2	C_{2v}	2.351	0.396	0.338	r_0(NF) = 1.35; ∠FNF = 103.3
NH_2	C_{2v}	23.72	12.94	8.16	r_0(NH) = 1.02; ∠HNH = 103.4
$NOBr^{79}$	C_s	2.780	0.125	0.120	r_0(NBr) = 2.14; r_0(NO) = 1.15; ∠BrNO = 114
NOCl	C_s	2.849	0.191	0.179	r_0(NCl) = 1.98; r_0(NO) = 1.14; ∠ClNO = 113.3
NOF	C_s	3.175	0.395	0.351	r_0(NO) = 1.13; r_0(NF) = 1.52; ∠ONF = 110
NO_2	C_{2v}	8.003	0.434	0.410	r_0(N) = 1.19; ∠ONO = 134.1
$NS^{32}F$	C_s	1.658	0.291	0.247	r_0(SF) = 1.65; r_0(SN) = 1.45; ∠NSF = 116.9
$NS^{34}F$	C_s	1.611	0.290	0.245	r_0(SF) = 1.65; r_0(SN) = 1.45; ∠NSF = 116.9
O_3	C_{2v}	3.553	0.445	0.395	r_0(O′O) = 1.28; ∠OO′O = 116.8
SF_2	C_{2v}	0.838	0.307	0.228	r_0(SF) = 1.59; ∠FSF = 98.3
SO_2	C_{2v}	2.027	0.344	0.294	r_0(SO) = 1.43; ∠OSO = 119.5
S_2O	C_s	1.398	0.169	0.150	r_0(SO) = 1.46; r_0(SS) = 1.88; ∠SSO = 118
SiF_2	C_{2v}	1.021	0.294	0.228	r_0(SiF) = 1.59; ∠FSiF = 101.0
SiH_2	C_{2v}	8.096	7.021	3.700	r_0(SiH) = 1.52; ∠HSiH = 92.1

Table 2.7. Geometrical parameters and rotational constants of four-atomic molecules

Molecule	Point group	A_0 [cm^{-1}]	B_0 [cm^{-1}]	C_0 [cm^{-1}]	Geometrical parameters
Linear molecules					
C_2H_2	$D_{\infty h}$	....	1.177	....	r_0(CC) = 1.21; r_0(CH) = 1.06
C_2D_2	$D_{\infty h}$	....	0.848	....	r_0(CC) = 1.21; r_0(CD) = 1.06
C_2N_2	$D_{\infty h}$	....	0.157	....	r_0(CC) = 1.39; r_0(CN) = 1.16
HC_2 Cl	$C_{\infty v}$	....	0.190	....	r_0(CH) = 1.06; r_0(CC) = 1.20; r_0(CCl)= 1.64
DC_2 Cl	$C_{\infty v}$	....	0.173	....	r_0(CD) = 1.06; r_0(CC) = 1.20; r_0(CCl) = 1.64
FCCH	$C_{\infty v}$	....	0.324	....	r_0(CH) = 1.05; r_0(CC) = 1.20; r_0(CF) = 1.28
Symmetric top molecules					
BF_3	D_{3h}	0.345	0.345	0.176	r_0(BF) = 1.31
NF_3	C_{3v}	0.356	0.356	0.195	r_0(NF) = 1.37; ∠FNF = 102.1
NH_3	C_{3v}	9.944	9.944	6.196	r_0(NH) = 1.02; ∠HNH = 107.8
ND_3	C_{3v}	5.142	5.142	3.117	r_0(ND) = 1.02; ∠DND = 107.8
Asymmetric top molecules					
CCl_2O	C_{2v}	0.264	0.116	0.080	r_0(CO) = 1.18; r_0(CCl) = 1.74; ∠ClCCl = 111.9
CF_2O^{16}	C_{2v}	0.394	0.392	0.196	r_0(CF) = 1.32; r_0(CO) = 1.17; ∠FCF = 108.0
CF_2O^{18}	C_{2v}	0.394	0.363	0.188	r_0(CF) = 1.32; r_0(CO) = 1.17; ∠FCF = 108.0
CHFO	C_s	3.041	0.392	0.347	r_0(CF) = 1.34; r_0(CO) = 1.18; r_0(CH) = 1.09; ∠FCO = 122.8; ∠HCO = 127.3;
CDFO	C_s	2.171	0.392	0.332	r_0(CF) = 1.34; r_0(CO) = 1.18; r_0(CD) = 1.09; ∠FCO = 122.8; ∠DCO = 127.3
CH_2O	C_{2v}	9.405	1.295	1.134	r_0(CH) = 1.12; r_0(CO) = 1.21; ∠HCH = 116.5
ClF_3	C_{2v}	0.459	0.154	0.115	ClF_2F': r_0(ClF) = 1.70; ∠FClF = 175.0; r_0(ClF′) = 1.60; ∠F′ClF = 87.5
HN_3	C_s	20.38	0.401	0.393	HN′N″N‴: r_0(N′H) = 1.0; ∠N′N″N‴ = 180; r_0(N′N″) = 1.24; r_0(N″N‴) = 1.13; ∠HN′N″ = 114.1

Table 2.8. Geometrical parameters and rotational constants of five-atomic molecules

Molecule	Point group	A_0 [cm^{-1}]	B_0 [cm^{-1}]	C_0 [cm^{-1}]	Geometrical parameters
Linear molecules					
C_3O_2	$D_{\infty h}$	....	0.076	....	$r_0(CO) = 1.16$; $r_0(CC) = 1.29$
HC_3N	$C_{\infty v}$	....	0.152	....	$r_0(CH) = 1.06$; $r_0(C{\equiv}C) = 1.20$; $r_0(C{-}C) = 1.38$; $r_0(CN) = 1.16$
DC_3N	$C_{\infty v}$	....	0.141	....	$r_0(CD) = 1.06$; $r_0(C{\equiv}C) = 1.20$; $r_0(C{-}C) = 1.38$; $r_0(CN) = 1.16$
Symmetric and spherical top molecules					
CF_3I	C_{3v}	0.191	0.051	0.051	$r_0(CF) = 1.33$; $r_0(CI) = 2.14$; $\angle FCF = 108$
$CHCl_3$	C_{3v}	0.110	0.110	0.056	$r_0(CH) = 1.09$; $r_0(CCl) = 1.76$; $\angle ClCCl = 111.3$
CHF_3	C_{3v}	0.345	0.345	0.189	$r_0(CH) = 1.10$; $r_0(CF) = 1.33$; $\angle FCF = 108.8$
CH_3Br	C_{3v}	5.18	0.319	0.319	$r_0(CBr) = 1.94$; $r_0(CH) = 1.10$; $\angle HCH = 111.2$
CH_3Cl	C_{3v}	5.20	0.443	0.443	$r_0(CCl) = 1.78$; $r_0(CH) = 1.11$; $\angle HCH = 110.8$
CH_3F	C_{3v}	5.18	0.852	0.852	$r_0(CH) = 1.10$; $r_0(CF) = 1.38$; $\angle HCH = 110.0$
CH_3I	C_{3v}	5.17	0.250	0.250	$r_0(CH) = 1.09$; $r_0(CI) = 2.14$; $\angle HCH = 111.5$
CH_4	T_d	....	5.241	....	$r_0(CH) = 1.09$
CD_4	T_d	....	2.633	....	$r_0(CD) = 1.09$
GeH_4	T_d	....	2.70	....	$r_0(GeH) = 1.52$
GeD_4	T_d	....	1.35	....	$r_0(GeD) = 1.52$
$Ge^{74}H_3Cl^{35}$	C_{3v}	2.603	0.145	0.145	$r_0(GeH) = 1.54$; $r_0(GeCl) = 2.15$; $\angle HGeH = 111.0$
$Ge^{74}H_3Cl^{37}$	C_{3v}	2.603	0.139	0.139	$r_0(GeH) = 1.54$; $r_0(GeCl) = 2.15$; $\angle HGeH = 111.0$
$POCl_3^{35}$	C_{3v}	....	0.067	....	$r_0(PCl) = 1.99$; $r_0(PO) = 1.45$; $\angle ClPCl = 103.3$
$POCl_3^{37}$	C_{3v}	....	0.064	....	$r_0(PCl) = 1.99$; $r_0(PO) = 1.45$; $\angle ClPCl = 103.3$
$PO^{16}F_3$	C_{3v}	....	0.153	....	$r_0(PF) = 1.52$; $r_0(PO) = 1.44$; $\angle FPF = 101.3$
$PO^{18}F_3$	C_{3v}	....	0.147	....	$r_0(PF) = 1.52$; $r_0(PO) = 1.44$; $\angle FPF = 101.3$
$PS^{32}F_3$	C_{3v}	....	0.089	....	$r_0(PF) = 1.53$; $r_0(PS) = 1.87$; $\angle FPF = 100.3$
$PS^{34}F_3$	C_{3v}	....	0.086	....	$r_0(PF) = 1.53$; $r_0(PS) = 1.87$; $\angle FPF = 100.3$
SiF_3Br^{79}	C_{3v}	....	0.052	....	$r_0(SiF) = 1.56$; $r_0(SiBr) = 2.15$; $\angle FSiF = 108.5$
SiF_3Cl^{35}	C_{3v}	....	0.083	....	$r_0(SiF) = 1.56$; $r_0(SiCl) = 1.99$; $\angle FSiF = 108.5$
$SiHCl_3^{35}$	C_{3v}	....	0.082	....	$r_0(SiH) = 1.47$; $r_0(SiCl) = 2.02$; $\angle ClSiCl = 109.4$
$Si^{28}HF_3$	C_{3v}	....	0.240	....	$r_e(SiF) = 1.56$; $r_e(SiH) = 1.45$; $\angle_e FSiF = 108.3$

Table 2.8. *Continued*

Molecule	Point group	A_0 [cm^{-1}]	B_0 [cm^{-1}]	C_0 [cm^{-1}]	Geometrical parameters
SiH_3Br	C_{3v}	....	0.144	....	r_0(SiH) = 1.48; r_0(SiBr) = 2.21; ∠HSiBr = 107.8
SiH_3Cl	C_{3v}	....	0.223	....	r_0(SiCl) = 2.05; r_0(SiH) = 1.48; ∠HSiCl = 107.9
SiH_3F	C_{3v}	....	0.478	....	r_0(SiH) = 1.48; r_0(SiF) = 1.59; ∠HSiH = 110.6
SiH_3I	C_{3v}	....	0.107	....	r_0(SiH) = 1.48; r_0(SiI) = 2.44; ∠HSiI = 107.8
$Si^{28}H_4$	T_d	....	2.864	....	r_0(SiH) = 1.48
SnH_4	T_d	....	2.16	....	r_0(SnH) = 1.70
Asymmetric top molecules					
CH_2CO	C_{2v}	9.37	0.343	0.331	r_0(CH) = 1.08; r_0(CO) = 1.16; r_0(CC) = 1.31; ∠HCH = 122
CH_2Cl_2	C_{2v}	1.067	0.111	0.102	r_e(CH) = 1.09; r_e(CCl) = 1.76; $∠_e$HCH = 111.5; $∠_e$ClCCl = 112.0
CD_2Cl_2	C_{2v}	0.790	0.110	0.100	r_e(CD) = 1.09; r_e(CCl) = 1.76; $∠_e$DCD = 111.5; $∠_e$ClCCl = 112.0
CH_2F_2	C_{2v}	1.639	0.354	0.308	r_0(CH) ≐ 1.09; r_0(CF) = 1.36; ∠HCH = 113.7; ∠FCF = 108.3
CH_2N_2	C_{2v}	9.112	0.377	0.362	r_0(NN) = 1.12; r_0(CN) = 1.32; r_0(CH) = 1.08; ∠HCH = 126
HCO_2H	C_s	2.585	0.402	0.347	r_0(CH) = 1.10; r_0(CO) = 1.23, 1.32; r_0(OH) = 0.97; ∠OCO = 124.9; ∠HCO = 124.5; ∠COH = 106.8
DCO_2H	C_s	1.925	0.402	0.332	r_0(CD) = 1.10; r_0(CO) = 1.23, 1.32; r_0(OH) = 0.97; ∠OCO = 124.9; ∠DCO = 124.5; ∠COH = 106.8
HCO_2D	C_s	2.205	0.392	0.333	r_0(CH) = 1.10; r_0(CO) = 1.23, 1.32; r_0(OD) = 0.97; ∠OCO = 124.9; ∠HCO = 124.5; ∠COD = 106.8
HNO_3	C_s	0.434	0.404	0.209	r_0(NO) = 1.20, 1.21, 1.41; r_0(OH) = 0.96; ∠NOH = 102.2; ∠ONO = 115.9, 113.9
DNO_3	C_s	0.433	0.377	0.201	r_0(NO) = 1.20, 1.21, 1.41; r_0(OD) = 0.96; ∠NOD = 102.2; ∠ONO = 115.9, 113.9
SO_2F_2	C_{2v}	0.171	0.169	0.169	r_0(SO) = 1.40; r_0(SF) = 1.53; ∠OSO = 123.0; ∠FSF = 97.0
SiH_2F_2	C_{2v}	0.824	0.260	0.213	r_0(SiH) = 1.47; r_0(SiF) = 1.58; ∠FSiF = 107.9
SiD_2F_2	C_{2v}	0.630	0.248	0.204	r_0(SiD) = 1.47; r_0(SiF) = 1.58; ∠FSiF = 107.9

Table 2.9. Geometrical parameters and rotational constants of six-atomic molecules

Molecule	Point group	A_0 [cm^{-1}]	B_0 [cm^{-1}]	C_0 [cm^{-1}]	Geometrical parameters
Linear molecules					
C_4H_2	$D_{\infty h}$	....	0.147	....	r_0(C–C) = 1.38; r_0(C≡C) = 1.20; r_0(CH) = 1.05
C_4D_2	$D_{\infty h}$	....	0.128	....	r_0(C–C) = 1.38; r_0(C≡C) = 1.20; r_0(CD) = 1.05
Symmetric top molecules					
CH_3HgCl^{35}	C_{3v}	....	0.069	....	r_0(CH) = 1.15; r_0(CHg) = 1.99; r_0(HgCl) = 2.28; ∠HCH = 109.2
CH_3HgCl^{37}	C_{3v}	....	0.067	....	r_0(CH) = 1.15; r_0(CHg) = 1.99; r_0(HgCl) = 2.28; ∠HCH = 109.2
CH_3NC^{12}	C_{3v}	....	0.335	....	r_0(CH) = 1.09; r_0(C–N) = 1.43; r_0(N=C) = 1.17; ∠NCH = 109.12
CH_3NC^{13}	C_{3v}	....	0.323	....	r_0(CH) = 1.09; r_0(C–N) = 1.43; r_0(N=C) = 1.17; ∠NCH = 109.12
CD_3NC^{12}	C_{3v}	....	0.286	....	r_0(CD) = 1.09; r_0(C–N) = 1.43; r_0(N=C) = 1.17; ∠NCD = 109.12
Asymmetric top molecules					
CH_2CFCl	C_s	0.356	0.170	0.115	r_0(CC) = 1.31; r_0(CH) = 1.07; r_0(CF) = 1.32; r_0(CCl) = 1.73; ∠CCF = 124.1; ∠CCCl = 123.0; ∠CCH = 121.0
CH_2CF_2	C_{2v}	0.367	0.348	0.178	r_0(CH) = 1.09; r_0(CF) = 1.32; r_0(CC) = 1.34; ∠CCF = 124.7 ∠CCH = 119.0
CD_2CF_2	C_{2v}	0.353	0.300	0.162	r_0(CD) = 1.09; r_0(CF) = 1.32; r_0(CC) = 1.34; ∠CCF = 124.7; ∠CCD = 119.0
CH_2CHCl	C_s	1.896	0.201	0.182	r_0(CH) = 1.09; r_0(CH′) = 1.09; r_0(CC) = 1.34; r_0(CCl) = 1.73; ∠CCH = 124.0, 120.0, 121.1; ∠CCCl = 122.5
CH_2CHF	C_s	2.154	0.355	0.304	r_0(CC) = 1.33; r_0(CF) = 1.35; r_0(CH) = 1.08; r_0(CH′) = 1.09; ∠CCH = 111.3, 121.4, 123.9; ∠FCC = 121.0
C_2H_4	V_h	4.865	1.001	0.828	r_0(CH) = 1.09; r_0(C=C) = 1.34; ∠CCH = 121.3
C_2D_4	V_h	2.432	0.737	0.563	r_0(CD) = 1.09; r_0(C=C) = 1.34; ∠CCD = 121.3
HC_2CHO	C_s	2.269	0.161	0.150	r_0(C=C) = 1.22; r_0(C≡C) = 1.21; r_0(C–C) = 1.44; r_0(CH) = 1.11, 1.06; ∠CCO = 123.7; ∠CCC = 178.4
DC_2CDO	C_s	1.704	0.148	0.136	r_0(C=C) = 1.22; r_0(C≡C) = 1.21; r_0(C–C) = 1.44; r_0(CD) = 1.11, 1.06; ∠CCO = 123.7; ∠CCC = 178.4

3 Energy Constants of Molecules

In this chapter the bond dissociation energies, ionization potentials, electron and proton affinities of diatomic and polyatomic molecules are given.

3.1 Bond Dissociation Energies of Molecules

The bond dissociation energy $D_T^0(R–X)$ or the strength of a chemical bond (R–X) is defined as the heat of reaction which leads to bond breaking:

$$RX \to R + X \ ,$$

and is given by

$$D_T^0(R–X) = \Delta H_{fT}^0(R) + \Delta H_{fT}^0(X) - \Delta H_{fT}^0(RX) \ ,$$

where $\Delta H_{fT}^0(R)$, $\Delta H_{fT}^0(X)$ and $\Delta H_{fT}^0(R\ X)$ are the enthalpies of formation of the molecules R, X and RX at the temperature T from the elements in their standard states.

Usually, the bond dissociation energies are tabulated at two temperatures: 0 K and 298 K. The exact conversion of D_0^0 to D_{298}^0 should be made by use of enthalpy values taken from thermochemical tables. The conversion can be performed by the following approximate relation:

$$D_{298}^0 \approx D_0^0 + (3/2)RT \ .$$

Table 3.1 lists the dissociation energies D_{298}^0 [kcal/mol] of diatomic molecules at a temperature of 298 K. The molecules in the table are arranged in alphabetical order of the constituent atoms. The bond dissociation energies D_{298}^0 for polyatomic molecules are given in Table 3.2. The molecules are arranged in accordance with the breaking bonds, as accepted in the chemical literature. The order of the breaking-bonds arrangement is as follows: H–C, H–Si, H–Ge, H–Sn, H–N, H–O, H–S, C–C, C–B, C–Si, C–N, C–O, C–S, C–Haloid, C–Metall, B–B, N–N, N–O, N–Haloid, O–O, O–S, O–Haloid, Si–Si,. . ., etc. The basic information on the bond dissociation-energy values and on the methods of their measurements can be found in [3.1–5]. The data of Tables 3.1, 2 are taken from [3.5] and recent publications.

Table 3.1. Dissociation energies of diatomic molecules

Molecule	D^0_{298}[kcal/mol]	Molecule	D^0_{298}[kcal/mol]
Ag–Ag	39 ± 2	Ar–Hg	1.47
Ag–Al	43.9 ± 2.2	Ar–I	2.4
Ag–Au	48.5 ± 2.2	Ar–K	1.0
Ag–Bi	46 ± 10	As–As	91.3 ± 2.5
Ag–Br	70 ± 7	As–Cl	107
Ag–Cl	81.6	As–D	64.6
Ag–Cu	41.6 ± 2.2	As–F	98
Ag–D	54.2	As–Ga	50.1 ± 0.3
Ag–Dy	31 ± 5	As–H	84
Ag–Eu	31.0 ± 3.0	As–I	70.9 ± 6.7
Ag–F	84.7 ± 3.9	As–In	48
Ag–Ga	43 ± 4	As–N	139 ± 30
Ag–Ge	41.7 ± 5.0	As–O	115 ± 2
Ag–H	51.4 ± 2	As–P	103.6 ± 3.0
Ag–Ho	29.5 ± 4.0	As–S	90.7 ± 1.5
Ag–I	56 ± 7	As–Sb	79.0 ± 1.3
Ag–In	39.9 ± 1.5	As–Se	23
Ag–Li	42.4 ± 1.5	As–Tl	47.4 ± 3.5
Ag–Mn	24 ± 5	At–At	~ 19
Ag–Na	33.0 ± 2.0	Au–Au	53.8 ± 0.5
Ag–Nd	< 50	Au–B	87.9 ± 2.5
Ag–O	52.6 ± 5.0	Au–Ba	60.9 ± 2.4
Ag–S	51.9	Au–Be	68 ± 2
Ag–Se	48.4	Au–Bi	71 ± 2.0
Ag–Si	42.5 ± 2.4	Au–Ca	58
Ag–Sn	32.5 ± 5.0	Au–Ce	81 ± 5
Ag–Te	46.8	Au–Cl	82 ± 2.3
Al–Al	44.5 ± 2.2	Au–Co	53 ± 4
Al–As	48.5 ± 1.7	Au–Cr	51 ± 4
Al–Au	77.9 ± 1.5	Au–Cu	54.5 ± 1.2
Al–Br	106 ± 2	Au–Cs	61 ± 0.8
Al–Cl	122.2 ± 0.2	Au–D	76.1
Al–Cu	51.8 ± 2.5	Au–Dy	62 ± 5
Al–D	69.5	Au–Eu	57.6 ± 2.5
Al–F	158.6 ± 1.5	Au–Fe	44.7 ± 4.0
Al–H	68.1 ± 1.5	Au–Ga	56 ± 9
Al–I	88.4 ± 0.5	Au–Ge	66.2 ± 3.5
Al–Li	42.0 ± 3.5	Au–H	69.8 ± 2
Al–N	71 ± 23	Au–Ho	63.9 ± 4.0
Al–P	51.8 ± 3.0	Au–La	80.4 ± 5.0
Al–Pd	60.8 ± 2.9	Au–Li	68.0 ± 1.6
Al–S	89.3 ± 1.9	Au–Lu	79.4 ± 4.0
Al–Sb	51.7 ± 1.4	Au–Mg	58 ± 10
Al–Se	80.7 ± 2.4	Au–Mn	44.3 ± 3.0
Al–Si	54.8 ± 7.2	Au–Na	51.4 ± 3.0
Al–O	122.4 ± 2.2	Au–Nd	71.5 ± 5.0
Al–Te	64.0 ± 2.4	Au–Ni	60 ± 5
Al–U	78 ± 7	Au–O	53.0 ± 5.0
Ar–Ar	1.13 ± 0.01	Au–Pb	31 ± 10
Ar–He	0.93	Au–Pd	37 ± 5

Table 3.1. *Continued*

Molecule	D^0_{298}[kcal/mol]	Molecule	D^0_{298}[kcal/mol]
Au–Pr	74 ± 5	Ba–S	95.6 ± 4.5
Au–Rb	58 ± 0.7	Be–Be	14
Au–Rh	55.2 ± 7	Be–Br	91 ± 20
Au–S	100 ± 6	Be–Cl	92.8 ± 2.2
Au–Sc	67.0 ± 4.0	Be–D	48.53
Au–Se	58.1	Be–F	138 ± 10
Au–Si	73.0 ± 1.4	Be–H	47.8 ± 0.3
Au–Sn	60.9 ± 1.7	Be–O	103.9 ± 3.2
Au–Sr	63 ± 10	Be–S	89 ± 14
Au–Tb	69.2 ± 8.0	Bi–Bi	47.9 ± 1.8
Au–Te	75.9	Bi–Br	63.9 ± 1.0
Au–U	76 ± 7	Bi–Cl	72 ± 1
Au–V	57.5 ± 2.9	Bi–D	67.8
Au–Y	73.4 ± 2.0	Bi–F	62 ± 7
B–B	71 ± 5	Bi–Ga	38 ± 4
B–Br	101 ± 5	Bi–H	≤ 67.7
B–C	107 ± 7	Bi–I	52.1 ± 1.1
B–Ce	73 ± 5	Bi–In	36.7 ± 0.4
B–Cl	128	Bi–Li	36.8 ± 1.2
B–D	81.5 ± 1.5	Bi–O	80.6 ± 3.0
B–F	181	Bi–P	67 ± 3
B–H	79.8	Bi–Pb	33.9 ± 3.5
B–I	45.8 ± 0.2	Bi–S	75.4 ± 1.1
B–Ir	122.9 ± 4.1	Bi–Sb	60 ± 1
B–La	81 ± 15	Bi–Se	67.0 ± 1.4
B–N	93 ± 5	Bi–Sn	50.2 ± 2.0
B–O	193.3 ± 5.0	Bi–Te	55.5 ± 2.7
B–P	82.9 ± 4.0	Bi–Tl	29 ± 3
B–Pd	78.7 ± 5.0	Br–Br	46.082
B–Pt	114.2 ± 4.0	Br–C	67 ± 5
B–Rh	113.7 ± 5.0	Br–Ca	74.3 ± 2.2
B–Ru	106.8 ± 5.0	Br–Cd	38 ± 23
B–S	138.8 ± 2.2	Br–Cl	51.99 ± 0.07
B–Sc	66 ± 15	Br–Co	79 ± 10
B–Se	110.4 ± 3.5	Br–Cr	78.4 ± 5.8
B–Si	68.9	Br–Cs	93.0 ± 1
B–Te	84.7 ± 4.8	Br–Cu	79 ± 6
B–Th	71	Br–D	88.61
B–Ti	66 ± 15	Br–F	59.8 ± 0.2
B–U	77 ± 8	Br–Fe	59 ± 23
B–Y	70 ± 15	Br–Ga	106 ± 4
Ba–Br	86.7 ± 2	Br–Ge	61 ± 7
Ba–Cl	104.2 ± 2.0	Br–H	87.56
Ba–D	≤ 46.3	Br–Hg	17.4 ± 1
Ba–F	140.3 ± 1.6	Br–I	42.8 ± 0.1
Ba–H	42 ± 3.5	Br–In	99 ± 5
Ba–I	73.8 ± 2	Br–K	90.8 ± 0.2
Ba–O	134.3 ± 3.2	Br–Li	100.1 ± 1
Ba–Pd	53.0 ± 1.2	Br–Mg	≤ 78.2
Ba–Rh	62.0 ± 6.0	Br–Mn	75.1 ± 2.3

Table 3.1. *Continued*

Molecule	D^0_{298}[kcal/mol]	Molecule	D^0_{298}[kcal/mol]
Br–N	66 ± 5	Ca–Ca	≤ 11
Br–Na	78.8 ± 0.2	Ca–Cl	95 ± 3
Br–Ni	86 ± 3	Ca–D	≤ 40.6
Br–O	56.2 ± 0.1	Ca–F	126 ± 5
Br–Pb	59 ± 9	Ca–H	40.1
Br–Rb	91.0 ± 1	Ca–I	63.0 ± 2.5
Br–Sb	75 ± 14	Ca–Li	20.3 ± 2.0
Br–Sc	106 ± 15	Ca–O	96.1 ± 4.0
Br–Se	71 ± 20	Ca–S	80.7 ± 4.5
Br–Si	87.9 ± 2.4	Cd–Cd	1.76
Br–Sn	≥ 132	Cd–Cl	49.8
Br–Sr	79.6 ± 2.2	Cd–F	73 ± 5
Br–Ti	105	Cd–H	16.5 ± 0.1
Br–Tl	79.8 ± 0.4	Cd–I	33 ± 5
Br–U	90.2 ± 1.5	Cd–In	33
Br–V	105 ± 10	Cd–O	56.3 ± 20.0
Br–W	78.7	Cd–S	49.8 ± 5.0
Br–Y	116 ± 20	Cd–Se	30.5 ± 6.0
Br–Zn	34 ± 7	Cd–Te	23.9 ± 3.6
C–C	145 ± 5	Ce–Ce	58.6
C–Ce	106 ± 3	Ce–F	139 ± 10
C–Cl	95 ± 7	Ce–Ir	140
C–D	81.6	Ce–N	124 ± 5
C–F	132	Ce–O	190 ± 2
C–Ge	110 ± 5	Ce–Os	121 ± 8
C–H	80.86	Ce–Pd	77.0
C–Hf	129 ± 6	Ce–Pt	133
C–I	50 ± 5	Ce–Rh	131
C–Ir	151 ± 1	Ce–Ru	127 ± 6
C–La	121 ± 15	Ce–S	136
C–Mo	115 ± 3.8	Ce–Se	118.2 ± 3.5
C–N	184 ± 1	Ce–Te	93 ± 10
C–Nb	136 ± 3.1	Cl–Cl	58.978 ± 0.001
C–O	257.3 ± 0.1	Cl–Co	93
C–Os	≥ 142	Cl–Cr	87.5 ± 5.8
C–P	122.7 ± 2	Cl–Cs	107 ± 2
C–Pt	143 ± 1.4	Cl–Cu	91.5 ± 1.1
C–Rh	139.5 ± 1.5	Cl–D	104.32
C–Ru	154.9 ± 3	Cl–Eu	~ 78
C–S	170.5 ± 0.3	Cl–F	61.24
C–Sc	≤ 106 ± 5	Cl–Fe	~ 84
C–Se	141.1 ± 1.4	Cl–Ga	115 ± 3
C–Si	107.9	Cl–Ge	~ 103
C–Tc	135 ± 7	Cl–H	103.16
C–Th	108.3 ± 4.1	Cl–Hg	24 ± 2
C–Ti	101 ± 7	Cl–I	50.5 ± 0.1
C–U	109 ± 4	Cl–In	105 ± 2
C–V	102 ± 5.7	Cl–K	103.5 ± 2
C–Y	100 ± 15	Cl–Li	112 ± 3
C–Zr	134 ± 6	Cl–Mg	78.3 ± 0.5

Table 3.1. *Continued*

Molecule	D^0_{298}[kcal/mol]	Molecule	D^0_{298}[kcal/mol]
Cl–Mn	86.2 ± 2.3	Cs–I	80.6 ± 0.5
Cl–N	79.8 ± 2.3	Cs–Na	15.1 ± 0.3
Cl–Na	98.5 ± 2	Cs–O	70.7 ± 15.0
Cl–Ni	89 ± 5	Cs–Rb	11.81 ± 0.01
Cl–O	65 ± 1	Cu–Cu	42.19 ± 0.57
Cl–P	69 ± 10	Cu–D	64.6
Cl–Pb	72 ± 7	Cu–Dy	34 ± 5
Cl–Ra	82 ± 18	Cu–F	98.8 ± 3
Cl–Rb	102.2 ± 2	Cu–Ga	51.6 ± 3.6
Cl–S	66.2	Cu–Ge	49.9 ± 5
Cl–Sb	86 ± 12	Cu–H	66.4
Cl–Sc	79	Cu–Ho	34 ± 5
Cl–Se	77	Cu–I	47 ± 5
Cl–Si	96	Cu–Li	46.1 ± 2.1
Cl–Sm	≥ 101 ± 3	Cu–Mn	37.9 ± 4
Cl–Sn	99 ± 44	Cu–Na	42.1 ± 4.0
Cl–Sr	97 ± 3	Cu–Ni	49.2 ± 4
Cl–Ta	130	Cu–O	64.3 ± 5.0
Cl–Ti	118	Cu–S	66
Cl–Tl	89.1 ± 0.5	Cu–Se	60
Cl–U	108 ± 2	Cu–Si	52.9 ± 1.5
Cl–V	114 ± 15	Cu–Sn	40.5 ± 1.6
Cl–W	101 ± 10	Cu–Tb	46 ± 5
Cl–Xe	1.6	Cu–Te	66.6
Cl–Y	126 ± 20	D–D	106.007
Cl–Yb	~ 77	D–F	137.8
Cl–Zn	54.7 ± 4.7	D–Ga	< 65.2
Cm–O	176	D–Ge	≤ 77
Co–Co	40 ± 6	D–H	105.027
Co–Cu	40 ± 4	D–Hg	10.05
Co–F	104 ± 15	D–In	58.8
Co–Ge	56 ± 5	D–Li	57.4066 ± 0.0011
Co–H	54 ± 10	D–Mg	32.3
Co–I	68 ± 5	D–Ni	≤ 72.4
Co–O	91.9 ± 3.2	D–Pt	≤ 83.7
Co–S	79 ± 5	D–S	84
Co–Si	66 ± 4	D–Si	72.3
Cr–Cr	37 ± 5	D–Sr	≥ 65.9
Cr–Cu	37 ± 5	D–Zn	21.2
Cr–F	106.3 ± 4.7	Dy–F	127
Cr–Ge	40.6 ± 7	Dy–O	145 ± 4
Cr–H	67 ± 12	Dy–S	99 ± 10
Cr–I	68.6 ± 5.8	Dy–Se	77 ± 10
Cr–N	90.3 ± 4.5	Dy–Te	56 ± 10
Cr–O	102.6 ± 7.0	Er–F	135 ± 4
Cr–S	79	Er–O	147 ± 3
Cs–Cs	10.701 ± 0.002	Er–S	100 ± 10
Cs–F	124 ± 2	Er–Se	78 ± 10
Cs–H	41.90 ± 0.02	Er–Te	57 ± 10
Cs–Hg	2	Eu–Eu	8.0 ± 4

Table 3.1. *Continued*

Molecule	D^0_{298}[kcal/mol]	Molecule	D^0_{298}[kcal/mol]
Eu–F	130	F–Xe	3.77
Eu–Li	16.0 ± 0.7	F–Y	144.6 ± 5.0
Eu–O	115 ± 2	F–Yb	⩾124.6 ± 2.3
Eu–Rh	55.9 ± 8	F–Zn	88 ± 15
Eu–S	86.6 ± 3.1	F–Zr	149 ± 15
Eu–Se	72 ± 3.5	Fe–Fe	24 ± 5
Eu–Te	58 ± 3.5	Fe–Ge	50.4 ± 7
F–F	37.95	Fe–H	43 ± 6
F–Ga	138 ± 3.5	Fe–O	93.3 ± 4.1
F–Gd	141.1 ± 6.5	Fe–S	77
F–Ge	116 ± 5	Fe–Si	71 ± 6
F–H	136.3	Ga–Ga	33 ± 5
F–Hg	∼ 43	Ga–H	< 65.5
F–Ho	129	Ga–I	81 ± 2.3
F–I	≤ 64.9	Ga–Li	31.8 ± 3.5
F–In	121 ± 3.5	Ga–O	84.5 ± 10.0
F–K	118.9 ± 0.6	Ga–P	54.9 ± 3.0
F–La	143 ± 10	Ga–Sb	45.9 ± 3.0
F–Li	138 ± 5	Ga–Te	60 ± 6
F–Lu	79.7	Gd–O	172 ± 2
F–Mg	110.4 ± 1.2	Gd–S	125.9 ± 2.5
F–Mn	101.2 ± 3.5	Gd–Se	103 ± 3.5
F–Mo	111.1	Gd–Te	82 ± 3.5
F–N	82	Ge–Ge	63.0 ± 1.7
F–Na	124	Ge–H	≤ 76.9
F–Nd	130.3 ± 3.0	Ge–Ni	67 ± 3
F–Ni	104	Ge–O	157.6 ± 3.0
F–O	53 ± 4	Ge–Pd	63.2
F–P	105 ± 23	Ge–S	131.7 ± 0.6
F–Pb	85 ± 2	Ge–Se	113.0 ± 2.0
F–Pm	129 ± 10	Ge–Si	72 ± 5
F–Pr	139 ± 11	Ge–Te	109 ± 3
F–Pu	128.7 ± 7	H–H	104.204
F–Rb	118 ± 5	H–Hg	9.523
F–S	81.9 ± 1.2	H–I	71.321
F–Sb	105 ± 23	H–In	58.1
F–Sc	140.8 ± 2	H–K	41.734 ± 0.011
F–Se	81 ± 10	H–Li	56.895 ± 0.001
F–Si	132.1 ± 0.5	H–Mg	30.2 ± 0.7
F–Sm	135	H–Mn	56 ± 7
F–Sn	111.5 ± 3	H–N	≤ 81
F–Sr	129.5 ± 1.6	H–Na	44.38 ± 0.06
F–Ta	137 ± 3	H–Ni	60.3 ± 2
F–Tb	134 ± 10	H–O	102.2
F–Ti	136 ± 8	H–P	71
F–Tl	106.4 ± 4.6	H–Pb	42 ± 5
F–Tm	122	H–Pd	56 ± 6
F–U	157.5 ± 2.5	H–Pt	≤ 80
F–V	141 ± 15	H–Rb	40 ± 5
F–W	131 ± 15	H–Rh	59 ± 5

Table 3.1. *Continued*

Molecule	D^0_{298}[kcal/mol]	Molecule	D^0_{298}[kcal/mol]
H–Ru	56 ± 5	I–Tl	65 ± 2
H–S	82.3 ± 2.9	I–Zn	33 ± 7
H–Sc	~ 43	I–Zr	73
H–Se	75.16 ± 0.23	In–In	24 ± 2
H–Si	≤ 71.5	In–Li	22.1 ± 3.5
H–Sn	63 ± 4	In–O	< 76.5 ± 10.0
H–Sr	39 ± 2	In–P	47.3 ± 2.0
H–Te	64 ± 0.5	In–S	69 ± 4
H–Ti	~ 38	In–Sb	36.3 ± 2.5
H–Tl	45 ± 2	In–Se	59 ± 4
H–Yb	38 ± 9	In–Te	52 ± 4
H–Zn	20.5 ± 0.5	Ir–La	138 ± 3
He–He	0.9	Ir–O	99.1 ± 10.1
He–Hg	1.58	Ir–Si	110.6 ± 5.0
Hf–C	131 ± 15	Ir–Th	137
Hf–N	128 ± 7	Ir–Y	109.0 ± 4.0
Hf–O	191.6 ± 3.2	K–K	13.7 ± 1.0
Hg–Hg	2 ± 0.5	K–Kr	1.1
Hg–I	8.29 ± 0.23	K–Li	19.6 ± 1.0
Hg–K	1.97 ± 0.05	K–Na	15.773 ± 0.002
Hg–Li	3.3	K–O	66.4 ± 5.0
Hg–Na	2.2	K–Xe	1.2
Hg–O	52.8 ± 7.9	Kr–Kr	1.25
Hg–Rb	2.0	Kr–O	< 2
Hg–S	51.9 ± 5.3	La–La	59 ± 5
Hg–Se	34.5 ± 7.2	La–N	124 ± 10
Hg–Te	≤ 34	La–O	191 ± 1
Hg–Tl	1	La–Pt	120 ± 5
Ho–Ho	20 ± 4	La–Rh	126 ± 4
Ho–O	146 ± 4	La–S	137.0 ± 0.4
Ho–S	102.4 ± 3.5	La–Se	114 ± 4
Ho–Se	80 ± 4	La–Te	91 ± 4
Ho–Te	62 ± 4	La–Y	48.3
I–I	36.111	Li–Li	26.34 ± 1
I–In	80	Li–Mg	16.1 ± 1.5
I–K	77.7 ± 0.2	Li–Na	21.10 ± 0.01
I–Li	82.5 ± 1.0	Li–O	79.7 ± 2.0
I–Mg	~ 68	Li–Pb	18.8 ± 1.9
I–Mn	67.6 ± 2.3	Li–S	74.7 ± 1.8
I–N	38 ± 4	Li–Sb	41.3 ± 2.4
I–Na	72.7 ± 0.5	Li–Sm	11.7 ± 1.0
I–Ni	70 ± 5	Li–Tm	16.5 ± 0.8
I–O	43	Li–Yb	8.9 ± 0.7
I–Pb	47 ± 9	Lu–Lu	34 ± 8
I–Rb	76.2 ± 0.5	Lu–O	162 ± 2
I–Si	70	Lu–Pt	96 ± 8
I–Sn	56 ± 10	Lu–S	121.2 ± 3.5
I–Sr	64.5 ± 1.4	Lu–Se	100 ± 4
I–Te	46 ± 10	Lu–Te	78 ± 4
I–Ti	74 ± 10	Mg–Mg	2.044 ± 0.001

Table 3.1. *Continued*

Molecule	D^0_{298}[kcal/mol]	Molecule	D^0_{298}[kcal/mol]
Mg–O	86.8 ± 3.0	O–Pu	171.1 ± 8.1
Mg–S	56	O–Rb	61 ± 20
Mn–Mn	6.2	O–Re	149.8 ± 20.0
Mn–O	96.3 ± 10.0	O–Rh	96.8 ± 10.0
Mn–S	72 ± 4	O–Ru	126.3 ± 10.0
Mn–Se	57.2 ± 2.2	O–S	124.7 ± 1.0
Mo–Mo	97 ± 5	O–Sb	103.8 ± 10.0
Mo–Nb	109 ± 6	O–Sc	162.9 ± 2.7
Mo–O	133.9 ± 5.0	O–Se	111.1 ± 5.1
N–N	225.94 ± 0.14	O–Si	191.1 ± 3.2
N–O	150.71 ± 0.03	O–Sm	135 ± 3
N–P	147.5 ± 5.0	O–Sn	127.1 ± 3.0
N–Pu	113 ± 15	O–Sr	101.7 ± 4.0
N–S	111 ± 5	O–Ta	191.0 ± 3.0
N–Sb	72 ± 12	O–Tb	170 ± 3
N–Sc	112 ± 20	O–Te	89.9 ± 5.0
N–Se	91 ± 15	O–Th	210.0 ± 2.9
N–Si	105 ± 9	O–Ti	160.7 ± 2.2
N–Ta	146 ± 20	O–Tm	120 ± 3
N–Th	138.0 ± 7.9	O–U	181.5 ± 3.2
N–Ti	113.8 ± 7.9	O–V	149.8 ± 4.5
N–U	127.0 ± 0.5	O–W	160.6 ± 10.0
N–V	114.1 ± 4.1	O–Xe	8.7
N–Xe	5.5	O–Y	172.0 ± 2.7
N–Y	115 ± 15	O–Yb	95 ± 4
N–Zr	135.0 ± 6.0	O–Zn	< 64.7 ± 10.0
Na–Na	17.59 ± 0.60	O–Zr	185.5 ± 3.2
Na–O	61.2 ± 4.0	P–P	117.0 ± 2.5
Na–Rb	14 ± 0.9	P–Pt	≤ 99.6 ± 4
Nb–Nb	122 ± 2.4	P–Rh	84.4 ± 4
Nb–O	184.4 ± 6.0	P–S	106 ± 2
Nd–Nd	< 39	P–Sb	85.3
Nd–O	168 ± 3	P–Se	86.9 ± 2.4
Nd–S	112.7	P–Si	86.9
Nd–Se	92 ± 4	P–Te	71.2 ± 2.4
Nd–Te	73 ± 4	P–Th	131.5 ± 10
Ne–Ne	0.94	P–Tl	50 ± 3
Ni–Ni	48.58 ± 0.23	P–U	71 ± 5
Ni–O	91.3 ± 4.0	P–W	73 ± 1
Ni–S	82.3	Pb–Pb	20.7 ± 0.2
Ni–Si	76 ± 4	Pb–S	82.7 ± 0.4
Np–O	171.7 ± 10.0	Pb–Sb	38.6 ± 2.5
O–O	119.11 ± 0.04	Pb–Se	72.4 ± 1
O–Os	143.0 ± 20.0	Pb–Te	60 ± 3
O–P	143.2 ± 3.0	Pd–Pd	17
O–Pa	188.4 ± 4.1	Pd–Si	74.9 ± 3.3
O–Pb	91.3 ± 3.0	Pd–Y	57 ± 4
O–Pd	91.0 ± 20.0	Pm–S	101 ± 15
O–Pm	161 ± 15	Pm–Se	81 ± 15
O–Pr	180 ± 3	Pm–Te	61 ± 15
O–Pt	93.6 ± 10.0	Po–Po	44.7

Table 3.1. *Continued*

Molecule	D^0_{298}[kcal/mol]	Molecule	D^0_{298}[kcal/mol]
Pr–S	117.1 ± 1.1	Sb–Te	66.3 ± 0.9
Pr–Sc	106.7 ± 5.5	Sb–Tl	30.3 ± 2.5
Pr–Te	78 ± 10	Sc–Sc	38.9 ± 5
Pt–Pt	85.4 ± 3.6	Sc–Se	92 ± 4
Pt–Si	119.8 ± 4.3	Sc–Te	69 ± 4
Pt–Th	132	Se–Se	79.5 ± 0.1
Pt–Ti	95 ± 3	Se–Si	131
Pt–Y	113.3 ± 2.9	Se–Sm	79.1 ± 3.5
Rb–Rb	10.9 ± 0.5	Se–Sn	95.9 ± 1.4
Rh–Rh	68.2 ± 5.0	Se–Sr	~ 68
Rh–Sc	106.1 ± 2.5	Se–Tb	101 ± 10
Rh–Si	94.4 ± 4.3	Se–Te	69.7 ± 1
Rh–Th	123 ± 5	Se–Ti	91 ± 10
Rh–Ti	93.4 ± 3.5	Se–Tm	66 ± 10
Rh–U	124 ± 4	Se–V	83 ± 5
Rh–V	87 ± 7	Se–Y	104 ± 3
Rh–Y	106.4 ± 2.5	Se–Zn	40.8 ± 6.2
Ru–Si	94.9 ± 5.0	Si–Si	78.1 ± 2.4
Ru–Th	141.4 ± 10	Si–Te	108
Ru–V	99 ± 7	Sm–Te	65.1 ± 3.5
S–S	101.65	Sn–Sn	46.7 ± 4
S–Sb	90.5	Sn–Te	86.0
S–Sc	114 ± 3	Sr–Sr	3.7 ± 0.1
S–Se	88.7 ± 1.6	Tb–Tb	31.4 ± 6.0
S–Si	149	Tb–Te	81 ± 10
S–Sm	93	Te–Te	62.1 ± 1.2
S–Sn	111 ± 0.8	Te–Ti	69 ± 4
S–Sr	81	Te–Tm	66 ± 10
S–Tb	123 ± 10	Te–Y	81 ± 3
S–Te	81 ± 5	Te–Zn	28.1 ± 4.3
S–Ti	99.9 ± 0.7	Th–Th	≤ 69
S–Tm	88 ± 10	Ti–Ti	33.8 ± 5
S–U	124.9 ± 2.3	Tl–Tl	15.4 ± 4
S–V	115 ± 2	U–U	53 ± 5
S–Y	126.3 ± 2.5	V–V	57.9 ± 5
S–Yb	40	Xe–Xe	15.6 ± 0.07
S–Zn	49 ± 3	Y–Y	38 ± 5
S–Zr	137.5 ± 4.0	Yb–Yb	4.9 ± 4
Sb–Sb	71.5 ± 1.5	Zn–Zn	7

Table 3.2. Bond dissociation energies of polyatomic molecules

Molecule and bond breaking	D^0_{298} [kcal/mol]	Molecule and bond breaking	D^0_{298} [kcal/mol]
H–CH	100.8	H–C_2H_5	100.3 ± 1.0
H–CH_2	111.1	H–CH_2CCH	89.4 ± 2
H–CH_3	104.8 ± 0.2	H–CH_2CHCH_2	86.3 ± 1.5
H–CCH	132 ± 5	H–n-C_3H_7	99.7
H–$CHCH_2$	106	H–i-C_3H_7	95.9 ± 0.5

Table 3.2. *Continued*

Molecule and bond breaking	D^0_{298} [kcal/mol]	Molecule and bond breaking	D^0_{298} [kcal/mol]
H–CH_2CCCH_3	87.2 ± 2	H–$CHClCF_3$	101.8 ± 1.5
H–s-C_4H_9	96.4	H–CClCFCl	105 ± 2
H–t-C_4H_9	93.3 ± 0.5	H–CClCHCl	104 ± 2
H–C_6H_5	110.9 ± 2.0	H–CCl_2CHCl_2	94 ± 2
H–CN	123.8 ± 2	H–C_2Cl_5	95 ± 2
H–CH_2CN	93 ± 2.5	H–$CClBrCF_3$	96.6 ± 1.5
H–$CH(CH_3)CN$	89.9 ± 2.3	H–n-C_3F_7	104 ± 2
H–$C(CH_3)_2CN$	86.5 ± 2.0	H–i-C_3F_7	103.6 ± 0.6
H–CH_2NH_2	93.3 ± 2.0	H–$CHClCHCH_2$	88.6 ± 1.4
H–CH_2NHCH_3	87 ± 2	H–C_6F_5	113.9
H–$CH_2N(CH_3)_2$	84 ± 2	H–$CH_2Si(CH_3)_3$	99.2 ± 1
H–CHO	87 ± 1	H–SiH	84
H–$COCH_3$	86.0 ± 0.8	H–SiH_2	64
H–$COCHCH_2$	87.1 ± 1.0	H–SiH_3	90.3
H–COC_2H_5	87.4 ± 1	H–SiH_2CH_3	89.6
H–COC_6H_5	86.9 ± 1	H–$SiH(CH_3)_2$	89.4
H–$COCF_3$	91.0 ± 2	H–$Si(CH_3)_3$	90.3
H–CH_2CHO	94.8 ± 2.0	D–$Si(CH_3)_3$	93 ± 1.7
H–CH_2COCH_3	98.3 ± 1.8	H–$SiH_2C_6H_5$	88.2
H–$CH(CH_3)COCH_3$	92.3 ± 1.4	H–SiF_3	100.1
H–CH_2OCH_3	93 ± 1	H–$SiCl_3$	91.3
H–$CH(CH_3)OC_2H_5$	91.7 ± 0.4	H–Si_2H_5	86.3
H–CH_2OH	94 ± 2	H–GeH_3	83 ± 2
H–$CH(CH_3)OH$	93 ± 1	H–GeH_2I	79 ± 2
H–$CH(OH)CHCH_2$	81.6 ± 1.8	H–$Ge(CH_3)_3$	81 ± 2
H–$C(CH_3)_2OH$	91 ± 1	H–Sn(n–$C_4H_9)_3$	73.7 ± 2.0
H–$CH_2OCOC_6H_5$	100.2 ± 1.3	H–NH_2	107.4 ± 1.1
H–$COOCH_3$	92.7 ± 1	H–$NHCH_3$	100.0 ± 2.5
H–CH_2SH	96 ± 1	H–$N(CH_3)_2$	91.5 ± 2
H–CH_2SCH_3	96.6 ± 1.0	H–NHC_6H_5	88.0 ± 2
H–CH_2F	101.2 ± 1	H–$N(CH_3)C_6H_5$	87.5 ± 2
H–CHF_2	103.2 ± 1	H–NO	≤ 49.5
H–CF_3	106.7 ± 1	H–NO_2	78.3 ± 0.5
H–CF_2Cl	101.6 ± 1.0	H–NF_2	75.7 ± 2.5
H–CH_2Cl	100.9 ± 2	H–$NHNH_2$	87.5
H–$CHCl_2$	100.6	H–N_3	92 ± 5
H–CCl_3	95.8 ± 1	H–OH	119 ± 1
H–CH_2Br	102.0 ± 2	H–OCH_3	104.4 ± 1
H–$CHBr_2$	103.7 ± 2	H–OC_2H_5	104.2 ± 1
H–CBr_3	96.0 ± 1.6	H–$OC(CH_3)_3$	105.1 ± 1
H–CH_2I	103 ± 2	H–$OCH_2C(CH_3)_3$	102.3 ± 1.5
H–CHI_2	103 ± 2	H–OC_6H_5	86.5 ± 2
H–$CHCF_2$	107 ± 2	H–O_2H	88.2 ± 1.0
H–CFCHF	107 ± 2	H–O_2CH_3	88.5 ± 0.5
H–$CFCF_2$	108 ± 2	H–O_2-t-C_4H_9	89.4 ± 0.2
H–CH_2CF_3	106.7 ± 1.1	H–$OCOCH_3$	105.8 ± 2
H–CF_2CH_3	99.5 ± 2.5	H–$OCOC_2H_5$	106.4 ± 2
H–C_2F_5	102.7 ± 0.5	H–OCO-n-C_3H_7	105.9 ± 2
H–CFCFCl	106 ± 2	H–ONO	78.3 ± 0.5

Table 3.2. *Continued*

Molecule and bond breaking	D^0_{298} [kcal/mol]	Molecule and bond breaking	D^0_{298} [kcal/mol]
$H–ONO_2$	101.2 ± 0.5	n-C_4H_9–N_2-n-C_4H_9	50.0
H–SH	91.1 ± 1	i-C_4H_9–N_2-i-C_4H_9	49.0
$H–SCH_3$	88.6 ± 1	s-C_4H_9–N_2-s-C_4H_9	46.7
$H–SC_6H_5$	83.3 ± 2	t-C_4H_9–N_2-t-C_4H_9	43.5
H–SO	41.3	$C_6H_5CH_2–N_2CH_2C_6H_5$	37.6
HC≡CH	230 ± 2	$CF_3–N_2CF_3$	55.2
$H_2C=CH_2$	172 ± 2	$CH_3–NO$	40.0 ± 0.8
$CH_3–CH_3$	89.8 ± 0.5	i-C_3H_7–NO	36.5 ± 3
$CH_3–CH_2CCH$	76.0 ± 2	t-C_4H_9–NO	39.5 ± 1.5
$CH_3–CH_2CCCH_3$	73.7 ± 1.5	$C_6H_5–NO$	50.8 ± 1
$CH_3–CH(CH_3)CCH$	73.0	NC–NO	28.8 ± 2.5
$CH_3–C_6H_5$	75.8 ± 1.5	$CF_3–NO$	42.8 ± 2
$CH_3–CH_2C_6H_5$	79.4 ± 1	$C_6F_5–NO$	49.8 ± 1
$C_2H_5–CH_2C_6H_5$	70.3 ± 1	$CCl_3–NO$	32 ± 3
$CHCCH_2–CH_2C_6H_5$	61.4 ± 2	$CH_3–NO_2$	60.8
n-C_3H_7–$CH_2C_6H_5$	70.0 ± 1	$CH_2C(CH_3)–NO_2$	58.6
$CH_3–CN$	121.8 ± 2	i-C_3H_7–NO_2	59.0
$C_2H–CN$	144 ± 1	t-C_4H_9–NO_2	58.5
$C_2H_5–CH_2NH_2$	79.4 ± 2	$C_6H_5–NO_2$	71.3 ± 1
$CH_3–CH_2CN$	80.4 ± 1	$C(NO_2)_3–NO_2$	40.5 ± 1
$C_2H_5–CH_2CN$	76.9 ± 1.7	$CH_3–OC(CH_3)CH_2$	66.3
$CH_3–CH(CH_3)CN$	78.8 ± 2	$CH_3–OC_6H_5$	57 ± 2
$C_2H_5–CH_2CN$	76.9 ± 1.7	$CH_3–OCH_2C_6H_5$	67.0
$CH_3–C(CH_3)_2CN$	74.7 ± 1.6	$C_2H_5–OC_6H_5$	63 ± 1.5
$CH_3–C(CH_3)(CN)C_6H_5$	59.9	$CH_2CHCH_2–OC_6H_5$	49.8 ± 2
$C_6H_5CH_2–CH_2NH_2$	68.0 ± 2	O=CO	127.2 ± 0.1
$C_6H_5CH_2–C_5H_4N$	86.7	$CH_3–O_2$	32.4 ± 0.7
CN–CN	128 ± 1	$C_2H_5–O_2$	35.2 ± 1.5
$CH_3CO–COCH_3$	67.4 ± 2.3	$CH_2CHCH_2–O_2$	17.2 ± 1.0
$C_6H_5CH_2–COOH$	67	i-C_3H_7–O_2	37.7 ± 1.8
$CF_3–COC_6H_5$	73.8 ± 2	t-C_4H_9–O_2	36.7 ± 1.9
$CF_2=CF_2$	76.3 ± 3	$CF_3–O_2$	48.8
$CH_2F–CH_2F$	88 ± 2	$CF_3–O_2CF_3$	86.4
$CH_3–CF_3$	101.2 ± 1.1	$CH_3–SCH_3$	77.2 ± 2
$CF_3–CF_3$	98.7 ± 2.5	t-C_4H_9–SH	68.4 ± 1.5
$C_6F_5–C_6F_5$	116.6 ± 5.9	$C_6H_5–SH$	86.5 ± 2
$CH_3–BF_2$	~ 113	$CH_3–SC_6H_5$	69.4 ± 2
$C_6H_5–BCl_2$	~ 122	$C_6H_5CH_2–SCH_3$	61.4 ± 2
$CH_2CHCH_2–Si(CH_3)_3$	70	S–CS	102.9 ± 3
s-C_4H_9–$Si(CH_3)_3$	90	$F–CH_3$	108 ± 3
$CH_3–NHC_6H_5$	71.4 ± 2	F–CN	112.3 ± 1.2
$C_6H_5CH_2–NH_2$	71.1 ± 1	$F–CF_2Cl$	117 ± 6
$CH_3–N(CH_3)C_6H_5$	70.8 ± 2	$F–CFCl_2$	110 ± 6
$C_6H_5CH_2–NHCH_3$	68.7 ± 2	$F–CF_2CH_3$	124.8 ± 2
$C_6H_5CH_2–N(CH_3)_2$	62.1 ± 2	$F–C_2F_5$	126.8 ± 1.8
$CH_2=N_2$	41.7 ± 1	Cl–CN	100.8 ± 1.2
$CH_3–N_2CH_3$	52.5	$Cl–COC_6H_5$	74 ± 3
$C_2H_5–N_2C_2H_5$	50.0	Cl–CSCl	63.4 ± 0.5
i-C_3H_7–N_2-i-C_3H_7	47.5	$Cl–CF_3$	86.1 ± 0.8

Table 3.2. *Continued*

Molecule and bond breaking	D^0_{298} [kcal/mol]	Molecule and bond breaking	D^0_{298} [kcal/mol]
$Cl–CF_2Cl$	76 ± 2	$C_2H_5–HgC_2H_5$	49 ± 4
$Cl–CCl_2F$	73 ± 2	$CH_3–Tl(CH_3)_2$	40 ± 4
$Cl–CCl_3$	73.1 ± 1.8	$CH_3–Pb(CH_3)_3$	57 ± 4
$Cl–C_2F_5$	82.7 ± 1.7	$C_2H_5–Pb(C_2H_5)_3$	55 ± 4
$Cl–CF_2CF_2Cl$	78 ± 2	$CH_3–Bi(CH_3)_2$	52 ± 4
$Cl–SiCl_3$	111	$BH_3–BH_3$	35
$Br–CH_3$	70.0 ± 1.2	$NH_2–NH_2$	65.8
$Br–C_6H_5$	80.5 ± 2	$NH_2–NHCH_3$	64.1 ± 2
$Br–CN$	87.8 ± 1.2	$NH_2–N(CH_3)_2$	59.0 ± 2
$Br–CH_2COCH_3$	62.5	$NH_2–NHC_6H_5$	52.3 ± 2
$Br–COC_6H_5$	64.2	$ON–NO_2$	9.7 ± 0.5
$Br–CHF_2$	69 ± 2	$O_2N–NO_2$	13.6
$Br–CF_3$	70.6 ± 3.0	$NF_2–NF_2$	21 ± 1
$Br–CF_2CH_3$	68.6 ± 1.3	$O–N_2$	40
$Br–C_2F_5$	68.7 ± 1.5	$O–NO$	73
$Br–n\text{-}C_3F_7$	66.5 ± 2.5	$HO–NO$	49.3
$Br–i\text{-}C_3F_7$	65.5 ± 1.1	$HO–NO_2$	49.4
$Br–CH_2C_6F_5$	94 ± 2.5	$HO_2–NO_2$	23 ± 2
$Br–CHClCF_3$	65.7 ± 1.5	$CH_3O–NO$	41.8 ± 0.9
$Br–CCl_3$	55.3 ± 1	$C_2H_5O–NO$	42.0 ± 1.3
$Br–CClBrCF_3$	60.0 ± 1.5	$n\text{-}C_3H_7O–NO$	40.1 ± 1.8
$Br–CBr_3$	56.2 ± 1.8	$i\text{-}C_3H_7O–NO$	41.0 ± 1.3
$Br–NF_2$	≤53	$n\text{-}C_4H_9O–NO$	42.5 ± 1.5
$I–n\text{-}C_4H_9$	49.0 ± 1	$i\text{-}C_4H_9O–NO$	42.0 ± 1.5
I–Norbornyl	62.5 ± 2.5	$s\text{-}C_4H_9O–NO$	41.5 ± 0.8
$I–CN$	73 ± 1	$t\text{-}C_4H_9O–NO$	40.9 ± 0.8
$I–CF_3$	53.5 ± 0.7	$HO–NCHCH_3$	49.7
$I–CF_2CH_3$	52.1 ± 1.0	$Cl–NF_2$	~ 32
$I–CH_2CF_3$	56.3 ± 1	$I–NO$	18.6 ± 0.1
$I–C_2F_5$	52.3 ± 0.7	$I–NO_2$	18.3 ± 1
$I–n\text{-}C_3F_7$	49.8 ± 1.0	$HO–OH$	51 ± 1
$I–i\text{-}C_3F_7$	51.4 ± 0.7	$HO–OCH_2C(CH_3)_3$	46.3 ± 1.9
$I–n\text{-}C_4F_9$	49.0 ± 1.0	$CH_3O–OCH_3$	37.6 ± 2
$I–C_6H_5$	65.4 ± 2	$C_2H_5O–OC_2H_5$	37.9 ± 1
$I–C_6F_5$	~ 66	$CF_3O–OCF_3$	46.2
$C_5H_5–FeC_5H_5$	91 ± 3	$(CF_3)_3CO–OC(CF_3)_3$	35.5 ± 1.1
$CH_3–ZnCH_3$	68 ± 4	$t\text{-}C_4H_9O–OSi(CH_3)_3$	47
$C_2H_5–ZnC_2H_5$	57 ± 4	$SF_5O–OSF_5$	37.2
$CH_3–Ga(CH_3)_2$	63 ± 4	$t\text{-}C_4H_9O–OGe(C_2H_5)_3$	46
$C_2H_5–Ga(C_2H_5)_2$	50 ± 4	$t\text{-}C_4H_9O–OSn(C_2H_5)_3$	46
$CH_3–Ge(CH_3)_3$	83 ± 4	$FClO_2–O$	58.4
$CH_3–As(CH_3)_2$	67 ± 4	$CF_3O–O_2CF_3$	30.3 ± 2
$CH_3–CdCH_3$	60 ± 4	$SF_5O–O_2SF_5$	30.3
$CH_3–In(CH_3)_2$	49 ± 4	$CH_3CO_2–O_2CCH_3$	30.4 ± 2
$CH_3–Sn(CH_3)_3$	71 ± 4	$C_2H_5CO_2–O_2CC_2H_5$	30.4 ± 2
$C_2H_5–Sn(C_2H_5)_3$	63 ± 4	$n\text{-}C_3H_7CO_2–O_2C\text{-}n\text{-}C_3H_7$	30.4 ± 2
$CH_3–Sb(CH_3)_2$	61 ± 4	$O–SO$	132 ± 2
$C_2H_5–Sb(C_2H_5)_2$	58 ± 4	$F–OCF_3$	43.5 ± 0.5
$CH_3–HgCH_3$	61 ± 4	$HO–Cl$	60 ± 3

Table 3.2. *Continued*

Molecule and bond breaking	D^0_{298} [kcal/mol]	Molecule and bond breaking	D^0_{298} [kcal/mol]
O–ClO	59 ± 3	F–SF_3	84.1 ± 3.0
HO–Br	56 ± 3	F–SF_2	63.1 ± 7.1
HO–I	56 ± 3	F–SF	91.7 ± 4.3
O=PF_3	130 ± 5	I–SH	49.4 ± 2
O=PCl_3	122 ± 5	I–SO	43
O–PBr_3	119 ± 5	I–SCH_3	49.3 ± 1.7
HO–$Si(CH_3)_3$	128	I–$Si(CH_3)_3$	77
HS–SH	66 ± 2	H_3Si–SiH_3	74
F–SF_5	91.1 ± 3.2	$(CH_3)_3Si$–$Si(CH_3)_3$	80.5
F–SF_4	53.1 ± 6.0	$(C_6H_5)_3Si$–$Si(C_6H_5)_3$	88 ± 7

3.2 Ionization Potentials

The minimum energy required to remove an electron from a molecule, atom or ion is called the *ionization potential* or *ionization energy*. The first ionization potential of molecules is defined as the minimum transition energy between the lowest vibrational levels of the ground-state molecule and of the ground-state positive molecular ion.

Table 3.3 lists the first Ionization Potential (IP) for diatomic and polyatomic molecules. The data are taken from reference data books and review articles [3.1,2,6,7]. The molecules are ordered alphabetically by their molecular formulae which are composed according to the Hill system [3.8]. This system specifies the chemical element symbols in molecular formulae:

(a) for molecules that do not contain carbon – alphabetically by element symbols,
(b) for carbon-containing molecules – C first, followed by H (if present), then the other element symbols alphabetically.
(c) the resulting complete molecular formulae are arranged in alphabetical order, each element symbol with its particular numerical subscript being considered as a separate unit.

The values in parentheses are not well-established.

3.3 Electron Affinities

The energy released by the addition of an electron to a neutral particle is called the *electron-affinity energy* of the particle or simply electron affinity. Electron affinity of a molecule is a measure of bond strength of an electron in a negative ion. It is defined as the transition energy between the lowest state of the neutral molecule and the lowest state of the corresponding negative ion.

Table 3.3. Ionization potentials of molecules [3.1, 2, 6, 7]

Molecular formula	Molecular name	IP [eV]
Compounds not containing carbon		
AgCl	Silver chloride	($\leq$ 10.08)
AgF	Silver fluoride	(11.0 $\pm$ 0.3)
AlBr	Aluminum bromide	(9.3)
$AlBr_3$	Aluminum tribromide	(10.4)
AlCl	Aluminum chloride	9.4
$AlCl_3$	Aluminum trichloride	(12.01)
AlF	Aluminum fluoride	9.73 $\pm$ 0.01
AlF_3	Aluminum trifluoride	$\leq$ 15.45
AlI	Aluminum iodide	9.3 $\pm$ 0.3
AlI_3	Aluminum triiodide	(9.1)
$AsCl_3$	Arsenic trichloride	(10.55 $\pm$ 0.025)
AsF_3	Arsenic trifluoride	(12.84 $\pm$ 0.05)
AsH_3	Arsine	9.89
BBr_3	Boron tribromide	(10.51 $\pm$ 0.02)
BCl_3	Boron trichloride	11.60 $\pm$ 0.02
BF	Fluroroborane	11.12 $\pm$ 0.01
BF_2	Difluoroborane	(9.4)
BF_3	Boron trifluoride	15.56 $\pm$ 0.03
BH	Borane	9.77 $\pm$ 0.05
BH_3	Borane	12.3 $\pm$ 0.1
BI_3	Boron triiodide	(9.25 $\pm$ 0.03)
BO_2	Boron oxide	(13.5 $\pm$ 0.3)
B_2H_6	Diborane	11.38 $\pm$ 0.03
B_2O_3	Boron oxide	13.5 $\pm$ 0.15
B_4H_{10}	Tetraborane	10.76 $\pm$ 0.04
B_5H_9	Pentaborane	9.90 $\pm$ 0.04
B_6H_{10}	Hexaborane	(9.0)
BaO	Barium oxide	6.91 $\pm$ 0.06
BrCl	Bromine chloride	11.01
BrF	Bromine fluoride	11.77 $\pm$ 0.01
BrF_5	Bromine pentafluoride	(13.17 $\pm$ 0.01)
BrH	Hydrogen bromide	11.66 $\pm$ 0.03
BrH_3Si	Bromosilane	10.6
BrI	Iodine bromide	9.790 $\pm$ 0.004
BrK	Potassium bromide	7.85 $\pm$ 0.1
BrLi	Lithium bromide	(8.7)
BrNO	Nitrosyl bromide	10.17 $\pm$ 0.03
BrNa	Sodium bromide	8.31 $\pm$ 0.1
BrO	Bromine oxide	(10.2)
BrRb	Rubidium bromide	7.94 $\pm$ 0.03
BrTl	Thallium bromide	9.14 $\pm$ 0.02
Br_2	Bromine	10.515 $\pm$ 0.005
Br_2Hg	Mercury bromide	10.560 $\pm$ 0.003
Br_2Sn	Tin bromide	9.0
Br_3Ga	Gallium bromide ($GaBr_3$)	10.40
Br_3P	Phosphorous tribromide	9.7
Br_4Hf	Hafnium bromide ($HfBr_4$)	(10.9)
Br_4Sn	Tin bromide ($SnBr_4$)	10.6
Br_4Ti	Titanium bromide($TiBr_4$)	10.3

Table 3.3. *Continued*

Molecular formula	Molecular name	IP [eV]
Br_4Zr	Zirconium bromide ($ZrBr_4$)	(10.7)
CaCl	Calcium chloride (CaCl)	5.61 ± 0.13
CaO	Calcium oxide	(6.9)
ClCs	Cesium chloride	(7.84 ± 0.05)
ClF	Chlorine fluoride	12.65 ± 0.01
$ClFO_3$	Perchloryl fluoride	(12.945 ± 0.005)
ClF_2	Chlorine difluoride	(12.77 ± 0.05)
ClF_3	Chlorine trifluoride	(12.65 ± 0.05)
ClF_5S	Sulfur chloride pentafluoride	(12.335 ± 0.005)
ClH	Hydrogen chloride	12.747
ClHO	Hypochlorous acid (HOCl)	(11.12 ± 0.01)
ClH_3Si	Chlorosilane	11.4
ClI	Iodine chloride	10.088 ± 0.01
ClIn	Indium chloride (InCl)	(9.51)
ClK	Potassium chloride	(8.0 ± 0.4)
ClLi	Lithium chloride	9.57
ClNO	Nitrosyl chloride	10.87 ± 0.01
$ClNO_2$	Nitryl chloride	(11.84)
ClNa	Sodium chloride	8.92 ± 0.06
ClO	Chlorine oxide	10.95
ClO_2	Chlorine dioxide	10.36 ± 0.02
ClRb	Rubidium chloride	(8.50 ± 0.03)
ClTl	Thallium chloride (TlCl)	9.70 ± 0.03
Cl_2	Chlorine	11.480 ± 0.005
Cl_2CrO_2	Chromyl chloride (CrO_2Cl_2)	11.6
Cl_2Ge	Germanium dichloride	(10.20 ± 0.05)
Cl_2H_2Si	Dichlorosilane	11.4
Cl_2Hg	Mercury chloride ($HgCl_2$)	11.380 ± 0.003
Cl_2O	Oxygen dichloride	10.94
Cl_2OS	Thionyl chloride	10.96
Cl_2O_2S	Sulfuryl chloride	12.05
Cl_2Pb	Lead chloride ($PbCl_2$)	(10.0)
Cl_2S	Sulfur chloride (SCl_2)	9.45 ± 0.03
Cl_2Si	Dichlorosilylene ($SiCl_2$)	(10.93 ± 0.10)
Cl_2Sn	Tin chloride ($SnCl_2$)	(10.0)
Cl_3Ga	Gallium chloride ($GaCl_3$)	11.52
Cl_3HSi	Trichlorosilane	(11.7)
Cl_3N	Nitrogen trichloride	(10.12 ± 0.1)
Cl_3OP	Phosphoryl chloride	11.36 ± 0.02
Cl_3OV	Vanadium oxytrichloride	(11.6)
Cl_3P	Phosphorous trichloride	9.91
Cl_3PS	Phosphorous thiochloride	9.71 ± 0.03
Cl_3Sb	Antimony trichloride	(10.1 ± 0.1)
Cl_4Ge	Germanium tetrachloride	11.68 ± 0.05
Cl_4Hf	Hafnium chloride ($HfCl_4$)	(11.7)
Cl_4Si	Silicon tetrachloride	11.79 ± 0.01
Cl_4Sn	Tin chloride ($SnCl_4$)	(11.88 ± 0.05)
Cl_4Ti	Titanium chloride ($TiCl_4$)	11.65 ± 0.15
Cl_4V	Vanadium chloride (VCl_4)	(9.2)
Cl_4Zr	Zirconium chloride ($ZrCl_4$)	(11.2)

Table 3.3. *Continued*

Molecular formula	Molecular name	IP [eV]
Cl_5Mo	Molybdenum pentachloride	(8.7)
Cl_5Nb	Niobium chloride ($NbCl_5$)	(10.97)
Cl_5P	Phosphorous pentachloride	10.7
Cl_5Ta	Tantalum chloride ($TaCl_5$)	11.08
Cl_6W	Tungsten chloride (WCl_6)	(9.5)
CsF	Cesium fluoride	(8.80 ± 0.10)
CsNa	Cesium sodium	(4.05 ± 0.04)
FGa	Gallium fluoride (GaF)	(9.6 ± 0.5)
FH	Hydrogen fluoride	16.044 ± 0.003
FHO	Hypofluorous acid (HOF)	12.71 ± 0.01
FH_3Si	Fluorosilane	11.7
FI	Iodine fluoride	10.62
FIn	Indium fluoride (InF)	(9.6 ± 0.5)
FNO	Nitrosyl fluoride	12.63 ± 0.03
FNO_2	Nitryl fluoride	(13.09)
FNS	Thionitrosyl fluoride (NSF)	11.51 ± 0.04
FO	Fluorine oxide (FO)	12.77
FO_2	Fluorine superoxide (FOO)	12.6 ± 0.2
FS	Sulfur fluoride (SF)	10.09
FTl	Thallium fluoride (TlF)	10.52
F_2	Fluorine	15.697 ± 0.003
F_2Ge	Germanium fluoride (GeF_2)	(11.65)
F_2HN	Difluoramine	(11.53 ± 0.08)
F_2H_2Si	Difluorosilane	12.2
F_2Mg	Magnesium fluoride	(13.4 ± 0.4)
F_2N	Difluoroamidogen (NF_2)	11.628 ± 0.01
F_2N_2	*trans*-Difluorodiazine	(12.8)
F_2O	Oxygen difluoride	13.11 ± 0.01
F_2OS	Thionyl fluoride	12.25
F_2O_2S	Sufuryl fluoride	13.04 ± 0.01
F_2Pb	Lead fluoride (PbF_2)	(11.5)
F_2S	Sulfur fluoride (SF_2)	(10.08)
F_2Si	Difluorosilylene (SiF_2)	10.78 ± 0.05
F_2Sn	Tin fluoride (SnF_2)	(11.1)
F_2Xe	Xenon difluoride	12.35 ± 0.01
F_3HSi	Trifluorosilane	(14.0)
F_3N	Nitrogen trifluoride	13.00 ± 0.02
F_3NO	Trifluoramine oxide	13.26 ± 0.01
F_3OP	Phosphoryl fluoride	12.76 ± 0.01
F_3P	Phosphorous trifluoride	11.44
F_3PS	Thiophosphoryl trifluoride	$\leq 11.05 \pm 0.035$
F_3Si	Trifluorosilyl (SiF_3)	(9.3)
F_4Ge	Germanium tetrafluoride	(15.5)
F_4N_2	Tetrafluorohydrazine (gauche)	11.94 ± 0.03
F_4S	Sulfur fluoride (SF_4)	12.03 ± 0.05
F_4Si	Silicon tetrafluoride	(15.7)
F_4Xe	Xenon tetrafluoride	12.65 ± 0.1
F_5I	Iodine pentafluoride	12.943 ± 0.005
F_5P	Phosphorous pentafluoride	(15.1)
F_5S	Sulfur pentafluoride	10.5 ± 0.1

Table 3.3. *Continued*

Molecular formula	Molecular name	IP [eV]
F_6Mo	Molybdenum fluoride (MoF_6)	(14.5 ± 0.1)
F_6S	Sulfur fluoride (SF_6)	15.33 ± 0.03
F_6U	Uranium fluoride (UF_6)	14.00 ± 0.10
GaI_3	Gallium iodide (GaI_3)	9.40
GeH_4	Germane	11.33
GeI_4	Germanium tetraiodide	(9.42)
GeO	Germanium oxide	11.25 ± 0.01
GeS	Germanium sulfide (GeS)	9.98 ± 0.02
HI	Hydrogen iodide	10.386 ± 0.001
HLi	Lithium hydride	7.7
HN	Imidogen (NH)	13.49 ± 0.01
HNO	Nitrosyl hydride	(10.1)
HNO_2	Nitrous acid (HONO)	≤ 11.3
HNO_3	Nitric acid	11.95 ± 0.01
HN_3	Hydrazoic acid	10.72 ± 0.025
HO	Hydroxyl (OH)	13.00
HO_2	Hydroperoxy (HOO)	11.35 ± 0.01
HS	Mercapto (SH)	10.37 ± 0.01
H_2	Hydrogen	15.42589 ± 0.00005
H_2N	Amidogen (NH_2)	11.14 ± 0.01
H_2O	Water	12.612 ± 0.010
H_2O_2	Hydrogen peroxide	10.54
H_2S	Hydrogen sulfide	10.453 ± 0.008
H_2Se	Hydrogen selenide	9.882 ± 0.001
H_2Si	Silylene (SiH_2)	8.92 ± 0.07
H_3N	Ammonia	10.16 ± 0.01
H_3NO	Hydroxylamine (NH_2OH)	10.00
H_3P	Phosphine	9.869 ± 0.002
H_3Sb	Stibine	9.54 ± 0.03
H_4N_2	Hydrazine	8.1 ± 0.15
H_4Si	Silane	11.65
H_4Sn	Stannane	(10.75)
H_6Si_2	Disilane	(9.7)
H_8Si_3	Trisilane	(9.2)
HgI_2	Mercury iodide (HgI_2)	9.5088 ± 0.0022
IK	Potassium iodide	(7.21 ± 0.3)
ILi	Lithium iodide	(7.5)
INa	Sodium iodide	7.64 ± 0.02
ITI	Thallium iodide	8.47 ± 0.02
I_2	Iodine	9.3995 ± 0.0012
I_4Ti	Titanium iodide (TiI_4)	(9.1)
I_4Zr	Zirconium iodide (ZrI_4)	(9.3)
KLi	Lithium potassium (LiK)	4.57 ± 0.04
KNa	Potassium sodium (KNa)	4.41636 ± 0.00017
K_2	Potassium	4.0637 ± 0.0002
LiNa	Lithium sodium	5.05 ± 0.04
LiO	Lithium oxide	(8.45 ± 0.20)
LiRb	Lithium rubidium	4.3 ± 0.1
Li_2	Lithium	5.1127 ± 0.0003
MgO	Magnesium oxide	9.7

Table 3.3. *Continued*

Molecular formula	Molecular name	IP [eV]
NO	Nitric oxide	9.26436 ± 0.00006
NO_2	Nitrogen dioxide	9.75 ± 0.01
NS	Nitrogen sulfide	8.87 ± 0.01
NP	Phosphorous nitride (PN)	11.85
N_2	Nitrogen	15.5808
N_2O	Nitrous oxide	12.886
N_2O_4	Nitrogen tetroxide	10.8 ± 0.2
N_2O_5	Nitrogen pentoxide	(11.9)
NaRb	Rubidium sodium (RbNa)	4.32 ± 0.04
Na_2	Sodium	4.88898 ± 0.00016
OPb	Lead oxide (PbO)	9.08 ± 0.10
OS	Sulfur oxide (SO)	10.32 ± 0.02
OS_2	Sulfur oxide (SSO)	10.54 ± 0.04
OSi	Silicon oxide (SiO)	11.43
OSn	Tin oxide (SnO)	9.60 ± 0.02
OSr	Strontium oxdie	7.0 ± 0.15
O_2	Oxygen	12.071 ± 0.001
O_2S	Sulfur dioxide	12.32 ± 0.02
O_2Th	Thorium oxide (ThO_2)	(8.7 ± 0.15)
O2Ti	Titanium oxide (TiO_2)	(9.54 ± 0.1)
O_2U	Uranium oxide (UO_2)	(5.4 ± 0.1)
O_3	Ozone	12.43
O_3S	Sulfur trioxide	12.80 ± 0.04
O_3U	Uranium oxide (UO_3)	(10.5 ± 0.5)
O_4Os	Osmium oxide (OsO_4)	12.320
O_4Ru	Ruthenium oxide (RuO_4)	12.15 ± 0.03
O_7Re_2	Rhenium oxide (Re_2O_7)	(12.7 ± 0.2)
P_2	Phosphorous	10.53
PbS	Lead sulfide (PbS)	(8.5 ± 0.5)
SSn	Tin sulfide (SnS)	(8.8)
S_2	Sulfur	9.356 ± 0.002
Compounds containing carbon		
$CBrClF_2$	Bromochlorodifluoromethane	(≤ 11.83)
$CBrCl_3$	Bromotrichloromethane	(10.6)
$CBrF_3$	Bromotrifluoromethane	11.4
CBr_2F_2	Dibromodifluoromethane	11.07 ± 0.03
CBr_4	Tetrabromomethane	(10.31 ± 0.02)
CCl	Chloromethylidene	(8.9 ± 0.2)
$CClF_3$	Chlorotrifluoromethane	12.39
CClN	Cyanogen chloride	12.34 ± 0.01
CCl_2	Dichloromethylene	10.36
CCl_2F_2	Dichlorodifluoromethane	11.75 ± 0.04
CCl_2O	Carbonyl chloride	(11.4)
CCl_3F	Trichlorofluoromethane	11.77 ± 0.02
CCl_4	Tetrachloromethane	11.47 ± 0.01
CF	Fluoromethylidene	9.11 ± 0.01
CFN	Cyanogen fluoride	13.32 ± 0.01
CF_2	Difluoromethylene	11.42 ± 0.01
CF_2O	Carbonyl fluoride	13.03

Table 3.3. *Continued*

Molecular formula	Molecular name	IP [eV]
CF_3	Trifluoromethyl	($\leq$ 8.9)
CF_3I	Trifluoroiodomethane	10.23
CN	Cyanide	(14.09)
CNO	Cyanate (NCO)	(11.76 $\pm$ 0.01)
CO	Carbon monoxide	14.0139
COS	Carbon oxysulfide	11.1736 $\pm$ 0.0015
COSe	Carbon oxyselenide	10.36 $\pm$ 0.01
CO_2	Carbon dioxide	13.773 $\pm$ 0.002
CS	Carbon sulfide	11.33 $\pm$ 0.01
CS_2	Carbon disulfide	10.0685 $\pm$ 0.0020
CH	Methylidyne	10.64 $\pm$ 0.01
$CHBrCl_2$	Bromodichloromethane	10.6
$CHBr_2Cl$	Chlorodibromomethane	10.59 $\pm$ 0.01
$CHBr_3$	Tribromomethane	10.48 $\pm$ 0.02
CHCl	Chloromethylene (HCCl)	9.84
$CHClF_2$	Chlorodifluoromethane	(12.2)
$CHCl_2F$	Dichlorofluoromethane	(11.5)
$CHCl_3$	Trichloromethane	11.37 $\pm$ 0.02
CHF	Fluoromethylene (HCF)	(10.49)
CHF_3	Trifluoromethane	13.86
CHI_3	Triiodomethane	9.25 $\pm$ 0.02
CHN	Hydrogen cyanide	13.60 $\pm$ 0.01
CHN	Hydrogen isocyanide (HNC)	(12.5 $\pm$ 0.1)
CHNO	Fulminic acid (HCNO)	(10.83)
CHNO	Isocyanic acid (HNCO)	11.61 $\pm$ 0.03
CHO	Oxomethyl (HCO)	8.10 $\pm$ 0.05
CH_2	Methylene	10.396 $\pm$ 0.003
CH_2BrCl	Bromochloromethane	10.77 $\pm$ 0.01
CH_2Br_2	Dibromomethane	10.50 $\pm$ 0.02
CH_2ClF	Chlorofluoromethane	11.71 $\pm$ 0.01
CH_2Cl_2	Dichloromethane	11.32 $\pm$ 0.01
CH_2F_2	Difluoromethane	12.71
CH_2I_2	Diiodomethane	9.46 $\pm$ 0.02
CH_2N_2	Cyanamide	(10.4)
CH_2N_2	Diazomethane	8.999 $\pm$ 0.001
CH_2O	Formaldehyde	10.874 $\pm$ 0.002
CH_2O_2	Formic acid	11.33 $\pm$ 0.01
CH_3	Methyl	9.84 $\pm$ 0.01
CH_3BO	Carbonyltrihydroboron (BH_3CO)	11.14 $\pm$ 0.02
CH_3Br	Bromomethane	10.541 $\pm$ 0.003
CH_3Cl	Chloromethane	11.22 $\pm$ 0.01
CH_3Cl_3Si	Methyltrichlorosilane	(11.36 $\pm$ 0.03)
CH_3F	Fluoromethane	12.47 $\pm$ 0.02
CH_3I	Iodomethane	9.538
CH_3NO	Formamide	10.16 $\pm$ 0.06
CH_3NO_2	Nitromethane	11.02 $\pm$ 0.04
CH_3N_3	Methyl azide	9.81 $\pm$ 0.02
CH_3O	Methoxy	(8.6)
CH_4	Methane	12.51
CH_4N_2O	Urea	9.7

Table 3.3. *Continued*

Molecular formula	Molecular name	IP [eV]
CH_4O	Methanol	10.85 ± 0.01
CH_4S	Methanethiol	9.44 ± 0.005
CH_5N	Methylamine	8.97 ± 0.02
CH_6N_2	Methylhydrazine	7.67 ± 0.02
CH_6Si	Methylsilane	10.7
C_2	Carbon	12.11
$C_2Br_2F_4$	1,2-Dibromotetrafluoroethane	(11.1)
C_2ClF_3	Chlorotrifluoroethylene	9.81 ± 0.03
C_2ClF_5	Chloropentafluoroethane	(12.6)
C_2Cl_2	Dichloroacetylene	10.09
$C_2Cl_2F_4$	1,2-Dichlorotetrafluoroethane	12.2
$C_2Cl_3F_3$	1,1,1-Trichlorotrifluoroethane	11.5
$C_2Cl_3F_3$	1,1,2-Trichlorotrifluoroethane	11.99 ± 0.02
C_2Cl_4	Tetrachloroethylene	9.32
$C_2Cl_4F_2$	Tetrachloro-1,2-difluoroethane	11.3
C_2Cl_4O	Trichloroacetyl chloride	(11.0)
C_2Cl_6	Hexachloroethane	11.1
C_2F_3N	Trifluoroacetonitrile	13.86
C_2F_4	Tetrafluoroethylene	10.12 ± 0.02
C_2F_6	Hexafluoroethane	(13.4)
C_2N_2	Cyanogen	13.37 ± 0.01
C_2H	Ethynyl (HCC)	(11.7)
C_2HBr	Bromoacetylene	10.31 ± 0.02
$C_2HBrClF_3$	1,1,1-Trifluoro-2-bromo-2-chloroethane	11.0
C_2HCl	Chloroacetylene	10.58 ± 0.02
C_2HClF_2	1-Chloro-2,2-difluoroethylene	9.80 ± 0.04
C_2HCl_3	Trichloroethylene	9.47 ± 0.01
C_2HCl_3O	Dichloroacetyl chloride	(11.0)
C_2HCl_5	Pentachloroethane	(11.0)
C_2HF	Fluoroacetylene	11.26
C_2HF_3	Trifluoroethylene	10.14
$C_2HF_3O_2$	Trifluoroacetic acid	11.46
C_2H_2	Acetylene	11.400 ± 0.002
$C_2H_2Cl_2$	1,1-Dichloroethylene	9.79 ± 0.04
$C_2H_2Cl_2$	*cis*-1,2-Dichloroethylene	9.66 ± 0.01
$C_2H_2Cl_2$	*trans*-1,2-Dichloroethylene	9.65 ± 0.02
$C_2H_2Cl_2O$	Chloroacetyl chloride	(11.0)
$C_2H_2Cl_4$	1,1,1,2-Tetrachloroethane	(11.1)
$C_2H_2Cl_4$	1,1,2,2-Tetrachloroethane	($\leq$ 11.62)
$C_2H_2F_2$	1,1-Difluoroethylene	10.29 ± 0.01
$C_2H_2F_2$	*cis*-1,2-Difluoroethylene	10.23
C_2H_2O	Ketene	9.61 ± 0.02
$C_2H_2O_2$	Glyoxal	10.1
C_2H_3Br	Bromoethylene	9.80 ± 0.02
C_2H_3Cl	Chloroethylene	9.99 ± 0.02
$C_2H_3ClF_2$	1-Chloro-1,1-difluoroethane	11.98 ± 0.01
C_2H_3ClO	Acetyl chloride	10.85 ± 0.05
C_2H_3ClO	Chloroacetaldehyde	10.48 ± 0.03
$C_2H_3ClO_2$	Chloroacetic acid	(10.7)
$C_2H_3Cl_3$	1,1,1-Trichloroethane	(11.0)
$C_2H_3Cl_3$	1,1,2-Trichloroethane	11.0

Table 3.3. *Continued*

Molecular formula	Molecular name	IP [eV]
C_2H_3F	Fluoroethylene	10.363 ± 0.015
C_2H_3FO	Acetyl fluoride	11.51 ± 0.02
$C_2H_3F_3$	1,1,1-Trifluoroethane	12.9 ± 0.1
C_2H_3N	Acetonitrile	12.194 ± 0.005
C_2H_3NO	Methylisocyanate	(10.67 ± 0.0_)
C_2H_4	Ethylene	10.507 ± 0.004
$C_2H_4Br_2$	1,2-Dibromoethane	10.37
$C_2H_4Cl_2$	1,1-Dichloroethane	11.06
$C_2H_4Cl_2$	1,2-Dichloroethane	11.04
$C_2H_4F_2$	1,1-Difluoroethane	11.87 ± 0.03
C_2H_4O	Acetaldehyde	10.229 ± 0.0007
C_2H_4O	Ethylene oxide	10.566 ± 0.01
$C_2H_4O_2$	Acetic acid	10.66 ± 0.02
$C_2H_4O_2$	Methyl formate	10.815 ± 0.005
C_2H_5Br	Bromoethane	10.28
C_2H_5Cl	Chloroethane	10.97 ± 0.02
C_2H_5ClO	2-Chloroethanol	(10.52)
C_2H_5F	Fluoroethane	(11.6)
C_2H_5I	Iodoethane	9.346
C_2H_5N	Ethyleneimine	9.2 ± 0.1
C_2H_5NO	Acetamide	9.65 ± 0.03
C_2H_5NO	*N*-Methylformamide	9.79
$C_2H_5NO_2$	Nitroethane	10.88 ± 0.05
C_2H_6	Ethane	11.52 ± 0.01
$C_2H_6Cl_2Si$	Dichlorodimethylsilane	(10.7)
C_2H_6O	Dimethyl ether	10.025 ± 0.025
C_2H_6O	Ethanol	10.47 ± 0.02
C_2H_6OS	Dimethyl sulfoxide	(9.01)
$C_2H_6O_2$	Ethylene glycol	10.16
C_2H_6S	Dimethyl sulfide	8.69 ± 0.01
C_2H_6S	Ethanethiol	9.285 ± 0.005
$C_2H_6S_2$	Dimethyl disulfide	(7.4 ± 0.3)
C_2H_7N	Dimethylamine	8.23 ± 0.08
C_2H_7N	Ethylamine	8.86 ± 0.02
C_2H_7NO	Ethanolamine	8.96
$C_2H_8N_2$	1,1-Dimethylhydrazine	7.28 ± 0.04
$C_2H_8N_2$	1,2-Ethanediamine	(8.6)
C_3F_6	Perfluoropropene	10.60 ± 0.03
C_3F_6O	Perfluoroacetone	(11.44)
C_3F_8	Perfluoropropane	13.38
C_3HN	Cyanoacetylene	11.64 ± 0.01
C_3H_2O	Propynal	(10.8)
$C_3H_3F_3$	3,3,3-Trifluoropropene	(10.9)
C_3H_3N	Propenenitrile	10.91 ± 0.01
C_3H_3NO	Oxazole	(9.6)
C_3H_3NO	Isoxazole	9.93 ± 0.05
C_3H_4	Propyne	10.36 ± 0.01
C_3H_4	Allene	9.69 ± 0.01
C_3H_4	Cyclopropene	9.67 ± 0.01
$C_3H_4N_2$	Imidazole	8.81 ± 0.01
C_3H_4O	Propenal	10.103 ± 0.006

Table 3.3. *Continued*

Molecular formula	Molecular name	IP [eV]
C_3H_4O	Propargyl alcohol	10.51
C_3H_4O	Cyclopropanone	(9.1 ± 0.1)
$C_3H_4O_2$	Propenoic acid	10.60
$C_3H_4O_2$	2-Oxetanone	(9.70 ± 0.01)
C_3H_5Br	3-Bromopropene	10.06
C_3H_5Cl	3-Chloropropene	9.9
C_3H_5ClO	Epichlorohydrin	(10.2)
$C_3H_5ClO_2$	Methyl chloroacetate	(10.3)
C_3H_5F	3-Fluoropropene	10.11
C_3H_5N	Propanenitrile	11.84 ± 0.02
C_3H_5NO	Propenamide	9.5
C_3H_6	Propene	9.73 ± 0.02
C_3H_6	Cyclopropane	9.86
$C_3H_6Br_2$	1,2-Dibromopropane	10.1
$C_3H_6Br_2$	1,3-Dibromopropane	≤ 10.26
$C_3H_6Cl_2$	1,2-Dichloropropane	(10.87 ± 0.05)
$C_3H_6Cl_2$	1,3-Dichloropropane	10.85 ± 0.05
C_3H_6O	Acetone	9.705
C_3H_6O	Allyl alcohol	9.67 ± 0.05
C_3H_6O	Propanal	9.953 ± 0.005
C_3H_6O	Methyloxirane	10.22 ± 0.02
C_3H_6O	Oxetane	9.668 ± 0.005
C_3H_6O	Methyl vinyl ether	(8.93 ± 0.02)
$C_3H_6O_2$	Ethyl formate	10.61 ± 0.01
$C_3H_6O_2$	Methyl acetate	10.27 ± 0.02
$C_3H_6O_2$	Propanoic acid	10.525 ± 0.003
$C_3H_6O_2$	1,3-Dioxolane	(9.9)
$C_3H_6O_3$	Trioxane	(10.3)
C_3H_6S	Thiacyclobutane	8.69
C_3H_7Br	1-Bromopropane	10.18 ± 0.01
C_3H_7Br	2-Bromopropane	10.07 ± 0.01
C_3H_7Cl	1-Chloropropane	10.82 ± 0.03
C_3H_7Cl	2-Chloropropane	10.78 ± 0.02
C_3H_7F	1-Fluoropropane	(11.3)
C_3H_7F	2-Fluoropropane	(11.08 ± 0.02)
C_3H_7I	1-Iodopropane	9.269
C_3H_7I	2-Iodopropane	9.175
C_3H_7N	Allylamine	8.76
C_3H_7N	Cyclopropylamine	(8.7)
C_3H_7N	Propyleneimine	(9.0)
C_3H_7NO	*N,N*-Dimethylformamide	9.13 ± 0.02
$C_3H_7NO_2$	1-Nitropropane	10.81 ± 0.03
$C_3H_7NO_2$	2-Nitropropane	10.71 ± 0.05
C_3H_8	Propane	10.95 ± 0.05
C_3H_8O	1-Propanol	10.22 ± 0.03
C_3H_8O	2-Propanol	10.12 ± 0.08
C_3H_8O	Ethyl methyl ether	9.72
$C_3H_8O_2$	2-Methoxyethanol	9.6
$C_3H_8O_2$	Dimethoxymethane	9.5
C_3H_8S	Ethyl methyl sulfide	8.54 ± 0.1

Table 3.4 lists the values of the electron affinities of diatomic and polyatomic molecules. Basic information about the measurement methods and values of electron affinities of molecules have been presented in [3.1, 6, 9–12].

Table 3.4. Electron affinities of molecules

Molecule	Electron affinity [eV]	Uncertainty [eV]	Molecule	Electron affinity [eV]	Uncertainty [eV]
Diatomic molecules			KRb	0.486	0.020
Ag_2	1.023	0.007	LiCl	0.593	0.010
Al_2	1.10	0.15	LiD	0.337	0.012
As_2	0	–	LiH	0.342	0.012
AsH	1.0	0.1	MgCl	1.589	0.011
Au_2	1.938	0.007	MgH	1.05	0.06
BO	3.12	0.09	MgI	1.899	0.018
BeH	0.7	0.1	MnD	0.866	0.010
Bi_2	1.271	0.008	MnH	0.869	0.010
Br_2	2.55	0.10	NH	0.370	0.004
BrO	2.353	0.006	NO	0.026	0.005
C_2	3.269	0.006	NS	1.194	0.011
CH	1.238	0.008	Na_2	0.430	0.015
CN	3.862	0.004	NaBr	0.788	0.010
CS	0.205	0.021	NaCl	0.727	0.010
CaH	0.93	0.05	NaF	0.520	0.010
Cl_2	2.38	0.10	NaI	0.865	0.010
ClO	2.275	0.006	NaK	0.465	0.030
Co_2	1.110	0.008	Ni_2	0.926	0.010
CoD	0.680	0.010	NiD	0.477	0.007
CoH	0.671	0.010	NiH	0.481	0.007
Cr_2	0.505	0.005	O_2	0.451	0.007
CrD	0.568	0.010	OD	1.825548	0.000037
CrH	0.563	0.010	OH	1.827670	0.000021
CrO	1.222	0.010	P_2	0.589	0.025
Cs_2	0.469	0.015	PH	1.028	0.010
CsCl	0.455	0.010	PO	1.092	0.010
Cu_2	0.836	0.006	Pb_2	1.366	0.010
CuO	1.777	0.006	PbO	0.722	0.006
F_2	3.08	0.10	Pd_2	1.685	0.008
FO	2.272	0.006	Pt_2	1.898	0.008
Fe_2	0.902	0.008	PtN	1.240	0.010
FeD	0.932	0.015	Rb_2	0.498	0.015
FeH	0.934	0.011	RbCl	0.544	0.010
FeO	1.493	0.005	RbCs	0.478	0.020
I_2	2.55	0.05	Re_2	1.571	0.008
IBr	2.55	0.10	S_2	1.670	0.015
IO	2.378	0.006	SD	2.315	0.002
K_2	0.497	0.012	SF	2.285	0.006
KBr	0.642	0.010	SH	2.314344	0.000004
KCl	0.582	0.010	SO	1.125	0.005
KCs	0.471	0.020	Sb_2	1.282	0.008
KI	0.728	0.010	Se_2	1.94	0.07

Table 3.4. *Continued*

Molecule	Electron affinity [eV]	Uncertainty [eV]
SeH	2.212520	0.000025
SeO	1.456	0.020
Si_2	2.201	0.010
SiH	1.277	0.009
Sn_2	1.962	0.010
SnPb	1.569	0.008
Te_2	1.92	0.07
TeH	2.102	0.015
TeO	1.697	0.022
ZnH	≤ 0.95	–
Triatomic molecules		
Ag_3	2.32	0.05
Al_3	1.4	0.15
AsH_2	1.27	0.03
Au_3	3.7	0.3
BO_2	3.57	0.13
BO_2	4.3	0.2
Bi_3	1.60	0.03
C_3	1.981	0.020
CCl_2	1.591	0.010
CD_2	0.645	0.006
CDF	0.535	0.005
CF_2	0.165	0.010
CH_2	0.652	0.006
CHBr	1.454	0.005
CHCl	1.210	0.005
CHF	0.542	0.005
CHI	1.42	0.17
C_2H	2.969	0.006
C_2O	1.848	0.027
COS	0.46	0.20
CS_2	0.895	0.020
CoD_2	1.465	0.013
CoH_2	1.450	0.014
CrH_2	> 2.5	–
Cr_2D	1.464	0.005
Cr_2H	1.474	0.005
Cs_3	0.864	0.030
Cu_3	2.11	0.05
DCO	0.301	0.005
DNO	0.330	0.015
DO_2	1.089	0.017
DS_2	1.912	0.015
HS_2	1.907	0.015
FeCO	1.157	0.005
FeD_2	1.038	0.013
FeH_2	1.049	0.014
GeH_2	1.097	0.015
HCO	0.313	0.005
HCl_2	4.896	0.005
HNO	0.338	0.015
HO_2	1.078	0.017
K_3	0.956	0.050
MnD_2	0.465	0.014
MnH_2	0.444	0.016
N_3	2.70	0.12
NCO	3.609	0.005
NCS	3.537	0.005
NH_2	0.771	0.005
N_2O	0.22	0.10
NO_2	2.273	0.005
Na_3	1.019	0.060
Ni_3	1.41	0.05
NiCO	0.804	0.012
NiD_2	1.926	0.007
NiH_2	1.934	0.008
O_3	2.1028	0.0025
O_2Ar	0.52	0.02
OClO	2.140	0.008
OIO	2.577	0.008
PH_2	1.271	0.010
PO_2	3.8	0.2
Pt_3	1.87	0.02
Pd_3	< 1.5	0.1
Rb_3	0.920	0.030
S_3	2.093	0.025
SO_2	1.107	0.008
S_2O	1.877	0.008
Sb_3	1.85	0.03
SeO_2	1.823	0.050
SiH_2	1.124	0.020
VO_2	2.3	0.2
Larger polyatomic molecules		
Al_3O	1.00	0.15
BD_3	0.027	0.014
BH_3	0.038	0.015
Bi_4	1.05	0.010
CF_3Br	0.91	0.2
CF_3I	1.57	0.2
$CO_3(H_2O)$	2.1	0.2
CHO_2	3.498	0.005
CH_2S	0.465	0.023
CH_3	0.08	0.03
CH_3I	0.2	0.1
CH_3NO_2	0.48	0.10
CH_3Si	0.852	0.010
CH_3Si	2.010	0.010

Table 3.4. *Continued*

Molecule	Electron affinity [eV]	Uncertainty [eV]	Molecule	Electron affinity [eV]	Uncertainty [eV]
CD_3O	1.552	0.022	$C_4H_2O_3$	1.44	0.10
CH_3O	1.570	0.022	C_4H_4N	2.39	0.13
CD_3S	1.856	0.006	C_4H_5O	1.801	0.008
CH_3S	1.861	0.004	$C_4H_6O_2$	0.69	0.10
CD_3S_2	1.748	0.022	C_4H_7O	1.67	0.05
CH_3S_2	1.757	0.022	C_4H_5DO	1.67	0.05
CH_3SiH_2	1.19	0.04	$C_4H_5D_2O$	1.75	0.06
CO_3	2.69	0.14	C_4H_9O	1.912	0.054
C_2F_2	2.255	0.006	C_4H_9S	2.03	0.02
C_2DO	2.350	0.020	C_4H_9S	2.07	0.02
C_2HO	2.350	0.020	C_4O	2.05	0.15
C_2D_2	0.492	0.006	C_4O_2	2.0	0.2
C_2HD	0.489	0.006	C_5	2.853	0.001
C_2HF	1.718	0.006	C_5F_5N	0.68	0.11
C_2H_2	0.490	0.006	$C_5F_6O_3$	1.5	0.2
C_2H_2FO	2.22	0.09	C_5D_5	1.790	0.008
C_2D_2N	1.538	0.012	C_5H_5	1.804	0.007
C_2D_2N	1.070	0.024	C_5H_7	0.91	0.03
C_2H_2N	1.543	0.014	C_5H_7O	1.598	0.007
C_2H_2N	1.059	0.024	CeF_4	3.8	0.4
C_2H_3	0.667	0.024	CoF_4	6.4	0.3
C_2D_3O	1.81897	0.00012	$Cr(CO)_3$	1.349	0.006
C_2H_3O	1.82476	0.00012	CrO_3	3.6	0.2
C_2H_5N	0.56	0.01	$Fe(CO)_2$	1.22	0.02
C_2D_5O	1.702	0.033	$Fe(CO)_3$	1.8	0.2
C_2H_5O	1.726	0.033	$Fe(CO)_4$	2.4	0.3
C_2H_5S	1.953	0.006	FeF_3	3.6	0.1
C_2H_5S	0.868	0.051	GeH_3	≤ 1.74	0.04
$C_2H_7O_2$	2.26	0.08	HNO_3	0.57	0.15
C_3H	1.858	0.023	IrF_4	4.7	0.3
C_3H_2	1.794	0.025	IrF_6	6.5	0.4
$C_3H_2F_3O$	2.625	0.010	MnF_4	5.5	0.2
C_3H_3	0.893	0.025	$Mo(CO)_3$	1.337	0.006
C_3H_2D	0.88	0.15	MoF_5	3.5	0.2
C_3D_2H	0.907	0.023	MoF_6	3.8	0.2
C_3H_3N	1.247	0.012	MoO_3	2.9	0.2
C_3D_5	0.381	0.025	NO_3	3.937	0.014
C_3H_5	0.362	0.019	$(NO)_2$	≥ 2.1	–
C_3H_4D	0.373	0.019	$Ni(CO)_2$	0.643	0.014
C_3H_5O	1.758	0.019	$Ni(CO)_3$	1.077	0.013
C_3H_5O	1.621	0.006	$OH(H_2O)$	< 2.95	0.15
$C_3H_5O_2$	1.80	0.06	OsF_4	3.9	0.3
C_3H_7O	1.789	0.033	OsF_6	6.0	0.3
C_3H_7O	1.839	0.029	PBr_3	1.59	0.15
C_3H_7S	2.00	0.02	PBr_2Cl	1.63	0.20
C_3H_7S	2.02	0.02	PCl_2Br	1.52	0.20
C_3O	1.34	0.15	PCl_3	0.82	0.10
C_3O_2	0.85	0.15	PO_3	4.5	0.5
$C_4F_4O_3$	0.5	0.2	$POCl_2$	3.83	0.25

Table 3.4. *Continued*

Molecule	Electron affinity [eV]	Uncertainty [eV]	Molecule	Electron affinity [eV]	Uncertainty [eV]
$POCl_3$	1.41	0.20	SiD_3	1.386	0.022
PtF_4	5.5	0.3	SiF_3	≤ 2.95	0.10
PtF_6	7.0	0.4	SiH_3	1.406	0.014
ReF_6	4.7	–	TeF_6	3.34	0.17
RhF_4	5.4	0.3	UF_5	3.7	0.2
RuF_4	4.8	0.3	UF_6	5.1	0.2
RuF_5	5.2	0.4	UO_3	<2.1	–
RuF_6	7.5	0.3	VF_4	3.5	0.2
SF_4	1.5	0.2	V_4O_{10}	4.2	0.6
SF_6	1.05	0.10	$W(CO)_3$	1.859	0.006
SO_3	≥ 1.70	0.15	WF_5	1.25	0.3
$(SO_2)_2$	0.6	0.2	WF_6	3.36	+0.04/ – 0.20
SeF_6	2.9	0.2	WO_3	3.33	+0.04/ – 0.15
Si_4	2.17	0.01	WO_3	3.9	0.2

3.4 Proton Affinities

The energy released by the addition of a proton (H^+) to a neutral molecule is called the proton-affinity energy of the molecule or simply proton affinity. The proton affinity of a molecule M is defined in terms of the hypothetical protonation reaction:

$$M + H^+ \rightarrow MH^+ \ .$$

The proton affinity is the negative of the enthalpy change associated with this reaction.

Values of proton affinities are listed in Table 3.5 for many diatomic and polyatomic molecules. The data are taken from [3.1, 13, 14]. The first column contains empirical formulae of molecules and the second the accepted chemical formulae of molecules or their names. The constituent atoms are given in the empirical formulae alphabetically, and the molecules in the table are arranged in alphabetical order of these empirical formulae.

The numerical data in Table 3.5 are listed to tenth of kcal/mol, as in the original papers. The real absolute uncertainties can exceed several kcal/mol [3.13].

Table 3.5. Proton affinities of molecules

Empirical formula	Molecular formula or name	Proton affinity [kcal/mol]
$[AsC_3H_9]$	$(CH_3)_3As$	213.4
$[AsF_3]$	AsF_3	155
$[AsH_3]$	AsH_3	179.2
$[B_2H_6]$	B_2H_6	~146
$[B_4H_8]$	B_4H_8	188
$[B_4H_{10}]$	B_4H_{10}	~144
$[B_5H_8]$	B_5H_8	184
$[B_5H_9]$	B_5H_9	169
$[BrCH_3]$	CH_3Br	165.7
[BrCN]	BrCN	178.3
$[BrC_6H_5]$	Bromobenzene	182.4
[BrH]	HBr	136
$[CClH_3]$	CH_3Cl	~168
[CClN]	ClCN	175.7
$[CCl_2]$	CCl_2	~200.0
$[CFH_3]$	CH_3F	150
$[CF_2]$	CF_2	171.9
$[CF_2H_2]$	CH_2F_2	147
$[CF_2O]$	F_2CO	160.5
$[CF_3H]$	CHF_3	147
$[CF_3HO_3S]$	CF_3SO_3H	~169
$[CF_3NO]$	CF_3NO	169
$[CF_4]$	CF_4	~126
[CHN]	HNC	190.2
[CHN]	HCN	171.4
[CHNO]	HNCO	173.3
[CHO]	HCO	152
$[CH_2N_2]$	CH_2N_2	205
$[CH_2O]$	HCOH	229
$[CH_2O]$	H_2CO	171.7
$[CH_2O_2]$	HCOOH	178.8
$[CH_2S]$	CH_2S	186
$[CH_3I]$	CH_3I	~171
$[CH_3NO]$	$HCONH_2$	198.4
$[CH_3NO_2]$	CH_3ONO	192.5
$[CH_3NO_2]$	CH_3NO_2	179.2
$[CH_4]$	CH_4	132.0
$[CH_4N]$	CH_2NH_2	199
$[CH_4O]$	CH_3OH	181.9
$[CH_4S]$	CH_3SH	187.4
$[CH_5N]$	CH_3NH_2	214.1
$[CH_5P]$	CH_3PH_2	204.1
$[CH_6N_2]$	CH_3NHNH_2	214.1
[CO]	CO	141.9
[COS]	COS	151
$[CO_2]$	CO_2	130.9
[CS]	CS	175
$[CS_2]$	CS_2	167.1
$[C_2BrH_5]$	C_2H_5Br	~171

Table 3.5. *Continued*

Empirical formula	Molecular formula or name	Proton affinity [kcal/mol]
[C_2ClH_2N]	$ClCH_2CN$	179.5
[$C_2ClH_3O_2$]	$CH_2ClCOOH$	182.4
[C_2ClH_5]	C_5H_5Cl	169
[$C_2Cl_3HO_2$]	CCl_3COOH	183.5
[$C_2Cl_3H_3O$]	Cl_3CCH_2OH	177.4
[C_2Cl_3N]	CCl_3CN	175.8
[C_2D_6O]	$(CD_3)_2O$	190.6
[C_2FH_3]	C_2H_3F	175
[$C_2FH_3O_2$]	CH_2FCOOH	183.5
[C_2FH_5]	C_2H_5F	165
[C_2FH_6N]	$CH_2FCH_2NH_2$	212.3
[$C_2F_2H_2$]	CH_2CF_2	176
[$C_2F_2H_2$]	(E)–CHFCHF	166
[$C_2F_2H_4O$]	CF_2HCH_2OH	176.2
[$C_2F_2H_5N$]	$CF_2HCH_2NH_2$	207.5
[C_2F_3H]	C_2F_3H	~169
[C_2F_3HO]	CF_3CHO	165.1
[$C_2F_3HO_2$]	CF_3COOH	169.0
[$C_2F_3H_3O$]	CF_3CH_2OH	169.0
[$C_2F_3H_4N$]	$CF_3CH_2NH_2$	202.5
[C_2F_3N]	CF_3CN	166.1
[C_2F_4O]	CF_3CFO	160.2
[C_2H_2]	C_2H_2	153.3
[C_2H_2O]	$CH_2C{=}O$	198.0
[C_2H_3]	C_2H_3 radical	~181
[C_2H_3N]	CH_3CN	188.4
[C_2H_3NS]	CH_3SCN	195.9
[C_2H_3NS]	CH_3NCS	195.9
[C_2H_4]	C_2H_4	162.6
[$C_2H_4N_2$]	$NCCH_2NH_2$	197.4
[C_2H_4O]	c–C_2H_4O (Oxirane)	187.9
[C_2H_4O]	CH_3CHO	186.6
[$C_2H_4O_2$]	CH_3COOH	190.2
[$C_2H_4O_2$]	HCO_2CH_3	188.9
[C_2H_4S]	c–C_2H_4S (Thiirane)	194.6
[C_2H_5I]	C_2H_5I	~176
[C_2H_5N]	Aziridine (Azirane)	215.7
[C_2H_5N]	$CH_2{=}CHNH_2$	219.1
[C_2H_5N]	$CH_3CH{=}NH$	213.9
[C_2H_5NO]	CH_3CONH_2	206.2
[C_2H_5NO]	$HCONHCH_3$	205.8
[$C_2H_5NO_2$]	NH_2CH_2COOH (Glycine)	211.6
[$C_2H_5NO_2$]	C_2H_5ONO	197.3
[$C_2H_5NO_2$]	$C_2H_5NO_2$	184.8
[C_2H_5P]	c–C_2H_4PH (Phosphirane)	191.4
[C_2H_6]	C_2H_6	143.6
[C_2H_6Hg]	CH_3HgCH_3	~186
[$C_2H_6N_2$]	(E)–$CH_3N{=}NCH_3$	206.9
[C_2H_6O]	$(CH_3)_2O$	192.1

Table 3.5. *Continued*

Empirical formula	Molecular formula or name	Proton affinity [kcal/mol]
[C_2H_6O]	C_2H_5OH	188.3
[C_2H_6OS]	$(CH_3)_2SO$	211.3
[C_2H_6S]	$(CH_3)_2S$	200.6
[C_2H_6S]	C_2H_5SH	190.8
[$C_2H_6S_2$]	CH_3SSCH_3	~196
[C_2H_7N]	$(CH_3)_2NH$	220.6
[C_2H_7N]	$C_2H_5NH_2$	217.0
[C_2H_7NO]	$NH_2(CH_2)_2OH$	221.3
[$C_2H_7O_3P$]	$(CH_3O)_2PHO$	207.2
[C_2H_7P]	$(CH_3)_2PH$	216.3
[$C_2H_8N_2$]	1,2–Diaminoethane	225.9
[$C_2H_8N_2$]	$(CH_3)_2NNH_2$	219.9
[C_2N_2]	NCCN	162
[C_3]	C_3	~185
[C_3ClH_4N]	Cl $(CH_2)_2CN$	187.5
[C_3FH_5O]	CH_3COCH_2F	192.0
[C_3FH_8N]	$FCH_2CH_2CH_2NH_2$	217.8
[$C_3F_2H_4O$]	CFH_2COCFH_2	187
[$C_3F_3H_3O$]	CH_3COCF_3	174.2
[$C_3F_3H_3O_2$]	$HCOOCH_2CF_3$	179.4
[$C_3F_3H_3O_2$]	CF_3COOCH_3	178.8
[$C_3F_3H_6N$]	$CF_3CH_2CH_2NH_2$	210.6
[$C_3F_3H_6N$]	$CF_3CH_2NHCH_3$	209.8
[$C_3F_3H_6N$]	$CF_3N(CH_3)_2$	193.8
[$C_3F_4H_2O$]	CF_2HCOCF_2H	170
[C_3F_5N]	C_2F_5CN	167.1
[$C_3F_6H_2O$]	$(CF_3)_2CHOH$	165.0
[C_3F_6O]	$(CF_3)_2CO$	161.5
[C_3GeH_8]	$(CH_3)_2Ge{=}CH_2$	204.9
[C_3HN]	HCCCN	184
[$C_3H_2N_2$]	$CH_2(CN)_2$	175.6
[C_3H_3]	c–C_3H_3 radical	175.8
[C_3H_3N]	$CH_2{=}CHCN$	189.7
[C_3H_3NO]	Oxazole	208.4
[C_3H_3NO]	Isooxazole	202.3
[C_3H_3NS]	Thiazole	213.2
[$C_3H_3N_3$]	1,3,5-Triazine	201.1
[C_3H_4]	Cyclopropene	198
[C_3H_4]	$H_2C{=}C{=}CH_2$	186.3
[C_3H_4]	CH_3 CCH	182
[$C_3H_4N_2$]	Imidazole	219.8
[$C_3H_4N_2$]	Pyrazole	209.8
[C_3H_4O]	$CH_3CH{=}CO$	199.4
C_3H_4O]	$CH_2{=}CHCHO$	193.9
[C_3H_5]	c-C_3H_5 radical	188
[C_3H_5]	$CH_2{=}CH{-}CH_2$ radical	175.8
[C_3H_5N]	l-Azabicyclo [1.1.0] butane	212
[C_3H_5N]	$HCCCH_2NH_2$	210.8
[C_3H_5N]	C_2H_5CN	192.6

Table 3.5. *Continued*

Empirical formula	Molecular formula or name	Proton affinity [kcal/mol]
[$C_3H_5O_3P$]	2,6,7-Trioxa-l-phosphabicyclol[2.2.1.] heptane	194.0
[C_3H_6]	c-C_3H_6	179.8
[C_3H_6]	$CH_3CH{=}CH_2$	179.5
[$C_3H_6N_2$]	$H_2N\ (CH_2)_2\ CN$	207.0
[$C_3H_6N_2$]	CH_3NHCH_2CN	206.0
[C_3H_6O]	$CH_2{=}CHOCH_3$	207.4
[C_3H_6O]	c-C_3H_6O (Oxetane)	196.9
[C_3H_6O]	$(CH_3)_2\ CO$	196.7
[C_3H_6O]	2-Methyloxirane	194.7
[C_3H_6O]	C_2H_5CHO	189.6
[$C_3H_6O_2$]	CH_3COOCH_3	197.8
[$C_3H_6O_2$]	$HCO_2C_2H_5$	193.1
[$C_3H_6O_2$]	C_2H_5COOH	191.8
[$C_3H_6O_3$]	$(CH_3O)_2CO$	200.2
[C_3H_6S]	Thietane	201.3
[C_3H_6S]	2-Methylthiirane	200.6
[C_3H_7]	i-C_3H_7	159.8
[C_3H_7N]	$CH_2{=}C(CH_3)\ NH_2$	226.3
[C_3H_7N]	Azetidine	223.5
[C_3H_7N]	N-Methylaziridine	221.6
[C_3H_7N]	$(CH_3)_2\ C{=}NH$	221
[C_3H_7N]	2-Methylaziridine	219.2
[C_3H_7N]	$H_2C{=}CHCH_2NH_2$	215.8
[C_3H_7N]	c-$C_3H_5NH_2$	215.0
[C_3H_7NO]	$(CH_3)_2\ NCHO$	211.4
[$C_3H_7NO_2$]	Sarcosine	218.7
[$C_3H_7NO_2$]	L-Alanine	214.8
[$C_3H_7NO_2$]	i-C_3H_7ONO	201.9
[$C_3H_7NO_2S$]	L-Cysteine	214.3
[$C_3H_7NO_3$]	L-Serine	216.8
[$C_3H_7O_3P$]	2-Methoxy-1,3,2-dioxaphospholane	212.7
[C_3H_8]	C_3H_8	150
[C_3H_8O]	$CH_3OC_2H_5$	196.4
[C_3H_8O]	i-C_3H_7OH	191.2
[C_3H_8O]	n-C_3H_7OH	190.8
[C_3H_8Pb]	$(CH_3)_2\ Pb{=}CH_2$	223.9
[C_3H_8S]	$CH_3SC_2H_5$	203.5
[C_3H_8S]	i-C_3H_7SH	194.1
[C_3H_8S]	n-C_3H_7SH	191.6
[C_3H_8Si]	$(CH_3)_2\ Si{=}CH_2$	226.4
[C_3H_8Sn]	$(CH_3)_2\ Sn{=}CH_2$	215.8
[C_3H_9N]	$(CH_3)_3N$	225.1
[C_3H_9N]	$(CH_3)\ (C_2H_5)\ NH$	222.8
[C_3H_9N]	i-$C_3H_7NH_2$	218.6
[C_3H_9N]	n-$C_3H_7NH_2$	217.9
[C_3H_9NO]	$NH_2(CH_2)_3\ OH$	228.6
[C_3H_9NO]	$CH_3OCH_2CH_2NH_2$	223.3
[$C_3H_9O_3P$]	$P(OCH_3)_3$	220.6

Table 3.5. *Continued*

Empirical formula	Molecular formula or name	Proton affinity [kcal/mol]
[$C_3H_9O_3PS$]	SP $(OCH_3)_3$	214.5
[$C_3H_9O_4P$]	OP $(OCH_3)_3$	212.0
[C_3H_9P]	$(CH_3)_3P$	227.1
[$C_3H_{10}N_2$]	1,3-Diaminopropane	234.1
[$C_4F_2H_7NO$]	CF_2 $HCON(CH_3)_2$	207.2
[$C_4F_3H_5O_2$]	$CF_3CO_2C_2H_5$	184.6
[$C_4F_3H_7O$]	$C_2H_5OCH_2CF_3$	186.4
[$C_4F_3H_8N$]	$CF_3CH_2N(CH_3)_2$	215.0
[$C_4F_3H_8N$]	$CF_3(CH_2)_3NH_2$	214.3
[$C_4F_4H_4O_2$]	$CF_3COOCH_2CH_2F$	178.6
[$C_4F_6H_4O$]	$(CF_3)_2C(CH_3)OH$	167.0
[C_4F_7N]	C_3F_7CN	167.4
[C_4F_9HO]	$(CF_3)_3COH$	163.1
[$C_4F_9H_2N$]	$(CF_3)_3CNH_2$	191.5
[C_4H_4O]	Furan	192.2
[C_4H_4S]	c-C_4H_4S (Thiophene)	196.5
[C_4H_5N]	Pyrrole	207.6
[C_4H_5N]	c-C_3H_5CN	195.4
[$C_4H_5NO_2$]	$NCCOOC_2H_5$	179.5
[$C_4H_5N_3O$]	Cytosine	223.8
[C_4H_6]	1-Methylcyclopropene	206
[C_4H_6]	(E)-$CH_2{=}CHCH{=}CH_2$	193
[C_4H_6]	Cyclobutene	191
[C_4H_6]	CH_3CCCH_3	187
[$C_4H_6N_2$]	1-Methylimidazole	228.9
[$C_4H_6N_2$]	4-Methylimidazole	224.4
[C_4H_6O]	$CH_2{=}CHCOCH_3$	200.2
[C_4H_6O]	$CH_3CH{=}CHCHO$	199.7
[C_4H_6O]	$CH_2{=}C(CH_3)CHO$	195.2
[$C_4H_6O_2$]	$CH_3COCOCH_3$	194.8
[C_4H_7N]	i-C_3H_7CN	194.3
[C_4H_7N]	n-C_3H_7CN	193.7
[C_4H_8]	$(CH_3)_2C{=}CH_2$	195.9
[C_4H_8]	E—$CH_3CH{=}CHCH_3$	179.4
[$C_4H_8N_2$]	$NCCH_2N(CH_3)_2$	211.1
[C_4H_8O]	$C_2H_5OCH{=}CH_2$	208.2
[C_4H_8O]	$CH_3COC_2H_5$	199.8
[C_4H_8O]	c-C_4H_8O (Tetrahydrofuran)	198.8
[C_4H_8O]	i-C_3H_7CHO	192.6
[C_4H_8O]	n-C_3H_7CHO	191.5
[$C_4H_8O_2$]	$C_2H_5COOCH_3$	200.2
[$C_4H_8O_2$]	$CH_3COOC_2H_5$	200.7
[$C_4H_8O_2$]	$HCOOCH(CH_3)_2$	196.0
[$C_4H_8O_2$]	$HCO_2(n\text{-}C_3H_7)$	194.2
[$C_4H_8O_2S$]	C_2H_5S $(OCH_3)CO$	201.0
[$C_4H_8O_3$]	$C_2H_5OCOOCH_3$	202.7
[$C_4H_8O_2S$]	$C_2H_5S(OCH_3)CO$	201.0
[C_4H_9N]	$(CH_3)_2NCH{=}CH_2$	227.8
[C_4H_9N]	Pyrrolidine	225.2

Table 3.5. *Continued*

Empirical formula	Molecular formula or name	Proton affinity [kcal/mol]
[C_4H_9N]	$CH_3CH{=}NC_2H_5$	222.7
[C_4H_9N]	$CH_2{=}C(CH_3)CH_2NH_2$	218.2
[ClH]	HCl	134.8
[FH]	HF	117
[F_2O_2S]	F_2SO_2	159.0
[F_3N]	NF_3	144
[F_3OP]	OPF_3	167.8
[F_3P]	PF_3	166.5
[HI]	HI	150
[HO_2]	HO_2	~158
[H_2]	H_2	101.3
[H_2O]	H_2O	166.5
[H_2O_2]	H_2O_2	162
[H_2O_4S]	H_2SO_4	~169
[H_2S]	H_2S	170.2
[H_2Se]	H_2Se	171.3
[H_2N]	NH_2	187
[H_3N]	NH_3	204.0
[H_3P]	PH_3	188.6
[H_4N_2]	H_2NNH_2	204.7
[H_4Si]	SiH_4	~155
[Mg_2]	Mg_2	~219
[NO]	NO	~127
[N_2]	N_2	118.2
[N_2O]	N_2O	136.5
[O_2]	O_2	100.9
[O_2S]	SO_2	161.6
[O_3S]	SO_3	~138

4 Electrical Properties of Molecules

The electrical structure of molecules which contain the charge particles (protons and electrons) manifests itself, for example, in the existence of molecular electric and magnetic moments, in the polarization of the molecules in an external electric field, in the molecular electric field, and in the electric field of a light wave.

In this chapter, data are presented for the values of the electric dipole and multi-pole moments and for the static average electric dipole polarizabilities of diatomic and polyatomic molecules.

4.1 Dipole Moments

The molecular electric dipole moment is defined as

$$\mu = \int \varrho(r) r \, \mathrm{d}r \ ,$$

where ϱ is the charge density formed by electrons and nuclei, and $\boldsymbol{r}$ is the radius vector. The electric dipole moment depends on the molecule's quantum state; for neutral molecules, μ is independent of the origin of the coordinate system. The comprehensive information on the molecular dipole moments and on the experimental techniques for their measurement has been given in [4.1–6].

Table 4.1, compiled mainly from [4.6], gives the electric dipole moments μ of molecules in their ground state. The molecules are listed by molecular formula in the modified Hill order (Sect. 3.2). The μ values are given in Debye units (D). The estimated accuracy of the dipole moments has the following notation:

A: $\pm 1\%$ for $\mu > 1.0$ D; ± 0.01 D for $\mu < 1.0$ D,
B: $\pm 2\%$ for $\mu > 1.0$ D; ± 0.02 D for $\mu < 1.0$ D,
C: $\pm 5\%$ for $\mu > 1.0$ D; ± 0.05 D for $\mu < 1.0$ D,
D: $\pm 10\%$ for $\mu > 1.0$ D; ± 0.10 D for $\mu < 1.0$ D;
Q: Questionable value;
i: indicates that the value may be ambiguous because of the presence of rotational isomers.

Table 4.1. Dipole moments of molecules

Molecular formula	Molecule name	μ [D]	Accuracy
Compounds not containing carbon			
AgCl	Silver chloride	5.70	C
AgF	Silver fluoride	6.22	C
AgI	Silver iodide	5.10	C
AlF	Aluminum fluoride	1.53	D
$AsCl_3$	Arsenic trichloride	1.59	C
AsF_3	Arsenic trifluoride	2.59	B
AsH_3	Arsine	0.20	C
BF	Fluoroborane	0.5	Q
B_4H_{10}	Tetraborane	0.486	A
B_5H_9	Pentaborane	2.13	B
B_6H_{10}	Hexaborane	2.50	B
BaO	Barium oxide	7.954	A
BaS	Barium sulfide	10.86	A
BrCl	Bromine chloride	0.519	A
BrF	Bromine fluoride	1.42	A
BrF_5	Bromine pentafluoride	1.51	D
BrH	Hydrogen bromide	0.827	A
BrH_3Si	Bromosilane	1.319	A
BrI	Iodine bromide	0.726	A
BrK	Potassium bromide	10.628	A
BrLi	Lithium bromide	7.268	A
BrNO	Nitrosyl bromide	1.8	Q
BrNa	Sodium bromide	9.118	A
BrO	Bromide oxide	1.76	B
BrTl	Thallium bromide	4.49	A
CaCl	Calcium monochloride	3.6	Q
ClCs	Cesium chloride	10.387	A
ClF	Chlorine fluoride	0.888	A
$ClFO_3$	Perchloryl fluoride	0.023	A
ClF_3	Chlorine trifluoride	0.6	D
$ClGeH_3$	Chlorogermane	2.13	A
ClH	Hydrogen chloride	1.109	A
ClHO	Hypochlorous acid (HOCl)	1.3	Q
ClH_3Si	Chlorosilane	1.31	A
ClI	Iodine chloride	1.24	B
ClIn	Indium chloride	3.79	C
ClK	Potassium chloride	10.269	A
ClLi	Lithium chloride	7.129	A
ClNO	Nitryl chloride	0.53	A
ClNS	Thionitrosyl chloride (NSCl)	1.87	A
ClNa	Sodium chloride	9.001	A
ClO	Chlorine oxide	1.239	A
ClRb	Rubidium chloride	10.510	A
ClTl	Thallium chloride	4.543	A
Cl_2H_2Si	Dichlorosilane	1.17	B
Cl_2OS	Thionyl chloride	1.45	B
Cl_2O_2S	Sulfuryl chloride	1.81	B
Cl_2S	Sulfur dichloride	0.36	A
Cl_3FSi	Trichlorofluorosilane	0.49	A

Table 4.1. *Continued*

Molecular formula	Molecule name	μ [D]	Accuracy
Cl_3HSi	Trichlorosilane	0.86	A
Cl_3N	Nitrogen trichloride	0.39	A
Cl_3OP	Phosphorous oxychloride	2.54	B
Cl_3P	Phosphorous trichloride	0.56	B
CsF	Cesium fluroide	7.884	A
CsNa	Cesium sodium	4.75	C
CuF	Copper fluoride	5.77	C
FGa	Gallium fluoride	2.45	B
$FGeH_3$	Fluorogermane	2.33	C
FH	Hydrogen fluoride	1.826	A
FHO	Hypofluorous acid (HOF)	2.23	C
FH_3Si	Fluorosilane	1.298	A
FI	Iodine fluoride	1.95	A
FIn	Indium fluoride	3.40	B
FK	Potassium fluoride	8.585	A
FLi	Lithium fluoride	6.326	A
FNO	Nitrosyl fluoride	1.730	A
FNO_2	Nitryl fluoride	0.466	A
FNS	Thionitrosyl fluoride (NSF)	1.902	A
FNa	Sodium fluoride	8.156	A
FRb	Rubidium fluoride	8.546	A
FS	Sulfur fluoride	0.794	B
FTl	Thallium fluoride	4.228	A
F_2Ge	Germanium fluoride	2.61	A
F_2HN	Difluoramine	1.92	A
F_2H_2Si	Difluorosilane	1.55	A
F_2N_2	*cis*-Difluorodiazine	0.16	A
F_2O	Oxygen difluoride	0.297	A
F_2OS	Thionyl fluoride	1.63	A
F_2O_2	Dioxygen difluoride (FOOF)	1.44	C
F_2O_2S	Sulfuryl fluoride	1.12	B
F_2S	Sulfur difluoride	1.05	C
F_2Si	Difluorosilylene (SiF_2)	1.23	B
F_3HSi	Trifluorosilane	1.27	B
$F_3H_3Si_2$	1,1,1-Trifluorodisilane	2.03	C
F_3N	Nitrogen trifluoride	0.235	A
F_3NO	Trifluoramine oxide	0.039	A
F_3OP	Phosphorous oxyfluoride	1.868	A
F_3P	Phosphorous trifluoride	1.03	A
F_3PS	Thiophosphoryl trifluoride	0.64	B
F_4N_2	Tetrafluorohydrazine (*gauche*)	0.26	B
F_4S	Sulfur tetrafluoride	0.632	A
F_4Se	Selenium tetrafluoride	1.78	C
F_5I	Iodine pentafluoride	2.18	C
GeO	Germanium oxide	3.282	A
GeS	Germanium sulfide	2.00	C
HI	Hydrogen iodide	0.448	A
HLi	Lithium hydride	5.884	A
HLiO	Lithium hydroxide	4.754	A
HN	Imidogen (NH)	1.39	C

Table 4.1. *Continued*

Molecular formula	Molecule name	μ [D]	Accuracy
HNO	Nitrosyl hydride	1.62	B
HNO_2	Nitrous acid (HONO) (*cis*)	1.423	A
	Nitrous acid (HONO) (*trans*)	1.855	A
HNO_3	Nitric acid	2.17	A
HN_3	Hydrazoic acid	1.70	C
HO	Hydroxyl (OH)	1.668	A
HS	Mercapto (SH)	0.758	A
H_2O	Water	1.854	A
H_2O_2	Hydrogen peroxide	1.573	A
H_2S	Hydrogen sulfide	0.97	A
H_3N	Ammonia	1.471	A
H_3NO	Hydroxylamine (NH_2OH)	0.59	C
H_3P	Phosphine	0.574	A
H_3Sb	Stibine	0.12	C
H_4N_2	Hydrazine	1.75	C
H_6OSi_2	Disiloxane	0.24	B
ILi	Lithium iodide	7.428	A
INa	Sodium iodide	9.236	A
IO	Iodine oxide	2.45	B
ITl	Thallium iodide	4.61	B
KLi	Lithium potassium	3.45	C
KNa	Potassium sodium	2.73	C
LiNa	Lithium sodium	0.463	A
LiO	Lithium oxide	6.84	A
LiRb	Lithium rubidium	4.00	C
NO	Nitric oxide	0.159	A
NO_2	Nitrogen dioxide	0.316	A
NS	Nitrogen sulfide	1.81	A
NP	Phosphorous nitride	2.747	A
N_2O	Nitrous oxide	0.161	A
N_2O_3	Nitrogen trioxide	2.122	A
NaRb	Rubidium sodium	3.1	D
OPb	Lead oxide	4.64	D
OS	Sulfur oxide (SO)	1.55	A
OS_2	Sulfur oxide (SSO)	1.47	B
OSi	Silicon monoxide	3.098	A
OSn	Tin oxide	4.32	C
OSr	Strontium oxide	8.900	A
O_2S	Sulfur dioxide	1.633	A
O_2Se	Selenium dioxide	2.62	B
O_3	Ozone	0.534	A
PbS	Lead sulfide	3.59	C
SSi	Silicon monosulfide	1.73	C
SSn	Tin sulfide	3.18	C
Compounds containing carbon			
$CBrF_3$	Bromotrifluoromethane	0.65	C
CBr_2F_2	Dibromodifluoromethane	0.66	C
$CClF_3$	Chlorotrifluoromethane	0.50	A
CClN	Cyanogen chloride	2.833	A

Table 4.1. *Continued*

Molecular formula	Molecule name	μ [D]	Accuracy
CCl_2F_2	Dichlorodifluoromethane	0.51	C
CCl_2O	Phosgene	1.17	A
CCl_3F	Trichlorofluoromethane	0.46	B
CF	Fluoromethylidene	0.645	A
CFN	Cyanogen fluoride	2.120	A
CF_2	Difluoromethylene	0.47	B
CF_2O	Carbonyl fluoride	0.95	A
CF_3I	Trifluoroiodomethane	1.048	A
CO	Carbon monoxide	0.110	A
COS	Carbon oxysulfide	0.715	A
COSe	Carbon oxyselenide	0.73	B
CS	Carbon sulfide	1.958	A
CSe	Carbon selenide	1.99	B
CH	Methylidyne	1.46	Q
$CHBr_3$	Tribromomethane (bromoform)	0.99	B
$CHClF_2$	Chlorodifluoromethane	1.42	B
$CHCl_2F$	Dichlorofluoromethane	1.29	B
$CHCl_3$	Trichloromethane (chloroform)	1.04	B
CHF_3	Trifluoromethane (fluoroform)	1.651	A
CHN	Hydrogen cyanide	2.984	A
CHN	Hydrogen isocyanide (HNC)	3.05	C
CHNO	Fulminic acid (HCNO)	3.099	A
CHNO	Isocyanic acid (HNCO)	1.6	Q
CH_2Br_2	Dibromomethane	1.43	B
CH_2ClF	Chlorofluoromethane	1.82	B
CH_2Cl_2	Dichloromethane	1.60	B
CH_2F_2	Difluoromethane	1.978	A
CH_2N_2	Cyanamide	4.27	C
CH_2N_2	Diazomethane	1.50	A
CH_2O	Formaldehyde	2.332	A
CH_2O_2	Formic acid	1.41	A
CH_3BO	Carbonyltrihydroboron (BH_3CO)	1.70	B
CH_3Br	Bromomethane (methyl bromide)	1.822	A
CH_3Cl	Chloromethane (methyl chloride)	1.892	A
CH_3F	Fluoromethane (methyl fluoride)	1.858	A
CH_3I	Iodomethane (methyl iodide)	1.62	B
CH_3NO	Formamide	3.73	B
CH_3NO_2	Nitromethane	3.46	A
CH_3N_3	Methyl azide	2.17	B
CH_4O	Methanol	1.70	A
CH_4S	Methanethiol (methyl mercaptan)	1.52	C
CH_5N	Methylamine	1.31	B
CH_6OSi	Methoxysilane	1.15	B
CH_6Si	Methylsilane	0.735	A
C_2ClF_3	Chlorotrifluoroethylene	0.40	D
C_2ClF_5	Chloropentafluoroethane	0.52	C
$C_2Cl_2F_4$	1,2-Dichlorotetrafluoroethane	0.5	Q
C_2F_3N	Trifluoroacetonitrile	1.262	A
C_2HBr	Bromoacetylene	0.234	A
C_2HCl	Chloroacetylene	0.44	A

Table 4.1. *Continued*

Molecular formula	Molecule name	μ [D]	Accuracy
C_2HCl_5	Pentachloroethane	0.92	C
C_2HF	Fluoroacetylene	0.721	A
C_2HF_3	Trifluoroethylene	1.32	B
$C_2HF_3O_2$	Trifluoroacetic acid	2.28	D
$C_2H_2Cl_2$	1,1-Dichloroethylene	1.34	A
$C_2H_2Cl_2$	*cis*-1,2-Dichloroethylene	1.90	B
$C_2H_2Cl_2O$	Chloroacetyl chloride	2.23	Ci
$C_2H_2Cl_4$	1,1,2,2-Tetrachloroethane	1.32	Ci
$C_2H_2F_2$	1,1-Difluoroethylene	1.368	A
$C_2H_2F_2$	*cis*-1,2-Difluoroethylene	2.42	A
C_2H_2O	Ketene	1.422	A
$C_2H_2O_2$	Glyoxal (*cis*)	4.8	C
C_2H_3Br	Bromoethylene (vinyl bromide)	1.42	B
C_2H_3Cl	Chloroethylene (vinyl chloride)	1.45	B
$C_2H_3ClF_2$	1-Chloro-1, 1-difluoroethane	2.14	B
C_2H_3ClO	Acetyl chloride	2.72	C
$C_2H_3Cl_3$	1,1,1-Trichloroethane	1.755	A
C_2H_3F	Fluoroethylene (vinyl fluoride)	1.468	A
C_2H_3FO	Acetyl fluoride	2.96	A
$C_2H_3F_3$	1,1,1-Trifluoroethane	2.347	A
C_2H_3N	Acetonitrile	3.924	A
C_2H_3NO	Methylisocyanate	2.8	Q
C_2H_4ClF	1-Chloro-1-fluoroethane	2.068	A
$C_2H_4Cl_2$	1,1-Dichloroethane	2.06	B
$C_2H_4F_2$	1,1-Difluoroethane	2.27	B
$C_2H_4F_2$	1,2-Difluoroethane (*gauche*)	2.67	C
C_2H_4O	Acetaldehyde	2.750	A
C_2H_4O	Ethylene oxide (oxirane)	1.89	A
$C_2H_4O_2$	Acetic acid	1.70	B
$C_2H_4O_2$	Methyl formate	1.77	B
$C_2H_4O_2$	Glycolaldehyde (hydroxyacetylene)	2.73	B
C_2H_5Br	Bromoethane (ethyl bromide)	2.03	A
C_2H_5Cl	Chloroethane (ethyl chloride)	2.05	A
C_2H_5ClO	2-Chloroethanol	1.78	Ci
C_2H_5F	Fluoroethane (ethyl fluoride)	1.94	B
C_2H_5I	Iodoethane (ethyl iodide)	1.91	B
C_2H_5N	Ethyleneimine	1.90	A
C_2H_5NO	Acetamide	3.76	Bi
C_2H_5NO	N-Methylformamide	3.83	Bi
$C_2H_5NO_2$	Nitroethane	3.23	A
C_2H_6O	Dimethyl ether	1.30	A
C_2H_6O	Ethanol (*average*)	1.69	Bi
	Ethanol (*trans*)	1.44	B
	Ethanol (*gauche*)	1.68	B
C_2H_6OS	Dimethyl sulfoxide	3.96	A
$C_2H_6O_2$	Ethylene glycol	2.28	Ci
C_2H_6S	Dimethyl sulfide	1.554	A
C_2H_6S	Ethanethiol (*trans*)	1.58	C
	Ethanethiol (*gauche*)	1.61	C
C_2H_7N	Dimethylamine	1.01	B

Table 4.1. *Continued*

Molecular formula	Molecule name	μ [D]	Accuracy
C_2H_7N	Ethylamine	1.22	Ci
$C_2H_8N_2$	1,2-Ethanediamine	1.99	Ci
C_3HN	Cyanoacetylene	3.724	A
C_3H_2O	Propynal (propargyl aldehyde)	2.47	B
$C_3H_3Cl_2F$	1,1-Dichloro-2-fluoropropene	2.43	A
$C_3H_3F_3$	3,3,3-Trifluoropropene	2.45	B
C_3H_3N	Propenenitrile (acrylonitrile)	3.87	B
C_3H_3NO	Oxazole	1.503	B
C_3H_3NO	Isoxazole	2.90	B
C_3H_4	Propyne (methylacetylene)	0.784	A
C_3H_4	Cyclopropene	0.45	A
$C_3H_4F_2$	1,1-Difluoro-1-propene	0.889	A
$C_3H_4N_2$	Imidazole	3.8	D
C_3H_4O	Propenal (*trans*)	3.12	B
C_3H_4O	Propargyl alcohol	1.13	C
C_3H_4O	Cyclopropanone	2.67	C
$C_3H_4O_2$	Vinyl formate	1.49	A
$C_3H_4O_2$	2-Oxetanone (beta-propiolactone)	4.18	A
$C_3H_4O_2$	3-Oxetanone	0.887	A
C_3H_5Br	3-Bromopropene (allyl bromide)	1.9	Q
C_3H_5Cl	*cis*-1-Chloropropene	1.67	C
C_3H_5Cl	*trans*-1-Chloropropene	1.97	C
C_3H_5Cl	2-Chloropropene	1.647	A
C_3H_5Cl	3-Chloropropene (allyl chloride)	1.94	Ci
C_3H_5F	*cis*-1-Fluoropropene	1.46	B
C_3H_5F	*trans*-1-Fluoropropene	1.9	Q
C_3H_5F	2-Fluoropropene	1.61	B
C_3H_5F	3-Fluoropropene (*cis*)	1.76	A
	3-Fluoropropene (*gauche*)	1.94	A
C_3H_5N	Propanenitrile	4.05	A
C_3H_6	Propene (propylene)	0.366	A
$C_3H_6Cl_2$	1,3-Dichloropropane	2.08	Bi
C_3H_6O	Acetone	2.88	A
C_3H_6O	Allyl alcohol (*average*)	1.60	Ci
C_3H_6O	Allyl alcohol (*gauche*)	1.55	C
C_3H_6O	Propanal (*cis*)	2.52	B
C_3H_6O	Methyloxirane	2.01	A
C_3H_6O	Oxetane (trimethylene oxide)	1.94	A
C_3H_6O	Methyl vinyl ether (*cis*)	0.96	B
$C_3H_6O_2$	Ethyl formate (*trans*)	1.98	A
$C_3H_6O_2$	Ethyl formate (*gauche*)	1.81	A
$C_3H_6O_2$	Methyl acetate	1.72	Ci
$C_3H_6O_2$	Propanoic acid (*average*)	1.75	Ci
$C_3H_6O_2$	Propanoic acid (*cis*)	1.46	C
$C_3H_6O_2$	1,3-Dioxolane	1.19	C
$C_3H_6O_3$	Trioxane	2.08	A
C_3H_6S	Thiacyclobutane (trimethylene sulfide)	1.85	C
C_3H_7Br	1-Bromopropane (propyl bromide)	2.18	Ci
C_3H_7Br	2-Bromopropane (isopropyl bromide)	2.21	C

4.2 Multipole Moments

The electric multipole moments of molecules are defined as [4.7–9]

$$\mu_\alpha = \int \varrho r_\alpha \, d\tau ,$$

$$Q_{\alpha\beta} = \tfrac{1}{2} \int \varrho (3r_\alpha r_\beta - r^2 \delta_{\alpha\beta}) \, d\tau ,$$

$$O_{\alpha\beta\gamma} = \tfrac{1}{2} \int \varrho \left[5r_\alpha r_\beta r_\gamma - r^2 (r_\alpha \delta_{\beta\gamma} + r_\beta \delta_{\gamma\alpha} + r_\gamma \delta_{\alpha\beta})\right] d\tau ,$$

$$H_{\alpha\beta\gamma\delta} = \tfrac{1}{8} \int \varrho \big[35 r_\alpha r_\beta r_\gamma r_\delta - 5r^2 (r_\alpha r_\beta \delta_{\gamma\delta} + r_\alpha r_\gamma \delta_{\beta\delta} + r_\alpha r_\delta \delta_{\beta\gamma}$$
$$+ r_\beta r_\gamma \delta_{\alpha\delta} + r_\beta r_\delta \delta_{\alpha\gamma} + r_\gamma r_\delta \delta_{\alpha\beta}) + r^4 (\delta_{\alpha\beta}\delta_{\gamma\delta} + \delta_{\alpha\gamma}\delta_{\beta\delta} + \delta_{\alpha\delta}\delta_{\beta\gamma})\big] d\tau$$

or, in general, as

$$M_{\alpha\beta\ldots\mu} = \frac{(-1)^n}{n!} \int \varrho r^{2n+1} \left[\frac{\partial^n}{\partial r_\alpha \partial r_\beta \cdots \partial r_\mu}\right] \frac{1}{r} \, d\tau ,$$

where μ_α, $Q_{\alpha\beta}$, $O_{\alpha\beta\gamma}$, $H_{\alpha\beta\gamma\delta}$ and $M_{\alpha\beta\ldots\mu}$ are components of the dipole, quadrupole, octopole, hexadecapole and 2^n-pole moment tensors, respectively. In these equations, ϱ is the charge density at the point (x, y, z); the subscripts r are x, y or z, depending on whether the subscripts have the value 1, 2 or 3; the δ's are Kronecker symbols; $d\tau = dx\,dy\,dz$.

The necessity of the definition of higher-order electric moments is connected with that for high symmetric molecules, the lower-order electric moments are vanishing. The multipole tensors are symmetric and have, in the general case, 3^n non-zero components, $2n+1$ components from which are independent. The numbers of non-zero and independent components can be decreased depending on the point symmetry group to which the molecule belongs. These numbers, defined by group-theory methods [4.10], are listed in Table 4.2 for the first four electric multipoles.

Table 4.3 lists the numbers of non-zero and independent components of the electric quadrupole moment $Q_{\alpha\beta}$ for the 51 point groups. It is seen from these tables, that for the highly symmetric molecules a single scalar quantity suffices to specify a multipole moment tensor. This scalar is referred to as the multipole moment and is denoted by μ, Q, O and H, respectively. If an axis of sufficient symmetry exists, we choose

$$\mu = \mu_z ,$$
$$Q = Q_{zz} = -2Q_{xx} = -2Q_{yy} ,$$
$$O = O_{zzz} = -2O_{xxz} = -2O_{yyz} ,$$
$$H = H_{zzzz} = -2H_{xxzz} = -2H_{yyzz} = 8H_{xxyy} = 8H_{xxxx}/3 = 8H_{yyyy}/3 .$$

Table 4.2. The number of non-zero (N) and independent (I) components of electric moment tensors: dipole μ_α, quadrupole $Q_{\alpha\beta}$, octopole $O_{\alpha\beta\gamma}$ and hexadecapole $H_{\alpha\beta\gamma\delta}$

Symmetry group	μ_α		$Q_{\alpha\beta}$		$O_{\alpha\beta\gamma}$		$H_{\alpha\beta\gamma\delta}$	
	I	N	I	N	I	N	I	N
C_1	3	3	5	9	7	27	9	81
$C_i (S_2)$	0	0	5	9	0	0	9	81
$C_s (C_{1h})$	2	2	3	5	4	14	5	41
C_2	1	1	3	5	3	13	5	41
C_{2h}	0	0	3	5	0	0	5	41
C_{2v}	1	1	2	3	2	7	3	21
$D_2 (V)$	0	0	2	3	1	6	3	21
$D_{2h} (V_h)$	0	0	2	3	0	0	3	21
C_4	1	1	1	3	1	7	3	29
S_4	0	0	1	3	2	12	3	29
C_{4h}	0	0	1	3	0	0	3	29
C_{4v}	1	1	1	3	1	7	2	21
D_{2d}	0	0	1	3	1	6	2	21
D_{4h}	0	0	1	3	0	0	2	21
D_4	0	0	1	3	0	0	2	21
C_3	1	1	1	3	3	15	3	53
S_6	0	0	1	3	0	0	3	53
C_{3v}	1	1	1	3	2	11	2	37
D_3	0	0	1	3	1	4	2	37
D_{3d}	0	0	1	3	0	0	2	37
C_{3h}	0	0	1	3	2	8	1	21
C_6	1	1	1	3	1	7	1	21
C_{6h}	0	0	1	3	0	0	1	21
D_{3h}	0	0	1	3	1	4	1	21
C_{6v}	1	1	1	3	1	7	1	21
D_6	0	0	1	3	0	0	1	21
D_{6h}	0	0	1	3	0	0	1	21
T	0	0	0	0	1	6	1	21
T_h	0	0	0	0	0	0	1	21
T_d	0	0	0	0	1	6	1	21
O	0	0	0	0	0	0	1	21
O_h	0	0	0	0	0	0	1	21
D_{4d}	0	0	1	3	0	0	1	21
D_{5h}	0	0	1	3	0	0	1	21
D_{5d}	0	0	1	3	0	0	1	21
D_{6d}	0	0	1	3	0	0	1	21
Y	0	0	0	0	0	0	0	0
Y_h	0	0	0	0	0	0	0	0
K	0	0	0	0	0	0	0	0
K_h	0	0	0	0	0	0	0	0
C_∞	1	1	1	3	1	7	1	21
$C_{\infty v}$	1	1	1	3	1	7	1	21
$C_{\infty h}$	0	0	1	3	0	0	1	21
$D_{\infty h}$	0	0	1	3	0	0	1	21

Table 4.3. The number of non-zero (N) and independent (I) components of electric quadrupole moments $Q_{\alpha\beta}$ for the 51 point symmetry groups (the components are designated by indexes x,y,z)

Symmetry group	N	I	Components of tensor $Q_{\alpha\beta}$
C_l, C_i	9	5	$A \equiv xx$, $B \equiv yy$, $C \equiv xy = yx$, $D \equiv xz = zx$, $E \equiv yz = zy$, $F \equiv zz = -(xx + yy)$
C_s, C_2, C_{2h}	5	3	Components A, B, C, F
C_{2v}, D_2, D_{2h}	3	2	Components A, B, F
C_4, S_4, C_{4h}, C_{4v}, D_{2d}, D_{4h} D_4, C_3, S_6, C_{3v}, D_3, D_{3d} C_{3h}, C_6, C_{6h}, D_{3h}, C_{6v}, D_6, D_{6h}, D_{4d}, C_5, C_{5v}, C_{5h}, D_5, D_{5h}, D_{5d}, D_{6d}, S_8, S_{10}, S_{12}, C_∞, $C_{\infty v}$, $C_{\infty h}$, $D_{\infty h}$	3	1	Components $A = B$, F
In the groups T, T_h, T_d, O, O_h, Y, Y_h, K, K_h all components are zero			

For molecules of the tetrahedral symmetry class T_d, all components of the dipole and quadrupole tensors are zero, while

$$O = O_{xyz}$$

and

$$H = H_{zzzz} = H_{xxxx} = H_{yyyy} = -16H_{zzxx} = -16H_{zzyy} = -16H_{xxyy} \; .$$

The first non-zero molecular multipole moment tensor for a given molecule is independent of the location of the origin of the coordinate system, but all higher-rank tensors depend on the location of the coordinate origin. In the Tables 4.2, 3, all multipole moments are presented with respect to the molecular centre of mass. The z-axis coincides with the axis of highest rotational symmetry of the molecule.

Table 4.4 lists the values of the electric quadrupole moments for diatomic, linear and symmetric top polyatomic molecules. The molecular quadrupole moments are characterized by the sign, but not all experimental methods allow one to determine this sign. Thus, the numerical values in Table 4.4 not preceded by a plus or minus sign refer to absolute values. The molecules are arranged in the Hill order (Sect. 3.2). The error of the values presented in most cases are given in parentheses as the uncertainty in the last significant digits. In the other cases, the error of the numerical values is estimated to be one or two in the last digits. Table 4.5 lists the values of electric octopole and hexadecapole moments for several molecules which belong to tetrahedral symmetry group T_d and have no dipole and quadrupole moments [4.11].

Table 4.4. Quadrupole moment of molecules [4.9, 11–14]

Molecule	Quadrupole moment Q [10^{-26} e.s.u. · cm^2]	Molecule	Quadrupole moment Q [10^{-26} e.s.u. · cm^2]
	Diatomic molecule		
AlF	−5.91 (48)	FLi	+5.82
^{79}BrF	0.91 (1.00)	FNa	−1.97 (72)
BrH	4.14 (13)	FRb	−12.49 (480)
BrLi	−48.0 (48)	F^{203}Tl	−13.11
CO	−1.94 (4)	F_2	+0.88
CS	+0.82 (139)	^{74}GeO	−4.51 (62)
CSe	−2.59 (158)	HI	+6
ClCs	−15.13 (278)	HLi	−5
ClF	1.54 (7)	H_2	+0.662
ClH	3.74 (12)	HD	+0.642 (24)
ClLi	13.06 (67)	D_2	+0.649
Cl^{205}Tl	11.77 (720)	Li_2	+13.8
Cl_2	+6.14	N_2	−1.52
CsF	−17.15 (82)	OPb	−7.40 (29)
CuF	−6.05 (82)	OSi	−4.56 (110)
FGa	−6.44 (24)		−1.10 (48)
FH	3.09	OSn	−8.17 (24)
FK	−9.32 (96)	O_2	−0.39
	Linear polyatomic molecules		
ArBrH	−0.5 (6)	CHNO	3.0 (3)
ArClH	−1.2 (6)	CHP	4.4 (12)
	−0.7 (6)	CI^{15}N	−7.33 (110)
ArHF	−3.1 (1.5)	COS	−0.786 (1)
ArDF	−3.4 (18)		−0.88 (15)
BHS	2.7 (6)	CO^{80}Se	−0.32 (24)
CBr^{15}N	−5.29 (110)	CO_2	−4.3 (2)
CCl^{15}N	−3.9 (10)	C_2HBr	8.5 (11)
CHClO	−2.3 (3)	C_2HCl	8.8 (5)
$CHClO_2$	−6.3 (2)	C_2HF	3.96 (14)
CHFO	−6.6 (3)	C_3HN	2.141 (77)
CHFOS	−7.4 (8)	C_4HD	17.4(7)
$CHFO_2$	−11.1 (9)	ClHKr	−2.44 (65)
$CDFO_2$	−12.0 (5)	FHN_2	−6.7 (2)
CH^{15}N	3.5 (9)	N_2O	−3.296 (15)
	Symmetric top molecules		
BrH_3 Si	−0.1 (4)	$C_3H_6O_3$	−6.5 (28)
CHF_3	3.867 (24)	F_3OP	−4.59 (33)
CH_3Br	3.55 (79)	F_3P	24.1 (31)
CH_3I	5.35 (85)	H_3N	−3.3 (4)

Table 4.5. Multipole moments of tetrahedral symmetry molecules

	Electric moments			
Molecule	Dipole μ [e.s.u. · cm]	Quadrupole Q [e.s.u. · cm^2]	Octopole O [10^{-34} e.s.u. · cm^3]	Hexadecapole H [10^{-42} e.s.u. · cm^4]
CCl_4	0	0	15.0	
			5.5	
CF_4	0	0	4.4	15.1
			4.8	
			5.7	
			13.7	
			3.31	
CH_4	0	0	4.5	−6.0
			5.9	−7.7
			5.0	4.8
			3.7	−4.6
			2.5	
SiH_4	0	0	6.6	−16.6
			2.4	7.0

4.3 Molecular Polarizabilities

The polarizability of a molecule is its property to acquire a dipole moment in an electric field. The polarizability (quantitative characteristic) describes the response of the electron cloud to an external electric field. The induced electric dipole moment is

$$\mu_a = \beta_{ab} E_b \ .$$

Thus, the electric dipole polarizability of a molecule is a tensor quantity.

The fundamental information on molecular polarizability is contained in [4.15, 16]. Table 4.6 gives the static average electric dipole polarizabilities for the ground-state molecules. The average polarizability is given by

$$\beta = \frac{\beta_1 + \beta_2 + \beta_3}{3} \ ,$$

where β_1, β_2 and β_3 are the principal magnitudes of the polarizability tensor.

Molecular polarizabilities are tabulated in $\mathrm{\AA}^3$ units. The conversion from $\mathrm{\AA}^3$ to atomic units a_0^3 (a_0 being the Bohr radius) is

$$\beta[\mathrm{\AA}^3] = 0.14818\, \beta[a_0^3]$$

The accuracy of most of the numerical data is better than several percent. However, one should treat many of the results with some caution and estimate the real accuracy of the data by studying the original paper.

Table 4.6. Average static electric dipole polarizabilities of ground-state molecules [4.1.7]

Molecule	Polarizability [Å^3]	Molecule	Polarizability [Å^3]
	Diatomic molecules		
Al_2	19	HgCl	7.4
BH	3.32	ICl	12.3
Br_2	7.02	K_2	77
CO	1.95		72
Cl_2	4.61	Li_2	34
Cs_2	104	LiCl	3.46
CsK	89	LiF	10.8
D_2	0.7954	LiH	3.68
DCl	2.84		3.84
F_2	1.38		3.88
H_2	0.804	N_2	1.7403
HBr	3.61	NO	1.70
HCl	2.63	Na_2	40
	2.77		38
HF	2.46	NaK	51
	0.80	NaLi	40
HI	5.44	O_2	1.5812
	5.35	Rb_2	79
	Triatomic molecules		
BeH_2	4.34	$HgCl_2$	11.6
CO_2	2.911	HgI_2	19.1
CS_2	8.74	N_2O	3.03
	8.86	NO_2	3.02
D_2O	1.26	Na_3	70
H_2O	1.45	O_3	3.21
H_2S	3.78	OCS	5.71
	3.95		5.2
HCN	2.59	SO_2	3.72
	2.46		4.28
$HgBr_2$	14.5		
	Inorganic polyatomic molecules		
$AsCl_3$	14.9	$(NaCl)_2$	23.4
AsN_3	5.75	$(NaF)_2$	20.7
BCl_3	9.47	$(NaI)_2$	26.9
	9.38	OsO_4	8.17
BF_3	3.31	PCl_3	12.8
$(BN_3)_2$	5.73	PF_3	6.10
$(BH_2N)_3$	8	PH_3	4.84
ClF_3	6.32	$(RbBr)_2$	48.2
$(CsBr)_2$	54.5	$(RbCl)_2$	43.2
$(CsCl)_2$	42.4	$(RbF)_2$	40.7
$(CsF)_2$	28.4	$(RbI)_2$	46.3
$(CsI)_2$	51.8	SF_6	6.54
$GeCl_4$	15.1	$(SF_5)_2$	13.2
GeH_3Cl	6.7	SO_3	4.84
$(HgCl)_2$	14.7	SO_2Cl_2	10.5
$(KBr)_2$	42.0	SeF_6	7.33
$(KCl)_2$	32.1	SiF_4	5.45

Table 4.6. *Continued*

Molecule	Polarizability [Å^3]	Molecule	Polarizability [Å^3]
$(KF)_2$	21.0	SiH_4	5.44
$(KI)_2$	36.3	$(SiH_3)_2$	11.1
$(LiBr)_2$	18.9	$SiHCl_3$	10.7
$(LiCl)_2$	13.1	SiH_2Cl_2	8.92
$(LiF)_2$	6.9	SiH_3Cl	7.02
$(LiI)_2$	23.4	$SnBr_4$	22.0
ND_3	1.70	$SnCl_4$	18.0
NF_3	3.62		13.8
NH_3	2.26	SnI_4	32.3
	2.10	TcF_6	9.00
	2.81	$TiCl_4$	16.4
$(NaBr)_2$	26.8	UF_6	12.5
		Hydrocarbon molecules	
CH_4		C_3H_8	
Methane	2.593	Propane	6.29
C_2H_2			6.37
Acetylene	3.33	C_4H_{10}	
	3.93	*n*-butane	8.20
C_2H_4		C_5H_{12}	
Ethylene	4.252	Pentane	9.99
C_2H_6		Neopentane	10.20
Ethane	4.47	C_6H_6	
	4.43	Benzene	10.32
C_3H_4		C_6H_{14}	
Propyne	6.18	*n*-hexane	11.9
C_3H_6		C_7H_{16}	
Propene	6.26	*n*-heptane	13.7
Cyclopropane	5.66	C_8H_{18}	
		n-octane	15.9
		Organic halide molecules	
CBr_2F_2		CCl_2S	
Dibromodifluoromethane	9	Thiophosgene	10.2
$CClF_3$		CCl_3F	
Chlorotrifluoromethane	5.59	Trichlorofluoromethane	9.47
	5.72	CCl_3NO_2	
CCl_2F_2		Trichloronitromethane	10.8
Dichlorodifluoromethane	7.81	CCl_4	
	7.93	Carbon tetrachloride	11.2
CCl_2O			10.5
Phosgene	7.29	CH_2ClNO_2	
CF_4		Chloronitromethane	6.9
Carbone tetrafluoride	3.838	CH_2Cl_2	
CF_2O		Dichloromethane	6.48
Carbonylfluoride	1.88	CH_3Br	
$CHBr_3$		Bromomethane	6.03
Bromoform	11.8		5.55
$CHBrF_2$			5.87
Bromodifluoromethane	5.7	CH_3Cl	
$CHClF_2$		Chloromethane	4.72

Table 4.6. *Continued*

Molecule	Polarizability [Å^3]	Molecule	Polarizability [Å^3]
Chlorodifluoromethane	5.91	CH_3F	5.35
	6.38	Fluoromethane	2.97
$CHCl_2F$		CH_3I	
Dichlorofluoromethane	6.82	Iodomethane	7.97
$CHCl_3$		C_2ClF_5	
Chloroform	9.5	Chloropentafluoroethane	6.3
CHF_3		C_2Cl_3N	
Fluoroform	3.57	Trichloroacetonitrile	6.10
	3.52	C_2F_6	
CHFO		Hexafluoroethane	6.82
Fluoroformaldehyde	1.76	C_2HBr	
CHI_3		Bromacetylene	7.39
Iodoform	18.0	C_2H_5Cl	
CH_2Br_2		Chloroethane	6.4
Dibromomethane	9.32		7.27
C_2HCl		C_2H_5F	
Chloroacetylene	6.07	Fluoroethane	4.96
C_2HCl_5		C_2H_5I	
Pentachloroethane	14.0	Iodoethane	10.0
$C_2H_2Cl_2O$		$C_3H_4Cl_2$	
Chloroacetyl chloride	8.92	Dichloropropene	10.1
C_2H_2ClN		C_3H_5Cl	
Chloroacetonitrile	6.10	Chloropropene	8.3
$C_2H_2F_2$		C_3H_5ClO	
1,1-Difluoroethylene	5.01	Chloroacetone	8.4
C_2H_3Br		$C_3H_5ClO_2$	
Bromoethylene	7.59	Ethylchloroformate	9.0
C_2H_3Cl		$C_3H_6Cl_2$	
Chloroethylene	6.41	Dichloropropane	10.9
C_2H_3ClO		C_3H_7Br	
Acetyl chloride	6.62	Bromopropane	9.5
C_2H_3I		C_3H_7Cl	
Iodoethylene	9.3	Chloropropane	10.0
C_2H_5Br		C_3H_7I	
Bromoethane	8.05	1-iodopropane	11.5
	8.29	C_4H_9I	
C_6H_5Br		1-iodobutane	13.3
Bromobenzene	14.7	C_6H_5F	
	14.1	Fluorobenzene	10.3
C_6H_5Cl		C_6H_5I	
Chlorobenzene	12.3	Iodobenzene	15.5
Other organic molecules			
CN_4O_8		C_2H_3N	
Tetranitromethane	15.3	Acetonitrile	4.40
CH_2O			4.48
Formaldehyde	2.8	C_2H_4O	
	2.45	Acetaldehyde	4.59
CH_2O_2		$C_2H_4O_2$	
Formic acid	3.4	Acetic acid	5.1

Table 4.6. *Continued*

Molecule	Polarizability [Å^3]	Molecule	Polarizability [Å^3]
CH_3NO		$C_2H_4O_4$	
Formamide	4.2	Formic acid dimer	12.7
	4.08	C_2H_5NO	
CH_3NO_2		Acetamide	5.67
Nitromethane	7.37	N-methyl formamide	5.91
CH_4O		$C_2H_5NO_2$	
Methanol	3.29	Nitroethane	9.63
	3.23	Ethylnitrite	7.0
	3.32	C_2H_5O	
CH_5N		Ethylene oxide	4.43
Methyl amine	4.7	C_2H_6O	
	4.01	Ethanol	5.11
C_2N_2			5.84
Cyanogen	7.99	C_3H_4O	
C_2H_2O		Propenal	6.38
Ketene	4.4	C_3H_6O	
C_2H_6O		Acetone	6.33
Methylether	5.16		6.4
	5.29		6.39
$C_2H_6O_2$		Allyl alcohol	7.65
Ethylene glycol	5.7	C_6H_6O	
C_2H_6S		Phenol	11.1
Ethanethiol	7.41	C_6H_7N	
C_2H_7N		Aniline	12.1
Ethylamine	7.10		
Dimethylamine	6.37		

5 Molecular Spectra

A quantitative interpretation of emission and absorption spectra of both hot gases and low-temperature plasmas requires the knowledge of radiative and collisional characteristics of the molecules such as oscillator strengths and transition probabilities, molecular dipole moments, lifetimes of exited states, Franck-Condon factors, cross sections, and rate coefficients of elementary processes. In this chapter the radiative characteristics of molecules are discussed.

5.1 Types of Transitions and Selection Rules

Molecular spectra – absorption, emission and scattering – arise when *transitions* of a molecule between quantum states take place [5.1–3]. Molecular spectra are defined by composition, structure and the nature of the chemical bonds in a molecule. Spectra of dilute (low-pressure) gases are the most characteristic because they consist of sharp lines broadened mainly by the Doppler effect [5.4, 5].

In accordance with three types of molecular energy levels – electronic, vibrational and rotational – molecular spectra are very complicated and cover a wide range of electromagnetic waves: from the radio frequency up to the X-ray spectral region [5.1]. The transition frequencies between rotational levels are in the 0.03–30 cm^{-1} microwave range, the frequencies of transitions between vibrational levels in the infrared region (400–10000 cm^{-1}), and the frequencies of transitions between electronic levels in visible and UltraViolet (UV) spectral region. As a rule, electronic transitions are also accompanied by a change of vibrational and rotational molecular energies. Therefore, the electronic spectra constitute a set of electronic-vibrational bands. If the spectral resolution of an apparatus is high enough, it is possible to resolve the rotational structure of the spectra.

Intensities of lines and bands in molecular spectra are defined by radiative transition probabilities of the corresponding quantum transitions [5.6, 7]. The strongest lines correspond to transitions allowed by *selection rules*.

Purely electronic molecular spectra arise due to transitions with a change of electronic energy when vibrational and rotational energies remain unchanged. The electronic spectra are observed in *absorption* as well as in *emission* (*luminescence*). The electric dipole moment of a transition between electronic states with symmetry types (representations) Γ' and Γ'' is non-zero if the direct matrix product $\Gamma' \times \Gamma''$ contains a symmetry type of one of the dipole moment $\boldsymbol{d}$ components. Because the electric dipole moment is independent of the molecular spin, the electronic transition takes place between states of the same multiplicity (i.e., transitions with a spin change, are forbidden). However, this rule is violated when the spin–orbital interaction is strong (the so-called *intercombination transitions*). For example, intercombination transitions are responsible for *phosphorescence spectra* when a molecular transition occurs from the excited triplet state into the ground singlet state [5.8].

Molecules in different electronic states have often different types of geometrical symmetry. In this case, the transition is possible if the condition $\Gamma' \times \Gamma'' \subset \Gamma_d$ is satisfied for the point group of low symmetry configuration; here, Γ_d is the symmetry type of one of the dipole moment projections.

The probability W_{mn} for an electric dipole transition from the electronic level m to the electronic level n, summed over all vibrational and rotational sublevels a and b of the electronic levels m and n, is defined by [5.6, 7]:

$$W_{mn} = \frac{2\omega^3}{3\hbar c^3}\frac{1}{g_m}\sum_{ab}|M_{ab}|^2 \ , \tag{5.1.1}$$

$$M_{ab} = \int \phi^*_{ema}\boldsymbol{d}\,\phi_{enb}\,\mathrm{d}\tau_e \ , \tag{5.1.2}$$

where ω is the transition frequency, g_m is the statistical weight of the initial state, M_{ab} is the dipole matrix element for transition ma–nb, and $\boldsymbol{d}$ is the operator of the dipole moment. The electronic transition is also characterized by the corresponding oscillator strength f_{nm} defined for absorption and related to the transition probability by

$$W_{mn} = \frac{2\omega^2 e^2}{mc^3}\frac{g_n}{g_m}|f_{nm}| \ . \tag{5.1.3}$$

The oscillator strength f_{mn} is a dimensionless quantity. For strong transitions, $f_{mn} \sim 1$.

The mean lifetime of the excited state τ_n is defined by

$$\tau_n = \left(\sum_{n<m} W_{mn}\right)^{-1} \ . \tag{5.1.4}$$

These formulae are also applicable to vibrational and rotational transitions but, in this case, one should redefine the matrix elements of dipole moment [5.9, 10].

5.2 Oscillator Strengths and Transition Probabilities

5.2.1 Rotational Spectra

Rotational spectra are usually observed in absorption by microwave-spectroscopy methods and more rarely in emission and Raman scattering [5.11–14]. For diatomic and linear polyatomic molecules, the rotational spectra arise due to dipole electric transitions (with a change of the rotational quantum number $\Delta J = \pm 1$) and consist of lines with characteristic frequencies ω. For example, for the R-branch one has

$$\omega = 2B(J+1) - 4D_v(J+1)^3, \quad D_v \ll B \ , \tag{5.2.1}$$

where B and D_v are rotational and centrifugal distortion constants, respectively. Absorption bands of diatomic molecules comprise two rotational branches R and P, which correspond to the transitions $\Delta J = +1$ and $\Delta J = -1$; the Q-branch ($\Delta J = 0$) is forbidden.

The rotational spectrum consists of almost equidistant lines with intervals of approximately $2B$. As to the rotational spectrum of the symmetric top molecules, it is formed by the lines with the frequencies (R-branch)

$$\omega = 2B(J+1) - 4D_v(J+1)^3 - 2D_{JK}(J+1)K^2 \tag{5.2.2}$$

in accordance with the selection rules $\Delta J = +1$, $\Delta K = 0$.

In contrast to the spectra of linear molecules, the lines of the symmetric top molecules exhibit the so-called K-structure corresponding to the last term in (5.2.2). The $\Delta K = 0$ selection rule is violated if the vibrational–rotational interaction, anharmonicity or non-rigidity of the molecule are taken into account.

In the rotational spectra of asymmetric top molecules, changes of the quantum numbers K_a and K_c are not restricted by selection rules. For this type of molecules all three components of the dipole moment can be non-zero in contrast to symmetric top molecules in which a single component of the dipole moment $\boldsymbol{d}$ is directed along the symmetry axis. The transitions arising due to interaction of the components d_a, d_b and d_c along the principal axes of inertia a, b, c with the electric vector of the radiation field are called the "a", "b" and "c" types of transitions, respectively. These transitions obey the following selection rules provided $K_a = K_c = J$ or $J+1$:

$$\text{“}a\text{” type: } \Delta K_a = 0, \pm 2, \pm 4, \ldots; \quad \Delta K_c = \pm 1, \pm 3, \pm 5, \ldots \ ;$$
$$\text{“}b\text{” type: } \Delta K_a = \pm 1, \pm 3, \pm 5, \ldots; \quad \Delta K_c = \pm 1, \pm 3, \pm 5, \ldots \ ;$$
$$\text{“}c\text{” type: } \Delta K_a = \pm 1, \pm 3, \pm 5, \ldots; \quad \Delta K_c = 0, \pm 2, \pm 4, \ldots \ .$$

The $\Delta J = 0, \pm 1$ rules are common for all three types of transitions. These selection rules are valid only for the rigid asymmetric top molecules as a result of the D_2-symmetry and break down for real non-rigid molecules.

Two methods are generally used for the classification of rotational transitions and corresponding rotational spectral lines [5.11–14]. According to the first method, one has to denote the quantum numbers J, K_a and K_c for both levels. The quantities for lower levels are always written at the right-handside. For example, a 1_{10}–1_{11} transition means that $J' = 1$, $K_a' = 1$, $K_c' = 0$ for the upper and $J'' = 1$, $K_a'' = 1$, $K_c'' = 1$ for the lower level. This type of designation is usually used for asymmetric top molecules. For the symmetric top molecules, the transitions with $\Delta J = -1, 0, +1$ are designated as P, Q and R, respectively. The J and K values are given in parentheses, for example, the rotational transition $P(2, 1)$. In Raman scattering transitions with $\Delta J = -2$ and $+2$ are also allowed and are denoted by O and S. The ΔK values are denoted by the lower-case letters o, p, q, r and s (in the upper left corner) for $\Delta K = -2, -1, 0, +1, +2$, respectively.

As a rule, the rotational spectra contain hundreds (or thousands) of lines. From the line frequencies, the molecular rotational and centrifugal constants can be obtained with high accuracy (up to 1 kHz). In turn, these constants can be used in constructing the *potential-energy surfaces* of the molecules. In the case of non-rigid molecules having several equilibrium configurations, the tunnel splitting of the rotational lines takes place and the height and form of the barrier of the potential surface are determined from these splittings [5.15].

5.2.2 Vibrational and Vibrational–Rotational Spectra

Vibrational spectra are observed when the vibrational energy is changed and the electronic and rotational energies remain unchanged [5.16]. Optically allowed vibrational transitions ($\Delta v_k = \pm 1$) correspond to the *fundamental vibrational bands* which are the most intense in the vibrational spectra. These bands are called *hot bands* if they are caused by transitions from some excited state to even higher states.

Transitions forbidden by the selection rules on v_k may be allowed if the nonlinear terms of anharmonicity in the expansion of the molecular dipole moment and dipole polarizability on the normal coordinates are taken into account. Transitions with a change of one of the numbers v_k on 2, 3, 4 etc. are called *overtone bands* ($\Delta v_k = 2$ is the first overtone, $\Delta v_k = 3$ is the second overtone, etc). If two or more v_k numbers are changed, a transition is called *combination* or *combined* transition (with all v_k numbers increasing) and *differential* transition (with some of the v_k numbers decreasing). The overtone bands are denoted by $2v_k, 3v_k, \ldots$, the combination bands by $v_k + v_l$, $2v_k + v_l$ etc., the differential bands are denoted by $v_k - v_l$, $2v_k - v_l$ and so forth.

For molecules having no symmetry elements, all vibrational transitions are allowed both in absorption and Raman spectra. For molecules having an inversion center, the transitions allowed in absorption are forbidden in Raman spectra and vice versa (this is the so-called *alternative interdiction*). Transitions between vibrational levels are allowed in absorption, if the direct product

$\Gamma_1 \times \Gamma_2$ for these levels contains a symmetry type (i.e., representation) of the dipole moment, and are allowed in Raman spectra, if the product $\Gamma_1 \times \Gamma_2$ has the symmetry type of the polarizability tensor. These selection rules are approximate, because they do not take into account the interactions of the vibration with the electronic and rotational movement.

Vibrational transitions of high-symmetry molecules are often forbidden by pure vibrational selection rules but their, rotational structure becomes allowed by the vibrational–rotational interaction. For example, in the absorption InfraRed (IR) spectrum of methane (CH_4) among four possible fundamental bands $\omega_1(A_1)$, $\omega_2(E)$, ω_3 and $\omega_4(F_2)$ only the ω_3 and ω_4 bands are allowed by vibrational selection rules. The Coriolis interaction of the ω_2 vibration with the ω_3 and ω_4 vibrations provokes the forming of the ω_2 band, while the effects of higher-order vibrational–rotational interactions are responsible for the activation of the ω_1 band, although the pure vibrational bands ω_1 and ω_2 $(J = 0)$ remain forbidden.

The rotational structure of vibrational bands is usually studied by *Fourier spectroscopy* and the *laser spectroscopy* methods. Molecular parameters are obtained from the transition frequencies and are then used for the construction of the potential energy surfaces and the calculation of line frequencies in other spectral ranges. The intensity of isolated band lines and the integral band intensity comprise information on molecular structure and are used in molecular spectral analysis.

5.2.3 Rovibronic Spectra

The potential energy surfaces and the respective systems of vibrational levels of different electronic states may differ essentially [5.3]. Therefore, the vibrational structure of electronic transitions follows quite complicated selection rules and the electronic–vibrational (vibronic) spectrum strongly differs from the pure vibrational one. The *Franck-Condon principle* [5.6, 7] is the theoretical background for the vibrational structure analysis and makes it possible to predict the distribution of vibrational band intensities.

The electronic-vibrational system of bands in diatomic molecule consists of *progressions* and *sequences*. If the initial molecular state is the ground electronic-vibrational state ($e'' = 0$, $v'' = 0$), the transition from this state to every excited state $e'v'$ can be observed in the absorption spectrum. A set of these bands is termed $v'' = 0$ *progression*. Similarly, transitions from the $v'' = 1$ state to higher states form a $v'' = 1$ progression and so on. If the vibrational frequencies ω of the upper (ω') and the lower (ω'') states do not differ essentially, the transitions with the same value $\Delta v = v' - v''$ give closely spaced bands making up a *sequence*.

As a rule, the electronic-vibrational band frequencies are listed in the *Delandre tables* [5.1–3], where the rows are labelled with the v' values and the columns with the v'' ones. The cells of the table containing the most intensive bands form a parabola symmetrical relative to the main diagonal of the table. This parabola is called the *Condon parabola* [5.1]. The wider the Condon parabola, the larger the difference between the potential curves of the upper and lower states.

In the case of electronic transitions in polyatomic molecules, the quantum numbers of several vibrational states can change simultaneously. Due to this fact, the bands are assigned to multidimensional progressions and sequences.

The main band wavelengths of N_2 and N_2^+ molecules, were calculated in [5.17, 18]. The review articles [5.19, 20] contain the data on wavelengths of the bound–bound transitions in N_2 and N_2^+.

Cold air is transparent for visible light. Absorption starts in the ultraviolet and is related to the system of the *Schumann-Runge bands* for the oxygen molecule [5.21]. The Schumann-Runge absorption bands due to transitions $B\,^3\Sigma_u^-$–$X\,^3\Sigma_g^-$ were studied by many researchers. Table 5.1 shows two examples [5.22, 23] of the oscillator strengths of absorption for Schumann-Runge bands. Another technique to determine the oscillator strengths was applied in [5.24]. The band oscillator strength depends on the gas temperature through the

Table 5.1. Absorption oscillator strengths for the Schuman-Runge bands of O_2 ($B^3\Sigma_u^-\,v'$–$X^3\Sigma_g^-\,v''=0$) [5.22, 24][a]

v'	λ [nm]	$f_{v'v''}$	
		[5.22]	[5.24]
0	202.593	$2.95 \pm 0.15(-10)$[b]	
1	199.817	$3.13 \pm 0.16(-9)$	$3.04 \pm 0.15(-9)$
2	197.197	$1.89 \pm 0.19(-8)$	$1.94 \pm 0.06(-8)$
3	194.733	$8.45 \pm 0.42(-8)$	$8.14 \pm 0.25(-8)$
4	192.419	$2.82 \pm 0.14(-7)$	$2.74 \pm 0.09(-7)$
5	190.254	$7.42 \pm 0.37(-7)$	$7.42 \pm 0.25(-7)$
6	188.243	$1.62 \pm 0.08(-6)$	$1.67 \pm 0.05(-6)$
7	186.372	$3.43 \pm 0.17(-6)$	$3.44 \pm 0.11(-6)$
8	184.651	$6.31 \pm 0.32(-6)$	$6.08 \pm 0.20(-6)$
9	183.076	$1.00 \pm 0.05(-5)$	$9.65 \pm 0.30(-6)$
10	181.650	$1.54 \pm 0.08(-5)$	$1.47 \pm 0.05(-5)$
11	180.379	$2.26 \pm 0.11(-5)$	$1.96 \pm 0.06(-5)$
12	179.261	$2.43 \pm 0.12(-5)$	$2.44 \pm 0.08(-5)$
13	178.3		$2.73 \pm 0.09(-5)$
14	177.5		$2.82 \pm 0.10(-5)$
15	176.8		$2.73 \pm 0.10(-5)$
16	176.3		$2.63 \pm 0.10(-5)$
17	175.9		
18	175.6		
19	175.4		
20	175.2		

[a] supplemented with the values for $v'=0$ and 1 [5.23]
[b] $2.95(-10) = 2.95 \cdot 10^{-10}$

Table 5.2. Absorption oscillator srengths for the Schuman-Runge bands of O_2 ($B^3\Sigma_u^- v'$–$X^3\Sigma_g^- v''$) from the excited vibrational states

	$v'' = 1$ [a]		$v'' = 2$ [b]	
v'	λ [nm]	$f_{v'v''}$	λ [nm]	$f_{v'v''}$
0				
1				
2				
3	200.7	$1.76 \pm 0.12(-6)$ [c]		
4	198.4	$5.03 \pm 0.25(-6)$		
5	196.1	$1.24 \pm 0.07(-5)$		
6	193.9	$2.68 \pm 0.14(-5)$	199.9	2.13(−4)
7	191.9	$5.03 \pm 0.25(-5)$	197.8	3.39(−4)
8	190.1	$8.57 \pm 0.40(-5)$	195.8	5.46(−4)
9	188.5	$1.20 \pm 0.10(-4)$	194.1	9.87(−4)
10	186.9	$1.71 \pm 0.10(-4)$	192.5	1.03(−3)
11	185.6	$2.05 \pm 0.10(-4)$	191.0	1.04(−3)
12	184.4	$2.61 \pm 0.14(-4)$	189.8	1.22(−3)
13	183.4	$2.58 \pm 0.30(-4)$	188.7	1.04(−3)
14	182.5	$2.93 \pm 0.30(-4)$		
15	181.8	$2.52 \pm 0.20(-4)$		
16	181.3	$2.09 \pm 0.25(-4)$		

[a] [5.24]
[b] [5.26]
[c] $1.76(-6) = 1.76 \cdot 10^{-6}$

population of the rotational states. The agreement between the two data sets given in Table 5.1 is encouraging. Measurements of the oscillator strengths at 79 K were described in [5.25]. The absorption from higher vibrational states has also been studied (Table 5.2).

The $B\ ^3\Sigma_u^-$ state of O_2 is well known to predissociate. The predissociation rate is related to the widths of rotational lines of the Schumann-Runge bands. The widths given in Table 5.3 are extrapolated to zero rotation ($J = 0$).

Another important absorption in O_2 is the Herzberg I band ($A^3\Sigma_u^+$–$X^3\Sigma_g^-$). Two sets of measurements of the absorption oscillator strengths are given in Table 5.4. The probabilities of transitions from the $a\ ^1\Delta_g$ and $b\ ^1\Sigma_g^+$ states in O_2 are summarized in [5.31].

The lifetimes of the excited electronic states are of special interest among the radiative characteristics of molecules [5.6, 32–34]. Table 5.5 gives the recommended experimental lifetimes of excited electronic states of some diatomic molecules.

Table 5.3. Predissociation linewidths [cm^{-1}] for Schumann-Runge bands of O_2 [5.27]

v	Linewidth
1	0.66 ± 0.10
2	0.39 ± 0.03
3	1.61 ± 0.08
4	2.99 ± 0.20
5	1.91 ± 0.10
6	1.38 ± 0.10
7	1.87 ± 0.10
8	1.61 ± 0.10
9	0.67 ± 0.05
10	0.95 ± 0.05
11	1.18 ± 0.10
12	0.42 ± 0.05
13	0.11 ± 0.01
14	0.18 ± 0.02
15	0.33 ± 0.03
16	0.35 ± 0.05
17	0.34 ± 0.03
18	0.22 ± 0.03
19	0.23 ± 0.04

Table 5.4. Absorption oscillator strengths for the Herzberg I bands of O_2 ($A\,^3\Sigma_u^+ v'$–$X\,^3\Sigma_g^- v'' = 0$)

v'	λ [nm]	[5.28][a]	[5.30]
0	(285.6)	8.47(−14)[b]	
1	279.4	7.39(−13)	
2	273.7	3.31(−12)	
3	268.5	1.02(−11)	
4	263.7	2.44(−11)	3.0(−11) ± 20%
5	259.3	4.84(−11)	5.38(−11) ± 15%
6	255.4	8.29(−11)	7.98(−11) ± 15%
7	251.9	1.23(−10)	1.24(−10) ± 15%
8	248.9	1.62(−10)	1.39(−10) ± 15%
9	246.3	1.84(−10)	1.4(−10) ± 20%
10	244.3	1.76(−10)	1.2(−10) ± 20%
11	242.9	1.25(−10)	1.0(−10) ± 30%

[a] The relative measurement by [5.28], normalized to the absolute value at $(v', v'') = (7, 0)$ obtained by [5.29]

[b] $8.47(-14) = 8.47 \cdot 10^{-14}$

Table 5.5. Values of radiative lifetimes of electronic excited states of some diatomic molecules recommended in [5.6]*

Molecule	Electronic state	$\tau_{nv'}$ [ns]	Molecule	Electronic state	$\tau_{nv'}$ [ns]
AlO	$B^2\Sigma^+$	110 ± 180(0)		$D^1\Delta$	97 000
Ar_2	$^3\Sigma^+$	3100 ± 400	CO^+	$B^2\Sigma^+$	5.20 ± 3.9(0)
	$^1\Sigma^+$	5 ± 11		$A^2\Pi$	3820(0)
BBr	$A^1\Pi$	25.6(0.1)			(3610 ± 360)(1)
BCl	$A^1\Pi$	19.1(0–2)	CS	$A^1\Pi$	215 ± 508(0)
BF	$A^1\Pi$	2.8(0–2)	CaBr	$A^2\Pi$	34. 2(1) $^2\Pi_{1/2}$
BH	$A^1\Pi$	159(0)			33.7(1) $^2\Pi_{3/2}$
BaBr	$C^2\Pi$	8		$B^2\Sigma^+$	42.9(0)
BaCl	$C^2\Pi$	20 ± 32		$C^2\Pi$	33.2(0) $^2\Pi_{1/2}$
BaF	$C^2\Pi$	23.6			31.8(1) $^2\Pi_{3/2}$
BaI	$C^2\Pi_{1/2}$	17.9 (25)			
	$C^2\Pi_{3/2}$	16.5	CaCl	$A^2\Pi$	29.4(2) $^2\Pi_{1/2}$
BaO	$A^1\Sigma$	356(0)			28.4(2) $^2\Pi_{3/2}$
	$A'^1\Pi$	9400		$B^2\Sigma$	38.2(0)
	$a^3\Pi$	~10 000		$C^2\Pi$	25.0(0)
BeO	$B^1\Sigma$	90(0)	CaF	$A^2\Pi$	21.9(0) $^2\Pi_{1/2}$
Br_2	$B^3\Pi^+_{0u}$	280(1)			18.4(0) $^2\Pi_{3/2}$
		500(5)		$B^2\Sigma$	25.1(0)
		420(20)	CaI	$A^2\Pi$	41.7(3) $^2\Pi_{1/2}$
		1200(27)			41.6(5) $^2\Pi_{3/2}$
		110(16)		$B^2\Sigma$	50.9(4)
		310(19)			
BrCl	$B^3\Pi$	$18.5 \cdot 10^3$	CdH	$A^2\Pi$	70(0)
C_2	$d^3\Pi_g$	122 ± 19(0)		$B^2\Sigma$	60–300(0)
	$C^1\Pi_g$	31.1(0)	Cl_2	$^3\Pi^+_{0u}$	10 000
	$D^1\Sigma^+_u$	16 ± 22(0−3)	FeO	$A^5\Sigma$, $B^5\Sigma$	450
CF	$A^2\Sigma^+$	19.0(1)	Cu_2	$A^1\Pi$	70(0)
	$B^2\Delta$	18.8(0)		$B^1\Pi$	30(0)
CH	$A^2\Delta$	497 ± 38 (0)			
	$B^2\Sigma^-$	364 ± 28(0)	GeO	$a^3\Pi$	$^c(420 - 2100) \cdot 10^3$
	$C^2\Sigma^+$	(6–26) depending on J	GeS	$a^3\Pi$	$^c(625 - 3000) \cdot 10^3$
CD	$A^2\Delta$	470 ± 50(0)	H_2	$B^1\Sigma^-_u$	0.8(3)
CH^+	$A^1\Pi$	360 ± 210(0)		$C^1\Pi_u$	0.6(0)
	$B^1\Delta$	230 ± 140(0)		$a^3\Sigma^+_g$	11.1 ± 1.7
	$b^3\Sigma$	480(0)		$d^3\Pi_u$	68
CN	$A^2\Pi$	8000(0)		$c^3\Pi_u$	1.10^6
	$B^2\Sigma$	61.7 ± 4.3(0)	HBr^-	$A^2\Sigma^+$	4400(0)
CO	$A^1\Pi$	10.4 ± 0.6(1)	HCl^+	$A^2\Sigma^+$	2600(0)
	$B^1\Sigma^+$	23.7 ± 1.8(0)	HgBr	$B^2\Sigma$	23
	$C^1\Sigma^+$	1.9 ± 1.0(0)	HgH	$A^2\Pi$	103(0)
	$a^3\Pi$	$(7.7 \pm 2.0) \cdot 10^6$	I_2	$B^3\Pi^+_{ou}$	for $v' < 13$
	$b^3\Sigma^+$	55.7 ± 5.0(0)			depends on J'
	$e^3\Sigma^-$	2000(4)			$10^3 \cdot 0.88(20)$
	$a'^3\Sigma^+$	10300 ± 760(4)			$10^3 \cdot 0.98(32)$
	$c^3\Pi$	16(0)			$10^3 \cdot 1.62(40)$
	$d^3\Delta$	7300(1)			$10^3 \cdot 9.0(62)$

Table 5.5. *Continued*

Molecule	Electronic state	$\tau_{nv'}$ [ns]	Molecule	Electronic state	$\tau_{nv'}$ [ns]
$E^3\Pi^+_{0g}$	27		OD	$A^2\Sigma^+$	$770 \pm 60(0)$
$D^1\Sigma^-_u$	15.5	ICl	OH^+	$A^3\Pi$	850(0)
$A^3\Pi_i$	$100 \cdot 10^3$	IF	OD^+	$A^3\Pi$	1010(0)
$B^3\Pi^+_{0u}$	10^6		PH	$A^3\Pi$	$440 \cdot 10^3$
		K_2	PN	$A^1\Pi$	227(0)
$B^1\Pi_u$	$11 \pm 17(0)$	Kr_2	PbO	B 1	1500
$^3\Sigma^+$	350	LaO	S_2	$B^3\Sigma^-_u$	$18.6 \pm 14.0(0–4)$
$B^2\Sigma$	34.8(0)		SCl	$^2\Pi$	10.2(?)
$C^2\Pi_{1/2}$	26.9(0)		SH	$A^2\Sigma^+$	550
$C^2\Pi_{3/2}$	28.3(0)	Li_2	SD	$A^2\Sigma^+$	280
$A^1\Sigma^+_u$	18(6)	LiH	SH^+	$A^3\Pi$	1080
$A^1\Sigma^+$		$(31.6 \pm 12.7)(5)$	SO	$A^3\Pi$	12.4
MgH	$A^2\Pi$	43.6(0)		$B^3\Sigma^-$	17.3(0)
N_2	$B^3\Pi_g$	$6500 \pm 1500(0)$	ScF	$E^1\Pi$	$1.1 \cdot 10^5$
	$C^3\Pi_u$	$41.0 \pm 2.9(0)$		$^3\Phi$	$1 \cdot 10^5$
	$a^1\Pi$	$(115 \pm 46) \cdot 10^3$	ScO	$A^2\Pi_{1/2}$	35.9(0)
	$D^3\Sigma^+_u$	14.1(0)		$A^2\Pi_{3/2}$	27.0(0)
	$c'^1\Sigma^+$	0.9(0)		$B^2\Sigma$	33.3(0)
	$E^3\Sigma^-_g$	$190 \cdot 10^3$	SiF	$A^2\Sigma^+$	230(0)
N_2^+	$A^2\Pi_u$	$13.9 \cdot 10^3(1)$	SiH	$A^2\Delta$	700(0)
		$(11.1 \pm 4.1) \cdot 10^3(2)$	SiD	$A^2\Delta$	680(0)
	$B^2\Sigma_u$	$63.1 \pm 1.7(0)$	SiO	$A^1\Pi$	9.6(0)
	$C^2\Sigma^+_u$	77(1)		$E^1\Sigma$	10.5(1–7)
NH	$A^3\Pi$	$434 \pm 28(0)$		$a^3\Pi$	$^c4.8 \cdot 10^6$
	$b^1\Sigma^+$	$17.8 \cdot 10^6$	SiS	$a^3\Pi$	$^c29.10^6$
	$c^1\Pi$	$441 \pm 92(0)$	SnO	$a^3\Pi$	$^c(110-320) \cdot 10^3$
	$d^1\Sigma^+$	46(0)	SnS	$a^3\Pi$	$^c(230-570) \cdot 10^3$
ND	$d^1\Sigma$	62(0)	SrBr	$A^2\Pi_{1/2}$	34.3(3)
NH^+	$A^2\Sigma^-$	1090(0)		$A^2\Pi_{3/2}$	33.2(2)
	$B^2\Delta$	980(0)		$B^2\Sigma$	42.2(3)
	$C^2\Sigma^+$	400(0)		$C^2\Pi_{1/2}$	30.3(1)
NO	$A^2\Sigma$	$170 \pm 44(0)$		$C^2\Pi_{3/2}$	28.1(1)
	$B^2\Pi$	3100(0)	SrCl	$A^2\Pi_{1/2}$	31.3(1)
	$C^2\Pi$	$25.7 \pm 15.1(0)$		$A^2\Pi_{3/2}$	30.4(1)
	$B'^2\Delta$	110(0)		$B^2\Sigma$	38.8(1)
	$D^2\Sigma^+$	$21.9 \pm 6.0(0)$		$C^2\Pi_{1/2}$	26.1(1)
	$F^2\Delta$	90(0)		$C^2\Pi_{3/2}$	26.0(1)
	$b^4\Sigma$	6430(1)	SrF	$A^2\Pi_{1/2}$	24.1(0)
	$a^4\Pi$	$^c160 \cdot 10^6$		$A^2\Pi_{3/2}$	22.6(0)
NO^+	$A^1\Pi$	$53.9 \pm 33.3(0)$	SrI	$A^2\Pi_{1/2}$	43.3(6)
Na_2	$A^1\Sigma^-_u$	12.2(1)		$A^2\Pi_{3/2}$	41.9(10)
	$B^1\Pi_u$	7.52(0)		$B^2\Sigma$	46.0
NaH	$A^1\Sigma^+$	24.0(3)		$C^2\Pi_{3/2}$	36.0(8)
Ne_2	$^3\Sigma^+_u$	$(5.9 \pm 9.1) \cdot 10^3$	TiO	$c^1\Phi$	17.5(0)
O_2	$d^1\Sigma^-_u$	$25 \cdot 10^3$		$C^3\Delta$	37(0)
O_2^+	$A^2\Pi_u$	$693 \pm 216(0)$	Xe_2	$a^3\Sigma^+_u(1u)$	60
	$b^4\Sigma^-$	$1128 \pm 98(0)$		$A^1\Sigma^+_u(0^+_u)$	2
OH	$A^2\Sigma^+$	$773 \pm 47(0)$		$b^3\Sigma^+_g(1g)$	150
	$C^2\Sigma^+$	6.1(0)	Xe_2	$B^1\Sigma^+_g(0^+_g)$	500

Table 5.5. *Continued*

Molecule	Electronic state	$\tau_{nv'}$ [ns]
XeF	$C(1/2)$	18 ± 13
XeO	$1S$	100
VO	$C^4\Sigma^-$	415
ZnH	$A^2\Pi_{1/2}$	77
	$A^2\Pi_{3/2}$	73
ZnD	$A^2\Pi_{1/2}$	76
	$A^2\Pi_{3/2}$	75
YO	$A^2\Pi_{1/2}$	33.0(0)
	$B^2\Sigma$	30.0(0)

* As a rule, Table 5.5 gives the lifetime values for the zero vibrational level. In the case that this level was not studied, the $\tau_{nv'}$ value for the most low-lying level being investigated is listed. Respective level number is specified in parentheses near $\tau_{nv'}$. The values of $\tau_{nv'}$ in low-temperature matrices are symbolized as c

5.2.4 Franck-Condon Factors

Relative line intensities in the electronic–vibrational (vibronic) spectra are determined by the Franck-Condon factors [5.3, 4]

$$q_{v'v''} = \left| \int \phi_{v'}(r)\, \phi_{v''}(r)\, \mathrm{d}r \right|^2 . \tag{5.2.3}$$

Here, $\phi_{v'}$ and $\phi_{v''}$ are the vibrational wave functions of the initial and final electronic states. The procedures of calculating the Franck-Condon factors are discussed in detail in [5.5–7]. The data ($q_{v'v''}$ factors) for 400 electronic transitions in 180 diatomic molecules are given in [5.7]; some data from [5.7] are presented in Tables 5.6–21.

In the case of symmetric polyatomic molecules, the Franck–Condon principle restricts possible transitions between the vibrational levels of the upper and lower electronic states. According to this principle, not only the e'–e'' electronic transition should be allowed, but also the Franck-Condon factors have to be invariant to all operations of the molecular symmetry group. In other words, the vibrational levels v' and v'' should have the same symmetry. In particular, if all molecules are in the totally symmetric vibronic ground state, only the progressions of bands of the totally symmetric vibrations should be observed in the absorption spectrum, the bands of the other vibrations being forbidden. For the antisymmetric vibrations of A_2, B_2, A'' types and so on, the levels with even v_k numbers are totally symmetric, while with the odd v_k levels are antisymmetric. Therefore, if the transition occurs from some level of such a vibration, the bands of sequences with even Δv_k ($\Delta v_k = 0, \pm 2, \pm 4, \ldots$) numbers will be allowed.

Table 5.6. Franck-Condon factors for transition $d\,^3\Pi$–$a\,^3\Pi$ in C_2 [5.7]

	0	1	2	3	4	5	6	7	8	9	10	11	12	13
0	0.7213	0.2206	0.0476	0.0088	0.0015	0.0002	0.0000	0.0002	0.0000					
1	0.2506	0.3371	0.2803	0.0999	0.0254	0.0054	0.0010	0.0027	0.0006	0.0001	0.0000			
2	0.0272	0.3742	0.1381	0.2621	0.1377	0.0453	0.0119	0.0202	0.0052	0.0012	0.0002	0.0000		
3	0.0008	0.0659	0.4255	0.0477	0.2112	0.1572	0.0647	0.0782	0.0279	0.0081	0.0020	0.0004	0.00001	0.0000
4	0.0000	0.0022	0.1055	0.4446	0.0143	0.1576	0.1592	0.1525	0.0837	0.0340	0.0108	0.0029	0.0006	0.0001
5		0.0000	0.0028	0.1341	0.4585	0.0046	0.1137	0.0804	0.1406	0.0858	0.0375	0.0126	0.0035	0.0008
6			0.0002	0.0019	0.1492	0.4831	0.0042	0.0103	0.0568	0.1322	0.0816	0.0389	0.0135	0.0039
7			0.0000	0.0008	0.0001	0.1413	0.5192	0.5602	0.0306	0.0367	0.1294	0.0727	0.0379	0.0132
8			0.0000	0.0000	0.0021	0.0017	0.1103	0.0585	0.5819	0.0796	0.0184	0.1348	0.0595	0.0355
9				0.0000	0.0000	0.0032	0.0121							

Table 5.7. Franck-Condon factors for transition $D\,^1\Sigma^+$–$X\,^1\Sigma_g^+$ in C_2 [5.7]

	0	1	2	3	4	5	6	7	8	9	10
0	9.97−1	2.51−3	1.62−4	3.64−6	1.76−8	8.36−9	5.43−9	1.59−9	3.32−10	5.91−11	1.01−11
1	2.59−3	9.93−1	3.94−3	4.21−4	1.18−5	4.90−8	4.97−8	3.23−8	1.06−8	2.79−9	5.52−10
2	9.14−5	4.18−3	9.90−1	4.51−3	7.28−4	2.40−5	8.34−8	1.73−7	1.19−7	4.01−8	1.03−8
3	3.75−8	2.65−4	4.92−3	9.89−1	4.43−3	1.05−3	3.87−5	9.31−8	4.39−7	3.13−7	1.11−7
4	1.07−7	3.12−7	5.09−4	4.96−5	9.89−1	3.90−3	1.36−3	5.46−5	7.23−8	9.15−7	6.61−7
5	1.94−8	4.66−7	1.27−6	8.04−4	4.47−3	9.90−1	3.09−3	1.66−3	7.04−5	4.06−8	1.62−6
6	2.25−9	1.16−7	1.22−6	3.68−6	1.13−3	3.62−3	9.91−1	2.16−3	1.92−3	8.62−5	2.15−8
7	2.88−10	1.67−8	3.92−7	2.42−6	8.62−8	1.47−3	2.59−3	9.92−1	1.27−3	2.16−3	1.03−4
8	3.65−11	2.22−9	6.50−8	9.91−7	4.02−6	1.76−5	1.80−3	1.57−3	9.94−1	5.57−4	2.38−3
9	3.94−12	2.81−10	9.62−9	1.93−7	2.07−6	5.77−6	3.24−5	2.10−3	7.17−4	9.94−1	1.20−4
10	6.22−13	3.46−11	1.51−9	3.11−8	4.58−7	3.75−6	7.23−6	5.62−5	2.37−3	1.75−4	9.94−1

Table 5.8. Franck-Condon factors for transition $C\,^1\Pi_g$–$A\,^1\Pi_u$ in C_2 [5.7]

	0	1	2	3	4	5	6	7	8	9	10
0	7.35−1	2.15−1	4.19−2	6.62−3	9.85−4	1.54−4	2.88−5	7.41−6	2.81−6	1.40−6	7.61−7
1	2.36−1	3.56−1	2.85−1	9.60−2	2.22−2	4.50−3	9.35−4	2.29−4	7.81−5	2.96−5	1.52−5
2	2.73−2	3.95−1	1.46−1	2.64−1	1.41−1	4.52−2	1.22−2	3.26−3	1.00−3	3.79−4	1.75−5
3	1.08−3	6.80−2	4.15−1	4.82−2	1.96−1	1.62−1	7.00−2	2.48−2	8.42−3	3.16−3	1.33−3
4	2.54−6	2.49−3	1.08−1	4.40−1	1.16−2	1.15−1	1.58−1	8.81−2	4.08−2	1.71−2	7.83−3
5	1.05−5	1.40−4	2.51−3	1.40−1	4.61−1	2.46−3	4.48−2	1.33−1	8.91−2	5.54−2	2.79−2
6	1.82−7	7.00−5	1.91−3	4.63−4	1.50−1	4.85−1	1.96−3	3.71−3	1.01−1	6.69−2	6.30−2
7	3.06−7	1.05−6	1.71−4	5.04−3	2.45−3	1.25−1	4.90−1	7.58−3	1.38−2	7.81−2	2.59−2
8	4.47−9	2.94−6	5.65−5	4.66−5	1.16−2	2.82−2	5.65−2	4.07−1	2.07−2	9.13−2	7.48−2
9	3.53−8	3.24−7	4.02−6	4.35−4	8.55−4	9.91−3	8.80−2	2.17−4	1.76−1	1.51−2	1.57−1
10	6.01−9	3.08−8	6.84−6	5.28−5	2.51−4	7.45−3	2.21−3	5.94−2	4.53−2	1.81−3	1.05−2

Table 5.9. Franck-Condon factors for transition $A\,^1\Pi_u$–$X\,^1\Sigma_g^+$ in C_2 [5.7]

	0	1	2	3	4	5	6	7	8	9	10
0	4.12−1	3.98−1	1.55−1	3.11−2	3.43−3	2.05−4	5.89−6	6.04−8	2.82−11	6.93−12	2.72−15
1	3.31−1	5.53−3	2.89−1	2.71−1	8.86−2	1.36−2	1.03−3	3.56−5	4.02−7	9.58−11	1.81−11
2	1.62−1	1.72−1	5.81−2	1.15−1	3.02−1	1.56−1	3.22−2	3.04−3	1.22−4	1.52−6	1.11−10
3	6.31−2	1.99−1	3.17−2	1.43−1	1.67−2	2.63−1	2.16−1	5.91−2	6.79−3	3.15−4	4.15−6
4	2.16−2	1.26−1	1.38−1	1.23−3	1.56−1	2.79−3	1.89−1	2.59−1	9.26−2	1.28−2	6.73−4
5	6.88−3	6.02−2	1.45−1	5.77−2	3.68−2	1.14−1	3.73−2	1.11−1	2.78−1	1.30−1	2.13−2
6	2.09−3	2.46−2	9.61−2	1.20−1	8.85−3	8.15−2	5.73−2	8.30−2	4.87−2	2.75−1	1.68−1
7	6.15−4	9.17−3	4.99−2	1.12−1	7.31−2	1.69−3	1.03−1	1.55−2	1.16−1	1.18−2	2.53−1
8	1.78−4	3.21−3	2.25−2	7.42−2	1.04−1	2.92−2	2.26−2	9.61−2	1.11−4	1.26−1	2.57−5
9	5.14−5	1.08−3	9.24−3	4.02−2	8.87−2	7.77−2	4.12−3	5.12−2	6.96−2	8.26−3	1.14−1
10	1.49−5	3.56−4	3.58−3	1.93−2	5.79−2	8.85−2	4.46−2	1.37−3	7.15−2	3.80−2	3.00−2

Table 5.10. Franck-Condon factors for transition $C\,^1\Pi_u$–$X\,^1\Sigma_g^+$ in H_2 [5.7]

	0	1	2	3	4	5	6	7	8	9	10	11	12	13	14
0	0.1248	0.3254	0.3318	0.1679	0.0443	0.0058	0.0003	0.0000							
1	0.1968	0.1397	0.0041	0.2138	0.2900	0.1306	0.0235	0.0014							
2	0.1936	0.0112	0.1157	0.0737	0.0406	0.2824	0.2255	0.0541	0.0032						
3	0.1539	0.0089	0.1044	0.0052	0.1319	0.0027	0.1934	0.3004	0.0942	0.0050					
4	0.1093	0.0435	0.0369	0.0633	0.0277	0.0797	0.0531	0.0962	0.3479	0.1379	0.1372	0.0049	0.0004	0.0000	0.0000
5	0.0736	0.0667	0.0028	0.0752	0.0031	0.0770	0.0153	0.1018	0.0328	0.3769	0.1718	0.0007	0.0011	0.0001	0.0000
6	0.0480	0.0714	0.0030	0.0450	0.0346	0.0160	0.0679	0.0014	0.1088	0.0060	0.4110	0.1827	0.0008	0.0008	0.0004
7	0.0310	0.0645	0.0153	0.0168	0.0505	0.0006	0.0493	0.0264	0.0246	0.0842	0.0009	0.4747	0.1408	0.0180	0.0011
8	0.0201	0.0532	0.0261	0.0026	0.0428	0.0150	0.0139	0.0485	0.0014	0.0476	0.0523	0.0020	0.5754	0.0340	0.0278
9	0.0130	0.0412	0.0309	0.0001	0.0268	0.0277	0.0003	0.0353	0.0218	0.0047	0.0518	0.0260	0.0243	0.6060	0.0284
10	0.0084	0.0306	0.0303	0.0025	0.0134	0.0294	0.0029	0.0151	0.0308	0.0022	0.0180	0.0411	0.0067	0.1569	0.2416
11	0.0054	0.0218	0.0261	0.0051	0.0054	0.0238	0.0083	0.0038	0.0245	0.0108	0.0018	0.235	0.0267	0.0050	0.3899
12	0.0034	0.0147	0.0200	0.0061	0.0018	0.0163	0.0102	0.0003	0.0146	0.0137	0.0002	0.0086	0.0182	0.0215	0.1561
13	0.0017	0.0016	0.0258	0.0041	0.0004	0.0081	0.0070	0.0000	0.0062	0.0089	0.0012	0.0021	0.0081	0.0075	0.0648

Table 5.11. Franck-Condon factors for transition $B'\,^1\Sigma_u^+$–$X\,^1\Sigma_g^+$ in H_2 [5.7]

	0	1	2	3	4	5	6	7	8	9	10	11	12	13	14
0	0.0332	0.1482	0.2825	0.2858	0.1712	0.0633	0.0140	0.0017	0.0001	0.0000					
1	0.0737	0.1674	0.0919	0.0015	0.1427	0.2616	0.1854	0.0642	0.0109	0.0007	0.0000				
2	0.1023	0.1053	0.0017	0.0866	0.0932	0.0008	0.1515	0.2660	0.1540	0.0360	0.0027	0.0000	0.0000		
3	0.114	0.0420	0.0162	0.0821	0.0009	0.0872	0.0612	0.0181	0.2264	0.2541	0.0885	0.0087	0.0001	0.0000	
4	0.1053	0.0081	0.0359	0.0302	0.0196	0.0596	0.0026	0.0919	0.0181	0.0931	0.3107	0.1927	0.0315	0.0008	0.0000
5	0.0674	0.0001	0.0267	0.0045	0.0259	0.0123	0.0228	0.0286	0.0119	0.0719	0.0000	0.1869	0.3558	0.1663	0.0188
6	0.0155	0.0000	0.0060	0.0003	0.0065	0.0012	0.0069	0.0033	0.0060	0.1010	0.0038	0.0364	0.0029	0.1106	0.5352
7	0.0021	0.0035	0.0140	0.0031	0.0018	0.0003	0.0035	0.0011	0.0034	0.0036	0.0030	0.0141	0.0001	0.0569	0.0312
8	0.0052	0.0000	0.0020	0.0001	0.0023	0.0003	0.0026	0.0008	0.0027	0.0025	0.0030	0.0097	0.0011	0.0442	0.0009

Table 5.12. Franck-Condon factors for transition $D\,^1\Pi_u^- - X\,^1\Sigma_g^+$ in H_2 [5.7]

	0	1	2	3	4	5	6	7	8	9	10	11	12	13	14
0	0.1084	0.3018	0.3339	0.1889	0.0573	0.0091	0.0006	0.0000							
1	0.1817	0.1557	0.0001	0.1727	0.2932	0.1580	0.0357	0.0029	0.0000						
2	0.1867	0.0217	0.0955	0.0954	0.0144	0.2455	0.2540	0.0797	0.0071	0.0000					
3	0.1537	0.0026	0.1112	0.0001	0.1233	0.0190	0.1313	0.3118	0.1350	0.0120	0.0000				
4	0.1122	0.0294	0.0534	0.0467	0.0462	0.0488	0.0856	0.0399	0.3302	0.1931	0.0141	0.0004	0.0000	0.0000	
5	0.0772	0.0526	0.0111	0.0757	0.0000	0.0811	0.0012	0.1153	0.0026	0.3287	0.2433	0.0089	0.0021	0.0001	0.0000
6	0.0514	0.0606	0.0000	0.0602	0.0209	0.0317	0.0483	0.0161	0.0933	0.0031	0.3389	0.2702	0.0002	0.0031	0.0010
7	0.0338	0.0571	0.0050	0.0325	0.0451	0.0014	0.0563	0.0076	0.0492	0.0525	0.0109	0.3931	0.2347	0.0162	0.0001
8	0.0223	0.0484	0.0130	0.0125	0.0491	0.0052	0.0280	0.0363	0.0022	0.0610	0.0212	0.0070	0.5149	0.0875	0.0457
9	0.0148	0.0386	00.181	0.0029	0.0395	0.0186	0.0060	0.0404	0.0072	0.0197	0.0493	0.0054	0.0012	0.6142	0.0139
10	0.0101	0.0301	0.0199	0.0001	0.0271	0.0270	0.0000	0.0270	0.0225	0.0006	0.0320	0.0310	0.0000	0.0838	0.2662
11	0.0069	0.0228	0.0191	0.0003	0.0166	0.0280	0.0026	0.0127	0.0269	0.0033	0.0100	0.0295	0.0185	0.0175	0.2718
12	0.0047	0.0170	0.0168	0.0013	0.0094	0.0244	0.0066	0.0042	0.0223	0.0100	0.0009	0.0168	0.0191	0.0240	0.1776
13	0.0033	0.0124	0.0138	0.0020	0.0051	0.0191	0.0008	0.0008	0.0152.	0.0129	0.0003	0.0067	0.0145	0.0071	0.1335
14	0.0025	0.0086	0.0104	0.0021	0.0026	0.0136	0.0087	0.0000	0.0091	0.0118	0.0018	0.0019	0.0084	0.0083	0.0083
15	0.0007	0.0125	0.0058	0.0017	0.0012	0.0085	0.0066	0.0001	0.0048	0.0083	0.0023	0.0004	0.0041	0.0049	0.0117

Table 5.13. Franck Condon factors for transition $a\,^1\Pi_g$–$X\,^1\Sigma_g^+$ in N_2 [5.7]

	0	1	2	3	4	5	6	7	8	9	10
0	0.4308−1	0.1526−0	0.2495−0	0.2502−0	0.1728−0	0.8679−1	0.3306−1	0.9684−2	0.2210−2	0.3933−3	0.5505−4
1	0.1255−0	0.1931−0	0.7983−1	0.5680−3	0.9050−1	0.1882−0	0.1755−0	0.1014−0	0.4104−1	0.1201−1	0.2674−2
2	0.1707−0	0.9710−1	0.3407−2	0.1084−0	0.8488−1	0.4928−3	0.6907−1	0.1681−0	0.1612−0	0.9090−1	0.3469−1
3	0.1832−0	0.1232−1	0.7583−1	0.6864−1	0.4177−2	0.9685−1	0.6357−1	0.5541−3	0.8533−1	0.1668−0	0.1405−0
4	0.1600−0	0.6199−2	0.9643−1	0.4799−3	0.7834−1	0.3555−1	0.1909−1	0.9785−1	0.3279−1	0.1244−1	0.1461−0
5	0.1217−0	0.4655−1	0.4658−1	0.3444−1	0.5591−1	0.9399−2	0.7899−1	0.6290−2	0.5127−1	0.8354−1	0.5291−2
6	0.8296−1	0.8456−1	0.4521−1	0.7273−1	0.2471−2	0.6401−1	0.1323−1	0.4320−1	0.5269−1	0.3557−2	0.8222−1
7	0.5262−1	0.9919−1	0.5751−2	0.5616−1	0.1783−1	0.4583−1	0.1448−1	0.5590−1	0.1899−2	0.6886−1	0.1194−1
8	0.3153−1	0.9235−1	0.3372−1	0.1805−1	0.5400−1	0.3650−2	0.5550−1	0.3734−2	0.5251−1	0.1475−1	0.3569−1
9	0.1784−1	0.7332−1	0.6077−1	0.1087−3	0.5440−1	0.1001−1	0.3844−1	0.1759−1	0.3658−1	0.1433−1	0.4834−1
10	0.9812−2	0.5294−1	0.7366−1	0.9022−2	0.2737−1	0.3999−1	0.4293−2	0.4821−1	0.4515−3	0.5006−1	0.8097−3
11	0.5187−2	0.3506−1	0.7087−1	0.3194−1	0.4209−2	0.4862−1	0.5673−2	0.3286−1	0.1914−1	0.2239−1	0.2554−1
12	0.2733−2	0.2236−1	0.6034−1	0.5118−1	0.1319−2	0.3228−1	0.2973−1	0.4797−2	0.4236−1	0.4316−1	0.4284−1
13	0.1408−2	0.1359−1	0.4631−1	0.5979−1	0.1448−1	0.1029−1	0.4277−1	0.3054−2	0.2914−1	0.1997−1	0.1397−1
14	0.7071−3	0.7888−2	0.3265−1	0.5754−1	0.3169−1	0.1516−3	0.3379−1	0.2170−1	0.5344−2	0.3729−1	0.5796−3
15	0.3585−3	0.4551−2	0.2223−1	0.5001−1	0.4488−1	0.4744−2	0.1578−1	0.3622−1	0.1354−2	0.2643−1	0.1922−1

Table 5.13. *Continued*

	11	12	13	14	15	16	17	18	19	20
0	0.7030−5	0.7162−6	0.1389−9	0.3433−8	0.1174−7	0.1391−7	0.1293−7	0.9650−7	0.1626−7	0.3021−7
1	0.4502−3	0.5839−4	0.7219−5	0.6019−6	0.1114−6	0.4265−8	0.1190−7	0.2830−7	0.1048−7	0.6822−9
2	0.9607−2	0.1958−2	0.2931−3	0.3332−4	0.2869−5	0.2305−6	0.3516−8	0.1265−7	0.3261−8	0.1027−7
3	07124−1	0.2434−1	0.5959−3	0.1056−2	0.1367−3	0.1435−4	0.1252−5	0.1937−6	0.4969−8	0.3211−8
4	0.1644−0	01136−0	0.4912−1	0.1433−1	0.2991−2	0.4451−3	0.4841−4	0.3776−5	0.2697−6	0.6947−9
5	0.4687−1	0.1472−0	0.1481−0	0.8134−1	0.2880−1	0.7705−2	0.1210−2	0.1451−3	0.1402−4	0.1068−5
6	0.4631−1	0.8962−2	0.9858−1	0.1602−0	0.1165−0	0.5072−1	0.1454−1	0.2881−2	0.3933−3	0.3832−4
7	0.3804−1	0.8192−1	0.7572−2	0.4170−1	0.1431−0	0.1453−0	0.7830−1	0.2636−1	0.6026−2	0.9357−3
8	0.5004−1	0.2884−2	0.7844−1	0.4214−1	0.5531−2	0.1027−0	0.1575−0	0.1077−0	0.4383−1	0.1146−1
9	0.2825−2	0.6449−1	0.8218−2	0.4236−1	0.7491−1	0.3162−2	0.5420−1	0.1496−0	0.1363−0	0.6649−1
10	0.5136−1	0.8016−2	0.4149−1	0.3926−1	0.8037−2	0.8054−1	0.2840−1	0.1526−1	0.1216−0	0.1544−0
11	0.2318−1	0.2276−1	0.3678−1	0.9269−2	0.6049−1	0.1508−2	0.5681−1	0.6110−1	0.5740−4	0.8120−1
12	0.1516−2	0.4472−1	0.8364−3	0.5113−1	0.9423−3	0.5228−1	0.2248−1	0.2239−1	0.7943−1	0.1047−1
13	0.1361−1	0.7767−2	0.3673−1	0.8838−2	0.3611−1	0.2005−1	0.2414−1	0.4884−1	0.1555−2	0.7303−1
14	0.3462−1	0.6902−2	0.3025−1	0.1166−1	0.3192−1	0.1017−1	0.4302−1	0.2421−2	0.5884−1	0.5302−2
15	0.8689−2	0.3280−1	0.1238−2	0.3827−1	0.3129−4	0.4212−1	0.1461−3	0.4644−1	0.3951−2	0.4449−1

Table 5.14. Franck-Condon factors for transition $\omega\,{}^1\Delta_u$–$X\,{}^1\Sigma_g^+$ in N_2 [5.7]

	0	1	2	3	4	5	6	7	8	9	10
0	0.3063−2	0.2129−1	0.6918−1	0.1399−0	0.1980−0	0.2073−0	0.1678−0	0.1067−0	0.5470−1	0.2247−1	0.7518−2
1	0.1409−1	0.6766−1	0.1343−0	0.1327−0	0.5356−1	0.1644−3	0.4318−0	0.1274−0	0.1646−0	0.1340−0	0.7848−1
2	0.3489−1	0.1130−0	0.1143−0	0.2702−1	0.9346−2	0.8176−1	0.8700−1	0.1714−1	0.1112−1	0.8948−1	0.1488−0
3	0.6176−1	0.1198−0	0.4489−1	0.4886−2	0.7366−1	0.5405−1	0.5298−9	0.5362−1	0.8503−1	0.2270−1	0.7664−2
4	0.8727−1	0.9314−1	0.2244−2	0.5014−1	0.5644−1	0.5824−6	0.5416−1	0.5282−1	0.5669−4	0.5087−1	0.7777−1
5	0.1058−0	0.5136−1	0.1023−1	0.6765−1	0.6363−2	0.3662−1	0.4922−1	0.1162−3	0.5402−1	0.4188−1	0.1010−2
6	0.1130−0	0.1645−1	0.4049−1	0.3890−1	0.8234−2	0.5516−1	0.2297−2	0.4057−1	0.3531−1	0.3903−2	0.5894−4

	11	12	13	14	15	16	17	18	19	20
0	0.2057−2	0.4566−3	0.8269−4	0.1220−4	0.1251−5	0.1300−6	0.2764−7	0.3901−7	0.6109−8	0.1311−7
1	0.3482−1	0.1193−1	0.3246−2	0.6911−3	0.1182−3	0.1437−4	0.1219−5	0.1311−7	0.4343−7	0.1592−7
2	0.1355−0	0.8202−1	0.3603−1	0.1182−1	0.2968−2	0.5865−3	0.8824−4	0.1108−4	0.6500−6	0.2100−7
3	0.8430−1	0.1441−0	0.1288−0	0.7388−1	0.3012−1	0.9033−2	0.2023−2	0.3364−3	0.4370−4	0.3952−5
4	0.1567−1	0.1425−1	0.9773−1	0.1452−0	0.1149−0	0.5935−1	0.2136−1	0.5535−2	0.1068−2	0.1507−3
5	0.6087−1	0.6556−1	0.4365−2	0.3284−1	0.1191−0	0.1417−0	0.9466−1	0.4134−1	0.1264−1	0.2740−2
6	0.2388−1	0.1030−1	0.7361−1	0.4362−1	0.5058−3	0.6630−1	0.1400−0	0.1267−0	0.6916−1	0.2488−1

Table 5.15. Franck-Condon factors for transition $C\,^3\Pi_u$–$X\,^1\Sigma_g^+$ in N_2 [5.7]

	0	1	2	3	4	5	6	7	8	9	10
0	0.5466–0	0.3470–0	0.9206–1	0.1370–1	0.1290–0	0.9019–4	0.6701–5	0.6867–6	0.4665–7	0.2808–7	0.2449–7
1	0.3050–0	0.8096–1	0.3612–0	0.1979–0	0.4741–1	0.6513–2	0.6425–3	0.5884–4	0.7036–5	0.1007–5	0.2039–8
2	0.1057–0	0.2679–0	0.2416–2	0.2351–0	0.2669–0	0.9899–1	0.1967–1	0.2779–2	0.3807–3	0.4507–4	0.1141–4
3	0.2963–1	0.1807–0	0.1283–0	0.7438–1	0.9156–1	0.2734–0	0.1629–0	0.4738–1	0.9458–2	0.1844–2	0.3140–3
4	0.7573–2	0.7749–1	0.1766–0	0.2155–1	0.1507–0	0.6433–2	0.2057–0	0.2167–0	0.9822–1	0.2865–1	0.8062–2

	11	12	13	14	15	16	17	18	19	20
0	0.3750–7	0.1681–7	0.7348–7	0.2133–7	0.1982–7	0.1192–7	0.1133–7	0.2974–7	0.1137–7	0.1298–8
1	0.5497–7	0.3093–7	0.6481–7	0.8725–9	0.2379–7	0.1382–7	0.1783–7	0.8937–7	0.1471–7	0.2954–7
2	0.6290–6	0.4449–7	0.2575–6	0.0001–9	0.2275–7	0.5902–8	0.1938–8	0.2832–7	0.1916–7	0.0973–9
3	0.7696–4	0.1254–4	0.1632–5	0.9500–6	0.2113–7	0.8046–8	0.1832–8	0.9178–7	0.7278–7	0.1719–7
4	0.2080–2	0.5658–3	0.1298–3	0.2400–4	0.9644–5	0.7741–6	0.2522–6	0.2233–6	0.3760–7	0.1250–6

Table 5.16. Franck-Condon factors for transition $b\,^1\Pi_u$–$X\,^1\Sigma_g^+$ in N_2 [5.7]

	0	1	2	3	4	5	6	7	8	9	10
0	1.07−2	3.36−2	6.40−2	9.34−2	1.15−1	1.27−1	1.27−1	1.16−1	9.67−2	7.49−2	5.43−2
1	4.64−2	9.88−2	1.21−1	1.00−1	5.66−2	1.76−2	3.21−4	8.44−3	3.41−2	6.31−2	8.35−2
2	9.47−2	1.21−1	7.29−2	1.47−2	1.76−3	2.95−2	5.99−2	6.45−2	4.24−2	1.36−2	4.65−5
3	1.42−1	8.55−2	8.91−3	1.17−2	5.43−2	6.12−2	2.87−2	1.54−3	9.13−3	3.82−2	5.66−2
4	1.81−1	2.80−2	1.07−2	6.20−2	5.25−2	7.76−3	6.39−3	3.97−2	5.17−2	2.66−2	1.71−3
5	2.01−1	4.77−4	6.82−2	5.37−2	1.67−3	2.50−2	5.59−2	3.12−2	7.85−4	1.67−2	4.73−2
6	1.63−1	5.39−2	7.54−2	1.13−3	3.63−2	5.73−2	9.43−3	9.39−3	4.54−2	3.91−2	4.19−3
7	8.69−2	1.19−1	1.58−2	2.88−2	4.98−2	4.60−3	2.02−2	4.56−2	1.30−2	3.07−3	3.41−2
8	3.96−2	1.20−1	7.53−4	5.19−2	1.18−2	9.15−3	3.87−2	1.05−2	5.73−3	3.06−2	1.98−2
9	2.26−2	1.36−1	3.31−2	5.42−2	2.11−2	4.23−2	1.93−2	5.15−3	4.07−2	1.74−2	1.59−3
10	6.59−3	7.38−2	5.96−2	1.28−2	2.49−2	2.37−2	3.15−4	2.60−2	1.59−2	1.67−3	2.33−2

	11	12	13	14	15	16	17	18	19	20	21
0	3.66−2	2.30−2	1.35−2	7.46−3	3.85−3	1.87−3	8.36−4	3.42−4	1.30−4	4.65−5	1.57−5
1	9.02−2	8.38−2	6.87−2	5.06−2	3.40−2	2.07−2	1.15−2	5.85−3	2.70−3	1.15−3	4.62−4
2	1.03−2	3.67−2	6.51−2	8.25−2	8.41−2	7.29−2	5.49−2	3.65−2	2.17−2	1.16−2	5.62−3
3	4.64−2	1.96−2	1.20−3	6.18−3	3.16−2	6.13−2	8.03−2	8.19−2	6.89−2	4.95−2	3.06−2
4	8.67−3	3.79−2	5.41−2	4.11−2	1.36−2	1.90−6	1.44−2	4.68−2	7.56−2	8.43−2	7.23−2
5	4.26−2	1.16−2	1.27−3	2.65−2	5.29−2	4.75−2	1.74−2	1.90−5	1.61−2	5.20−2	7.70−2
6	9.79−3	4.23−2	4.36−2	1.16−2	1.87−3	3.05−2	5.45−2	3.85−2	6.69−3	4.09−3	3.71−2
7	3.51−2	4.96−3	7.14−3	3.64−2	3.52−2	6.43−3	4.81−3	3.51−2	4.60−2	1.95−2	1.56−5
8	3.75−6	1.86−2	3.09−2	1.00−2	1.67−3	2.46−2	3.10−2	8.54−3	2.10−3	2.55−2	3.44−2
9	3.07−2	2.87−2	8.85−4	1.62−2	3.68−2	1.45−2	1.37−3	2.88−2	3.76−2	8.96−3	4.23−3
10	1.55−2	1.12−4	1.89−2	2.13−2	1.44−3	9.88−3	2.61−2	1.00−2	1.20−3	2.12−2	2.34−2

Table 5.16. *Continued*

	0	1	2	3	4	5	6	7	8	9	10
11	3.38−3	6.35−2	9.98−2	4.10−4	5.43−2	6.91−3	1.77−2	3.12−2	3.56−4	2.53−2	2.41−2
12	8.22−4	3.00−2	8.92−2	9.43−3	4.34−2	2.50−3	3.18−2	4.74−3	1.31−2	2.60−2	2.75−4
13	2.68−4	1.81−2	8.32−2	3.11−2	2.65−2	2.02−2	2.62−2	1.27−3	2.81−2	8.46−3	9.75−3
14	5.39−5	7.81−3	5.50−2	4.53−2	5.75−3	3.41−2	6.80−3	1.41−2	1.81−2	3.92−4	2.26−2
15	1.42−5	4.72−3	4.79−2	6.70−2	9.60−9	4.60−2	8.93−7	3.00−2	5.55−3	1.24−2	2.23−2
16	3.04−6	2.10−3	2.91−2	6.06−2	4.61−3	3.21−2	5.50−3	2.46−2	1.39−4	2.00−2	5.82−3
17	1.37−6	1.09−3	1.99−2	5.82−2	1.61−2	1.87−2	1.80−2	1.40−2	7.59−3	1.78−2	5.04−5
18	1.70−6	6.71−4	1.48−2	5.59−2	3.04−2	7.71−3	2.99−2	4.24−3	1.92−2	8.91−3	6.09−3
19	2.42−6	3.55−4	8.66−3	4.05−2	3.50−2	8.07−4	2.84−2	7.05−6	2.11−2	9.68−4	1.25−2
20	3.79−6	2.27−4	5.59−3	3.12−2	3.84−2	4.76−4	2.37−2	2.41−3	1.79−2	6.81−4	1.43−2
21	6.00−6	1.86−4	4.25−3	2.71−2	4.37−2	4.34−3	1.88−2	8.40−3	1.32−2	5.46−3	1.24−2
22	7.53−6	1.48−4	2.95−3	2.07−2	4.14−2	9.58−3	1.12−2	1.34−2	6.58−3	1.07−2	6.77−3
23	7.49−6	1.06−4	1.78−3	1.35−2	3.22−2	1.26−2	4.57−3	1.40−2	1.77−3	1.20−2	1.94−3
24	6.75−6	7.78−5	1.10−3	8.74−3	2.39−2	1.32−2	1.40−3	1.20−2	1.71−4	1.05−2	1.95−4

Table 5.17. Franck-Condon factors for transition $A\,^1\Pi$–$X\,^1\Sigma^+$ in NO [5.7]

	0	1	2	3	4	5	6
0	2.65–2	9.99–2	1.99–1	2.48–1	2.12–1	1.29–1	5.89–2
1	7.50–5	1.71–1	1.28–1	1.59–2	2.50–2	1.32–1	1.84–1
2	1.31–1	1.33–1	8.55–3	4.98–2	1.12–1	3.50–2	7.24–3
3	1.65–1	4.80–2	2.42–2	9.46–2	1.21–2	3.88–2	9.92–2
4	1.67–1	1.57–3	7.77–2	2.90–2	2.61–2	7.62–2	3.74–3
5	1.47–1	1.52–2	7.17–2	1.24–3	6.86–2	7.64–3	4.41–2
6	1.14–1	5.98–2	2.56–2	3.54–2	3.25–2	1.76–2	5.38–2
7	8.01–2	9.90–2	1.21–4	5.50–2	2.40–4	5.01–2	4.32–3
8	5.03–2	1.15–1	1.84–2	3.27–2	1.75–2	2.83–2	1.37–2
9	2.65–2	9.89–2	5.80–2	3.71–3	3.85–2	9.43–4	3.68–2
10	1.20–2	6.87–2	8.13–2	3.81–3	2.91–2	8.10–3	2.24–2

These rules are valid only in the framework of the Franck–Condon approximation. In some cases they may be violated and a lot of forbidden transitions can be observed. The main reason for the violation of the Franck-Condon principle is the presence of electronic–vibrational interactions. When this factor is included, the electronic dipole moment of transitions begins to depend on normal coordinates. This leads to the fact that the selection rule on the symmetry types of the vibronic levels becomes less rigid. Namely, dipole transitions between vibronic states of the Γ'_{ev} and Γ''_{ev} symmetry types are allowed if the direct product $\Gamma'_{ev} \times \Gamma''_{ev}$ has the symmetry type of at least one of the components of the electronic dipole moment of the transition:

$$\Gamma'_{ev} \times \Gamma''_{ev} \subset \Gamma_d \ . \tag{5.2.4}$$

In other words, the product of the wave functions $\phi^*_{v'}\phi_{v''}$ of the initial and final states may not be totally symmetric relative to those symmetry operations for which the product $\phi^*_{v'}\boldsymbol{d}\phi_{v''}$ is not totally symmetric, i.e., a forbidden electronic or vibronic transition becomes active if the symmetry types $\phi^*_{v'}\boldsymbol{d}\phi_{v''}$ and $\phi^*_{v'}\phi_{v''}$ coincide. For example, this is the case for the electronic transition B_{2u}–A_{1g} in the benzene molecule (point group of D_{6h} symmetry), which is forbidden according to the pure electronic selection rule.

For allowed electronic transitions, the rotational structure of the vibronic band is determined mainly by the type of electronic transition. In particular, the rotational structure of the electronic transition $^1\Sigma$–$^1\Sigma$ of a diatomic or linear polyatomic molecule consists, as in the case of a pure vibrational spectrum, of P and R branches, corresponding to the rotational transitions with $\Delta J = -1, +1$, respectively. If the spin–orbital interaction is small and doublet, triplet and other splittings are negligible, the rotational structure of these transitions is similar to the $^1\Sigma$–$^1\Sigma$ transition. All these transitions are related to the projection of the electronic dipole moment on the z-axis and do not contain Q branches because of the condition $K = \Lambda = 0$. However, the transitions with a change of Λ(Π–Σ, Δ–Π etc.) are related to projections d_x and d_y and have intensive Q branches. The Λ-doubling, spin splitting and other

Table 5.18. Franck-Condon factors for transition $A\,^3\Sigma_u^+ - X\,^3\Sigma_g^-$ in O_2 [5.7]

	0	1	2	3	4	5	6	7	8	9	10	11
0	1.81−6	3.34−5	2.92−4	1.60−3	6.25−3	1.84−2	4.26−2	7.93−2	1.21−1	1.55−1	1.65−1	1.50−1
1	1.49−5	2.38−4	1.76−3	8.01−3	2.48−2	5.52−2	8.98−2	1.05−1	8.30−2	3.50−2	1.51−3	1.50−2
2	6.30−5	8.76−4	5.49−3	2.04−2	4.92−2	7.92−2	8.16−2	4.46−2	4.49−3	1.05−2	5.49−2	7.59−2
3	1.84−4	2.22−3	1.18−2	3.55−2	6.52−2	7.05−2	3.59−2	1.44−3	1.70−2	5.48−2	4.68−2	6.82−3
4	4.20−4	4.44−3	1.99−2	4.82−2	6.47−2	4.10−2	3.90−3	1.16−2	4.59−2	3.56−2	1.70−3	1.85−2
5	7.97−4	7.38−3	2.79−2	5.37−2	5.00−2	1.34−2	2.71−3	3.37−2	3.54−2	3.17−3	1.41−2	4.23−2
6	1.31−3	1.07−2	3.41−2	5.13−2	3.04−2	8.58−4	1.67−2	3.61−2	1.03−2	4.85−3	3.36−2	1.96−2
7	1.87−3	1.35−2	3.68−2	4.28−2	1.39−2	1.72−3	2.67−2	2.22−2	3.75−5	2.01−2	2.52−2	6.21−4
8	2.37−3	1.54−2	3.57−2	3.17−2	4.15−3	7.86−3	2.64−2	8.09−3	4.34−3	2.45−2	8.21−3	4.75−3
9	2.62−3	1.54−2	3.10−2	2.09−2	4.59−4	1.23−2	1.95−2	1.18−3	1.07−2	1.77−2	4.56−4	1.31−2
10	2.46−3	1.33−2	2.37−2	1.23−2	4.93−5	1.25−2	1.15−2	2.01−5	1.20−2	9.05−3	6.29−4	1.39−2

	12	13	14	15	16	17	18	19	20	21	22	23	24
0	1.14−1	7.40−2	4.12−2	1.95−2	7.92−3	2.74−3	7.96−4	1.98−4	4.31−5	7.85−6	1.22−6	2.14−7	2.19−8
1	6.77−2	1.19−1	1.36−1	1.14−1	7.58−2	4.07−2	1.78−2	6.46−3	1.97−3	5.15−4	1.16−4	2.10−5	2.62−6
2	4.34−2	3.53−3	1.38−2	7.00−2	1.20−1	1.28−1	9.68−2	5.65−2	2.65−2	1.02−2	3.21−3	8.24−4	1.73−4
3	9.76−3	5.48−2	6.57−2	2.43−2	3.64−4	3.86−2	1.01−1	1.28−1	1.09−1	6.86−2	3.34−2	1.30−2	4.13−3
4	5.16−2	2.97−2	2.93−5	2.96−2	6.59−2	4.16−2	1.93−3	2.12−2	8.47−2	1.26−1	1.16−1	7.60−2	3.86−2
5	1.83−2	1.70−3	3.69−2	4.24−2	5.23−3	1.40−2	5.87−2	5.14−2	7.26−3	1.17−2	7.31−2	1.23−1	1:23−1
6	6.87−4	3.02−2	3.08−2	7.71−4	2.23−2	4.58−2	1.34−2	5.42−3	4.97−2	5.70−2	1.33−2	6.11−3	6.51−2
7	1.83−2	2.85−2	1.77−3	1.66−2	3.47−2	5.83−3	1.14−2	4.35−2	2.13−2	1.09−3	3.99−2	5.93−2	1.98−2
8	2.55−2	7.53−2	6.10−3	2.81−2	8.13−3	6.54−3	3.27−2	1.30−2	3.72−3	3.71−2	2.91−2	1.36−4	2.82−2
9	1.61−2	9.23−8	1.70−2	1.47−2	3.91−4	2.11−2	1.52−2	7.20−4	2.50−2	2.05−2	3.66−5	2.52−2	3.47−2
10	5.93−3	2.85−3	1.58−2	3.06−3	6.42−3	1.70−2	1.40−3	1.01−2	1.92−2	1.20−3	1.21−2	2.44−2	3.51−3
11	1.34−3	4.20−3	8.42−3	1.15−4	7.30−3	7.28−3	1.75−4	1.03−2	6.64−3	7.74−4	1.30−2	7.84−3	6.66−4

Table 5.19. Franck-Condon factors for transition $B\,^3\Sigma_u^- - X\,^3\Sigma_g^-$ in O_2 [5.7]

	0	1	2	3	4	5	6	7	8	9	10	11
0	3.547−9	1.021−7	1.430−6	1.283−5	8.264−5	4.087−4	1.608−3	5.146−3	1.362−2	3.038−2	5.765−2	9.277−2
1	3.489−8	9.090−7	1.142−5	9.071−5	5.105−4	2.167−3	7.164−3	1.875−2	3.910−2	6.537−2	8.637−2	8.593−2
2	2.143−7	5.059−6	5.697−5	4.005−4	1.963−3	7.108−3	1.951−2	4.087−2	6.457−2	7.475−2	5.774−2	2.167−2
3	9.135−7	1.958−5	1.979−4	1.231−3	5.243−3	1.610−2	3.623−2	5.900−2	6.621−2	4.494−2	1.080−2	1.697−3
4	2.914−6	5.680−5	5.162−4	2.842−3	1.049−2	2.710−2	4.904−2	5.936−2	4.172−2	9.656−3	2.093−3	2.861−2
5	7.663−6	1.361−4	1.113−3	5.419−3	1.726−2	3.713−2	5.256−2	4.372−2	1.390−2	4.503−4	2.222−2	3.847−2
6	1.733−5	2.811−4	2.071−3	8.921−3	2.445−2	4.323−2	4.603−2	2.261−2	5.668−4	1.264−2	3.386−2	1.985−2
7	3.498−5	5.198−4	3.458−3	1.317−2	3.092−2	4.420−2	3.326−2	6.546−3	3.409−3	2.597−2	2.552−2	2.222−3
8	6.299−5	8.594−4	5.173−3	1.744−2	3.488−2	3.942−2	1.881−2	1.396−4	1.355−2	2.764−2	9.543−3	2.026−3
9	1.008−4	1.269−3	6.938−3	2.074−2	3.527−2	3.072−2	7.566−3	2.089−3	2.092−2	1.906−2	6.325−4	1.153−2

	12	13	14	15	16	17	18	19	20	21	22	23
0	1.275−1	1.504−1	1.519−1	1.320−1	9.974−2	6.595−2	3.823−2	1.934−2	8.476−3	3.238−3	1.131−3	3.809−4
1	5.850−2	1.979−2	1.965−5	1.946−2	6.821−2	1.143−1	1.311−1	1.150−1	8.112−2	4.730−2	2.381−2	1.090−2
2	9.873−5	1.816−2	5.757−2	7.123−2	4.128−2	4.670−3	8.838−3	5.537−2	1.035−1	1.185−1	1.016−1	7.286−2
3	2.948−2	5.375−2	3.585−2	3.265−3	1.161−2	5.294−2	6.424−2	2.839−2	1.855−4	2.270−2	7.705−2	1.247−1
4	4.399−2	1.959−2	9.664−5	2.553−2	5.000−2	2.562−2	1.648−6	2.713−2	6.181−2	4.458−2	5.579−3	1.261−2
5	1.650−2	4.180−4	2.567−2	3.905−2	1.042−2	4.610−3	3.960−2	4.045−2	5.218−3	1.083−2	5.024−2	5.055−2
6	1.098−5	1.925−2	3.309−2	7.831−3	5.859−3	3.550−2	2.465−2	2.003−6	2.607−2	4.437−2	1.291−2	4.858−3
7	1.018−2	2.951−2	1.079−2	2.827−3	2.883−2	2.053−2	9.678−5	2.601−2	3.260−2	2.565−3	1.487−2	4.147−2
8	2.289−2	1.683−2	5.783−5	2.006−2	2.179−2	1.948−4	1.907−2	2.695−2	1.271−3	1.693−2	3.355−2	5.264−3
9	2.132−2	2.657−3	8.799−3	2.279−2	3.439−3	9.290−3	2.484−2	3.387−3	1.111−2	2.883−2	5.038−3	1.004−2

Table 5.19. *Continued*

	0	1	2	3	4	5	6	7	8	9	10	11	12
10	1.477−4	1.724−3	8.608−3	2.292−2	3.306−2	2.145−2	1.590−3	7.528−3	2.197−2	8.797−3	1.391−3	1.793−2	1.125−2
11	1.963−4	2.136−3	9.789−3	2.331−2	2.847−2	1.314−2	1.057−7	1.218−2	1.776−2	2.101−3	6.439−3	1.687−2	2.768−3
12	2.399−4	2.449−3	1.038−2	2.225−2	2.307−2	7.126−3	9.339−4	1.417−2	1.190−2	2.047−5	1.045−2	1.161−2	1.355−5
13	2.699−4	2.604−3	1.028−2	2.002−2	1.768−2	3.355−3	2.570−3	1.361−2	6.770−3	6.809−4	1.152−2	6.186−3	1.061−3
14	2.807−4	2.580−3	9.579−3	1.711−2	1.299−2	1.339−3	3.813−3	1.155−2	3.330−3	2.051−3	1.027−2	2.572−3	2.952−3
15	2.723−4	2.403−3	8.468−3	1.405−2	9.288−3	4.292−4	4.325−3	9.058−3	1.430−3	3.014−3	8.083−3	7.935−4	4.109−3
16	2.481−4	2.118−3	7.151−3	1.115−2	6.523−3	9.268−5	4.214−3	6.749−3	5.290−4	3.323−3	5.902−3	1.417−4	4.315−3
17	2.125−4	1.766−3	5.764−3	8.542−3	4.505−3	5.822−6	3.694−3	4.838−3	1.609−4	3.110−3	4.101−3	1.667−6	3.866−3
18	1.736−4	1.414−3	4.495−3	6.402−3	3.101−3	3.486−6	3.029−3	3.410−3	3.630−5	2.643−3	2.793−3	2.395−5	3.166−3
19	1.349−4	1.082−3	3.371−3	4.658−3	2.110−3	1.686−5	2.342−3	2.354−3	3.884−6	2.085−3	1.867−3	6.360−5	2.424−3
20	8.805−5	6.991−4	2.147−3	2.905−3	1.253−3	2.134−5	1.519−3	1.410−3	3.432−8	1.367−3	1.090−3	6.998−5	1.555−3

	13	14	15	16	17	18	19	20	21	22	23
10	6.030−4	1.759−2	1.120−2	1.162−3	2.006−2	9.571−3	3.071−3	2.386−2	8.444−3	4.573−3	2.537−2
11	6.118−3	1.647−2	1.611−3	9.065−3	1.609−2	1.835−4	1.430−2	1.432−2	2.637−4	1.898−2	1.171−2
12	1.085−2	9.787−3	3.256−4	1.375−2	6.591−3	2.856−3	1.642−2	2.548−3	8.232−3	1.718−2	3.825−4
13	1.161−2	3.783−3	3.265−3	1.222−2	9.511−4	8.039−3	1.040−2	1.833−4	1.378−2	6.924−3	2.688−3
14	9.549−3	7.701−4	5.831−3	7.986−3	8.035−5	9.805−3	4.176−3	3.015−3	1.200−2	8.901−4	7.740−3
15	6.738−3	7.226−6	6.552−3	4.254−3	1.163−3	8.507−3	9.585−4	5.448−3	7.475−3	1.145−4	9.114−3
16	4.341−3	1.808−4	5.920−3	1.939−3	2.188−3	6.180−3	4.037−5	5.980−3	3.763−3	1.228−3	7.607−3
17	2.647−3	4.926−4	4.716−3	7.722−4	2.555−3	4.056−3	8.117−5	5.216−3	1.618−3	2.128−3	5.319−3
18	1.588−3	6.641−4	3.517−3	2.737−4	2.414−3	2.549−3	2.938−4	4.060−3	6.205−4	2.371−3	3.425−3
19	9.462−4	6.741−4	2.502−3	8.452−5	2.010−3	1.565−3	4.135−4	2.942−3	2.130−4	2.142−3	2.112−3
20	5.029−4	5.107−4	1.521−3	2.048−5	1.354−3	8.478−4	3.637−4	1.801−3	6.051−5	1.506−3	1.138−3

Table 5.20. Franck-Condon factors for transition $c\,^1\Sigma_u^- - X\,^3\Sigma_g^-$ in O_2 [5.7]

	0	1	2	3	4	5	6
0	4.9034–9	1.4332–7	1.9938–6	1.7584–5	1.1051–4	5.2734–4	1.9884–3
1	4.1810–8	1.1242–6	1.4242–5	1.1301–4	6.2969–4	2.6165–3	8.3984–3
2	1.8850–7	4.6806–6	5.4202–5	3.8821–4	1.9220–3	6.9550–3	1.8931–2
3	5.9754–7	1.3757–5	1.4616–4	9.4806–4	4.1804–3	1.3174–2	3.0257–2
4	1.4943–6	3.2023–5	3.1340–4	1.8477–3	7.2739–3	1.9960–2	3.8429–2
5	3.1363–6	6.2815–5	5.6860–4	3.0584–3	1.0779–2	2.5749–2	4.1214–2
6	5.7386–6	1.0787–4	9.0699–4	4.4689–3	1.4143–2	2.9405–2	3.8721–2

Table 5.21. Franck-Condon factors for transition $A'\,^3\Delta_u - a\,^1\Delta_g^-$ in O_2 [5.7]

	0	1	2	3	4	5
0	8.8377–5	1.0910–3	6.3151–3	2.2808–2	5.7690–2	1.0874–1
1	5.1497–4	5.1932–3	2.3502–2	6.2355–2	1.0543–1	1.1336–1
2	1.6268–3	1.3371–2	4.6693–2	8.7441–2	8.7958–2	3.5630–2
3	3.7032–3	2.4739–2	6.5508–2	8.1288–2	3.7238–2	5.2831–6
4	6.8150–3	3.6873–2	7.2326–2	5.3304–2	4.0037–3	1.9620–2
5	1.0785–2	4.7068–2	6.6223–2	2.3231–2	2.8569–3	4.2661–2
6	1.5248–2	5.3428–2	5.1505–2	4.6199–3	1.8481–2	4.1175–2

effects are observed in spectra and cause the splitting of the *P*, *Q* and *R* branches into sub-branches. The splitting depends on the predominant interactions or to which of the Hund cases the state under discussion belongs to. For example, if the $^2\Pi$ state refers to the case (b), then in the band of transition $^2\Pi - ^2\Pi$ a simple doubling of the *P*, *Q* and *R* branches is observed. If the state $^2\Pi$ belongs to the case (a), this band consists of two subbands, $^2\Pi_{1/2} - ^2\Sigma$ and $^2\Pi_{3/2} - ^2\Sigma$, each consisting of three bands.

As the molecular constants for different electronic states may vary considerably, the structure of the *P*, *Q* and *R* electronic branches may strongly differ from the structure of these branches in purely vibrational bands. This is the reason why a larger convergence of lines and sharp band edges occur in electronic spectra. A red shading of the *R* band occurs when $B' < B''$, and when $B' > B''$ a violet shading of the *P* band occurs (*B* is the rotational constant).

The study of the rotational structure of vibronic bands of nonlinear polyatomic molecules is much more complicated. Practically, for each band type it is necessary to introduce a specific interaction model.

5.3 Photoionization

Molecular photoionization processes are of importance for fundamental research and different applications in many fields [5.8, 35, 36] including aeronomy [5.37, 38], astrophysics [5.39] and others. The reliable data on cross sections for the total and partial photoionization processes as well as for fragmentation processes in a wide spectral range are required for modeling purposes [5.40].

Although the molecular photoionization has long been studied, quantitatively reliable and absolute partial photoionization cross sections have been obtained only quite recently [5.41, 42]. One-photon, single-electron ionization and subsequent fragmentation events

$$AB + \hbar\omega \rightarrow [AB^+]^j + e^- \ , \tag{5.3.1}$$

$$[AB^+]^j \rightarrow A + B^+ \tag{5.3.2}$$

are most important. Here, $\hbar\omega$ is the incident photon energy, AB is a target molecule, $[AB^+]^j$ is a molecular ion in the electronic excited state j, and e^- is the ejected electron carrying the kinetic energy.

Three regions of the photon energy are distinguished:

(i) the discrete absorption region, extending from the lowest electronic excitation energy to the lowest (first) ionization potential,
(ii) the "structured" ionization region extending from the first ionization potential to an energy of about 20 eV. In this region, the ionization efficiency varies and is less than unity,
(iii) the smooth cross-section region from 20 eV to a few hundred eV, where inner-shell processes give a significant contribution [5.43].

The partial photoionization cross section σ^j_{ion} of the electronic j level is related to the corresponding asymmetry parameter β^j. These two quantities define the differential partial photoionization cross section. In the case of a non-oriented molecular target this relation has the form

$$d\sigma^j_{\text{ion}}(\Theta) = (\sigma^j_{\text{ion}}/4\pi)[1 + \beta^j \mathrm{P}_2(\cos\Theta)]d\Omega \ , \tag{5.3.3}$$

where $\mathrm{P}_2(\cos\Theta)$ is the Legendre polynomial, and Θ is the angle between the vector of the photon (linear) polarization and the momentum vector of the outgoing electron.

Photoionization cross section σ is related to the differential optical oscillator strength df/dE by the expression [5.44]:

$$\sigma[\mathrm{Mb}] = 109.75\frac{df}{dE}[\mathrm{eV}^{-1}] \ , \tag{5.3.4}$$

where $1\,\mathrm{Mb} = 10^{-18}\mathrm{cm}^2$.

The dipole excitation or ionization of atoms and molecules can be described by electron collisions with a small momentum transfer [5.45]. The quantitative relation between differential electron scattering $(d\sigma/dE)$ and photoabsorption cross sections is given by the *Bethe-Born relationship* [5.46]:

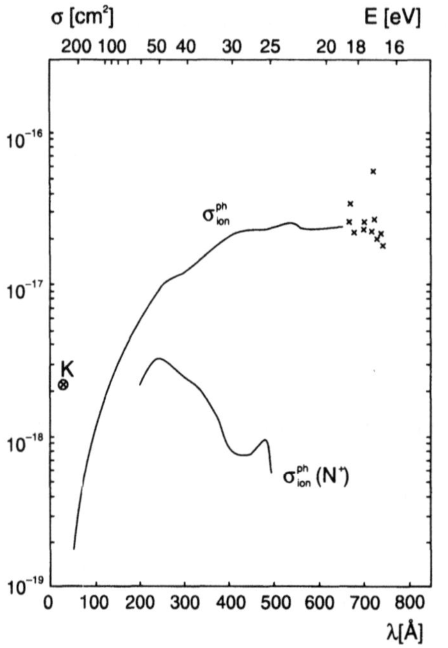

Fig. 5.1. The total (σ_{ion}^{ph}) and dissociative [$\sigma_{ion}^{ph}(N^+)$] photoionization cross sections for N_2 [5.49]

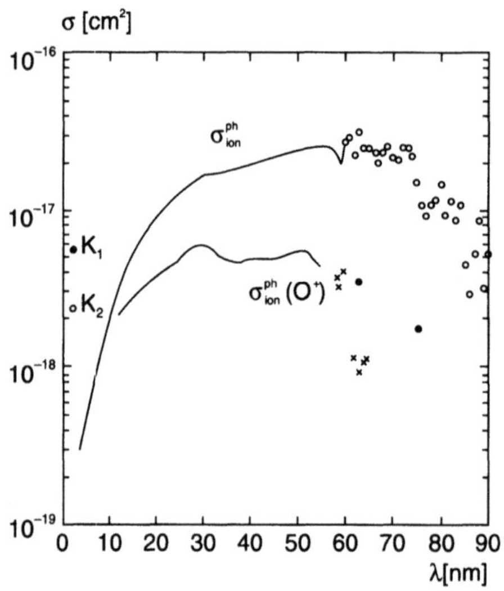

Fig. 5.2. Total (σ_{ion}^{ph}) and dissociative [$\sigma_{ion}^{ph}(O^+)$] photoionization cross sections for O_2 [5.54]

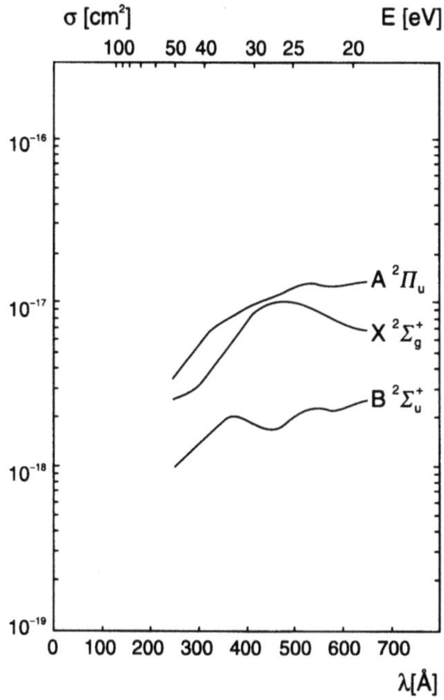

Fig. 5.3. Partial photoionization cross section for the production of N_2^+ in the states $X\ ^2\Sigma_g^+$, $A\ ^2\Pi_u$ and $B\ ^2\Sigma_u^+$ [5.49, 50, 53]

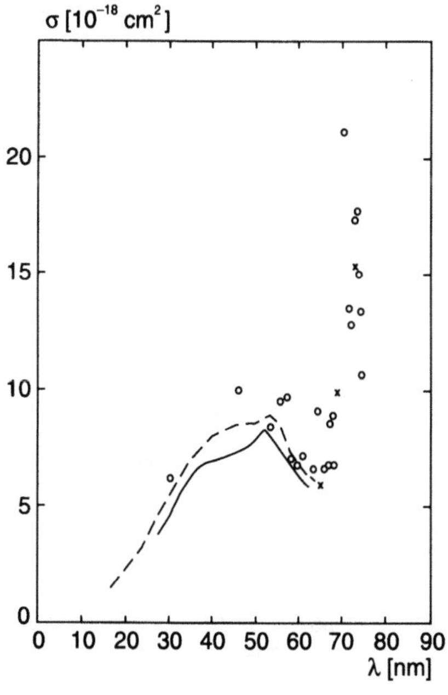

Fig. 5.4. Partial photoionization cross section for the production of O_2^+ in the states $X\ ^2\Pi_g^+$: - - [5.56]; - - - and xxx [5.57]; ooo [5.55]

$$\frac{d\sigma(K)}{dE} = \frac{2}{E}\frac{k_f}{k_i}\frac{1}{K^2}\frac{df(K)}{dE} , \tag{5.3.5}$$

where k_i, k_f and K are the incident, scattered and transferred electron momenta, respectively, E is the energy loss of the incident electron, and $df(K)/dE$ is the so-called *generalized differential* (momentum-transfer depen-

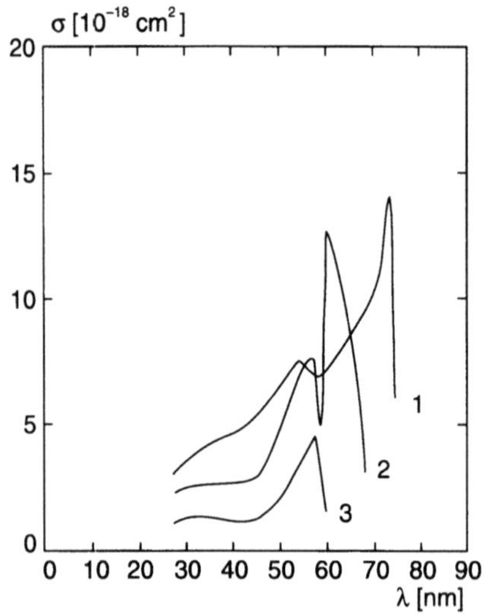

Fig. 5.5. Partial photoionization cross section for the production of O_2^+ in the states: (*1*) $A\,^2\Pi_u + a\,^4\Pi_u$; (*2*) $b\,^4\Sigma_g^-$ and (*3*) $B\,^2\Sigma_g^-$

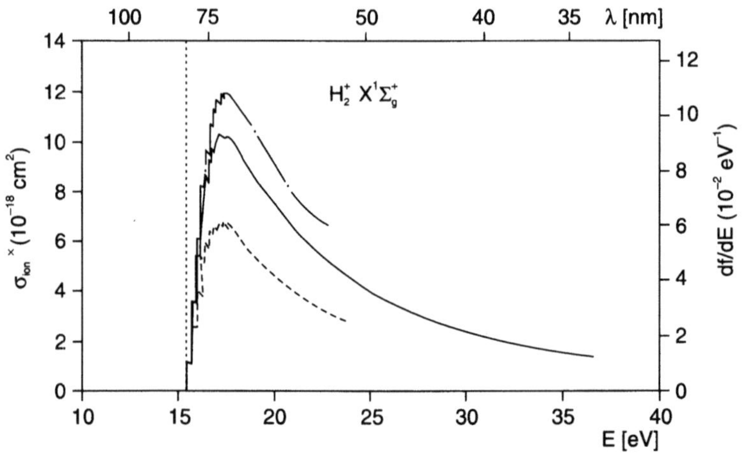

Fig. 5.6. Partial photoionization cross section for H_2; final ionic state $= X\,^1\Sigma_g^+$: -·-·- [5.59]; — [5.60]; - - - [5.61]

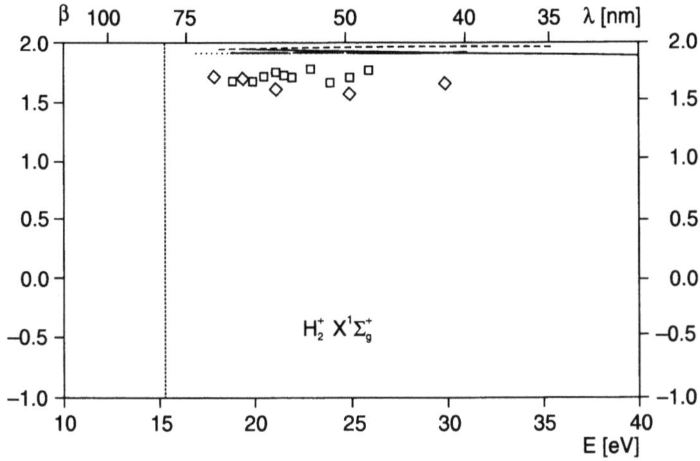

Fig. 5.7. Photoelectron asymmetry parameter β for photoionization of H_2; final ionic state $= X\ ^1\Sigma_g^+$: □ [5.62]; x [5.63]; — [5.64]; ... [5.65]; -·-· [5.66]; and -- [5.67]

dent) *oscillator strength.* The latter can be expanded in a series of terms in which the first one corresponds to the differential optical or dipole oscillator strength df/dE in (5.3.4); higher terms contain even powers of the momentum transfer K.

A number of measurements have been reported on photoionization and photoabsorption of N_2 in [5.47, 48]. In [5.49] these data are redetermined with some new results taken into account.

In [5.50] the absolute absorption cross section and the photoionization yield (the ratio of the photoionization to the photoabsorption cross sections)

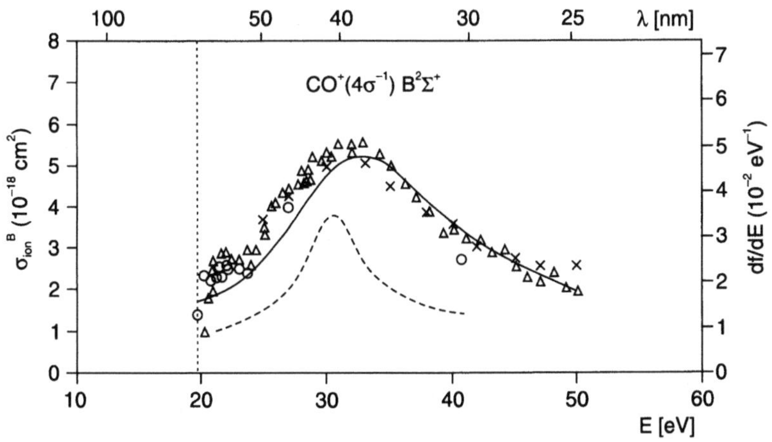

Fig. 5.8. Partial photoionization cross section for CO; final ionic state $= B\ ^2\Sigma^+$: — [5.68]; - - - [5.69]; xxx [5.70]; ooo [5.71]; and △△△ [5.72]

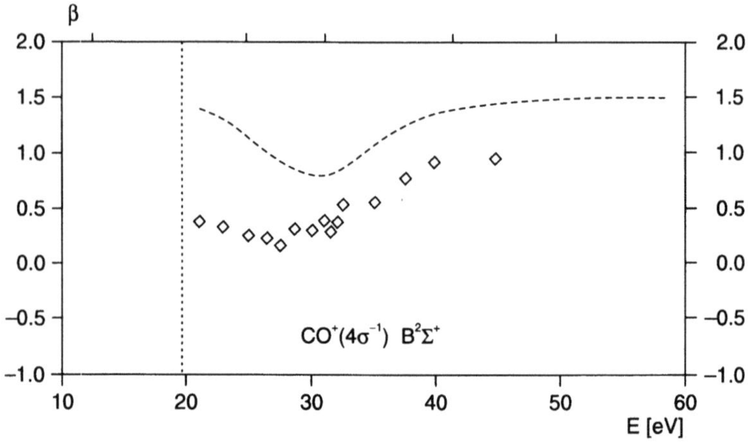

Fig. 5.9. Photoelectron asymmetry parameter β for photoionization of CO; final ionic state = $B\,^2\Sigma^+$: - - - [5.73] and xxx [5.74]

have been measured in the wavelength range from 800 to 1000 Å. These data are in quite good (within 1%) agreement with those given in [5.51]. The measurements were reported in [5.52] for the shorter-wavelength region. In this energy range, the ionization yield has been confirmed to be unity so that the photoabsorption cross section coincides with the photoionization cross section. The total $\sigma_{\text{ion}}^{\text{ph}}$ and dissociative $\sigma_{\text{ion}}^{\text{ph}}$ (N^+) photoionization cross sections for N_2 are shown in Fig. 5.1 [5.49]. In the spectral region above 650 Å, only the typical data are shown. The peak cross section at the K absorption shape resonance is indicated by the letter K.

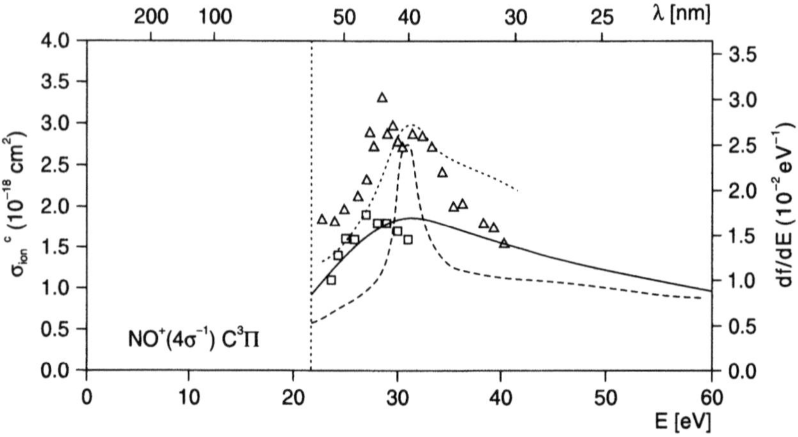

Fig. 5.10. The partial photoionization cross section for NO; the final ionic state = $c\,^3\Pi$: — [5.75]; ... [5.76]; - - - [5.77]; □□□ [5.62]; and △△△ [5.78]

The total and dissociative photionization cross sections of O_2 [5.54] are shown in Fig. 5.2. At wavelengths longer than 60 nm a complicated structure due to autoionization and predissociation is apparent. In those regions only typical data are shown. Peaks in the K absorption region are indicated by K_1 and K_2.

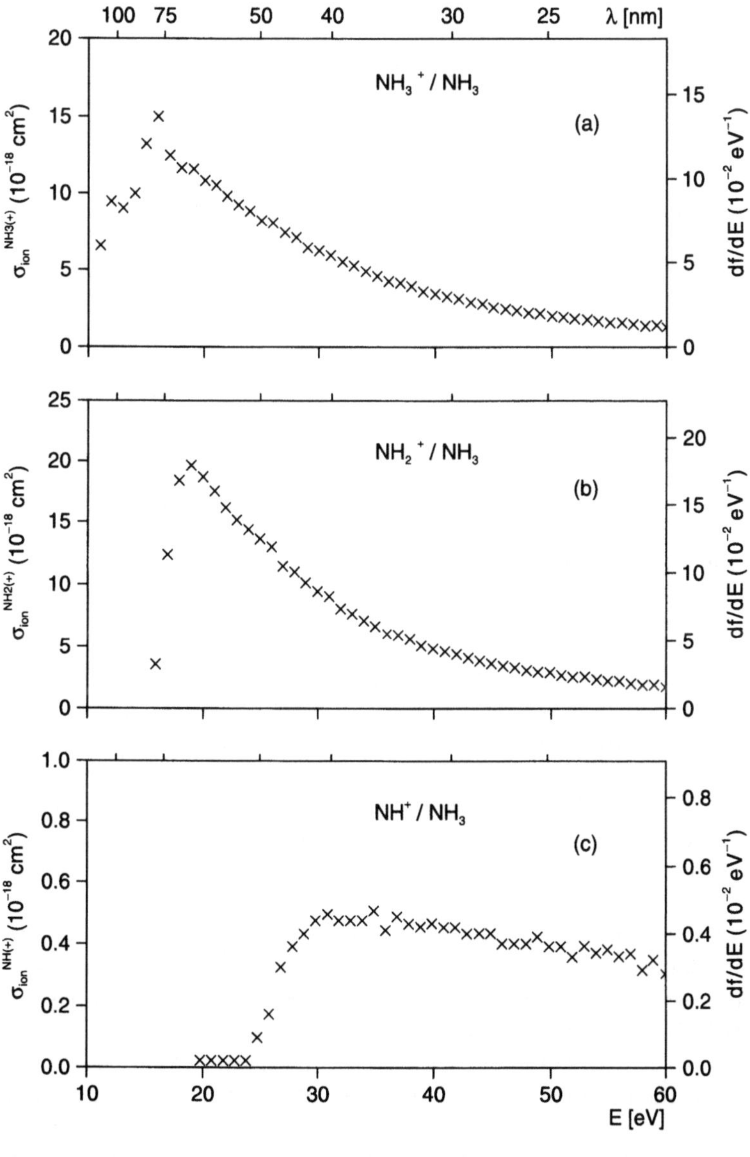

Fig. 5.11a–c. Partial ionic photofragmentation cross section for NH_3 [5.79]: (a) product ion $= NH_3^+$, (b) product ion $= NH_2^+$, (c) product ion $= NH^+$

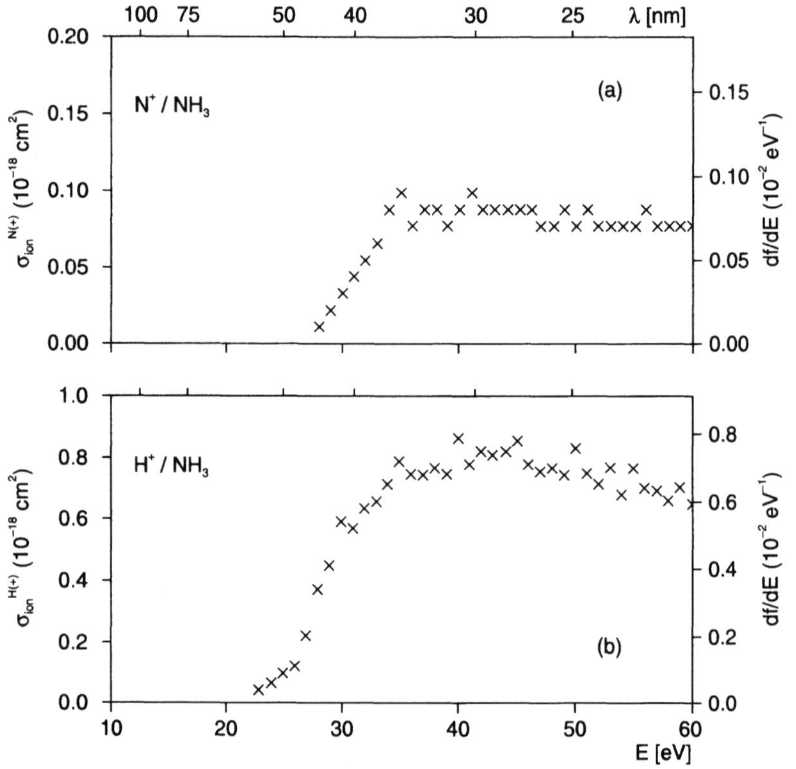

Fig. 5.12a, b Partial ionic photofragmentation cross section for NH_3 [5.79]: (**a**) product ion = N^+, (**b**) product ion = H^+

In [5.19] the contribution of N_2^{++} to the ionization cross section was estimated. It was concluded that the fraction of N_2^{++} in the total yield of ions is around 7% at 180–210 Å and decreases to about 2% at a wavelength longer than 260 Å. This is consistent with another estimate [5.52].

In [5.50] the partial photoionization cross sections for the production of N_2^+ in its states $X\,^2\Sigma_g^+$, $A\,^2\Pi_u$ and $B\,^2\Sigma_u^+$ are also reported. Another method has been applied in [5.53] and a general agreement has been obtained (Fig. 5.3).

Using photoelectron spectroscopy, the branching ratio for the production of the specific states of O_2^+ was determined [5.55, 56]. Independently from these experiments, the partial cross sections using the (e, 2e) technique were determined [5.57]. Figure 5.4 exhibits the partial ionization cross section for the production of $O_2^+(X\,^2\Pi_g)$. Figure 5.5 displays the results of the above three experiments for the production of O_2^+ in the $B\,^2\Sigma_g^-$, $b\,^4\Sigma_g^-$ and $A\,^2\Pi_u + a\,^4\Pi_u$ states.

A compilation of absolute total photoabsorption, photoionization, partial channel photoionization cross sections and partial cross sections for ionic photofragmentation (i.e., dissociative photoionization) for approximately twenty molecules is given in [5.58].

Figure 5.6 depicts the partial photoionization cross section of the H_2 molecule when the final ionic state of H_2^+ is $X\ {}^1\Sigma_g^+$. The angle dependence of the same cross section is shown in Fig. 5.7.

Figure 5.8 exhibits a partial photoionization cross section of CO molecule with the final ionic $B\ {}^2\Sigma^+$ state. The final vacancy is produced in the 4σ orbital. The asymmetry parameter for this process is given in Fig. 5.9.

The partial photoionization cross section of the NO molecule is presented in Fig. 5.10; the final ionic state is $c^3\Pi$. The cross sections for different channels of NH_3 photofragmentation taken from [5.79] are displayed in Figs. 5.11, 12.

6 Collisions of Molecules with Electrons

The collisions of molecules with electrons are considered in this chapter. Numerical data on the effective cross sections are given for hydrogen, nitrogen, oxygen and other diatomic and triatomic molecules.

Collisions of molecules with electrons play an important role in many physical processes such as energy deposition by incident particles, production of excited and charged species and others. The cross sectional data are mostly based on experimental results. Theoretical calculations are quite complicated because of the inherent complexity of molecular structure. Therefore, accurate calculations are very scarce. Table 6.1 illustrates the fields of application of the cross-section information in electron–molecule collisions in various areas of science [6.1].

Figure 6.1 depicts the cross sections of various processes arising from collisions of neutral H_2 molecules with electrons. It illustrates the relative order of magnitude of different channels in electron–molecule scattering.

6.1 Basic Approaches

A majority of calculations for cross-sections have been performed in the quantum-mechanical approximations, which are briefly discussed.

The differential cross section of *elastic* scattering is expressed through the scattering amplitude $f(\theta)$ by

$$\frac{d\sigma_{el}}{d\Omega} = |f(\theta)|^2 , \tag{6.1.1}$$

where θ is the scattering angle.

The differential cross section of *inelastic* scattering can be written with the inelastic scattering amplitude $f_{if}(\theta)$ as

$$\frac{d\sigma_{if}}{d\Omega} = \frac{k_f}{k_i}|f_{if}(\theta)|^2 , \tag{6.1.2}$$

where k_i and k_f are the initial and final momenta of the incident and scattered electrons. The $f(\theta)$ and $f_{if}(\theta)$ values are found from a solution of the Schrödinger equation for the electron plus molecule system with the boundary conditions derived from scattering theory [6.3, 4].

Table 6.1. Fields of application of various molecules in electron–molecule collisions [6.1]. (** indicates "very important")

Molecule	Plasma process	Gaseous dielectrics	Discharge switch	Gas laser	Space science	Radiation research
H_2	*				**	*
N_2	*	*		**	**	*
O_2	*	*			**	*
F_2	**			**		*
Cl_2	**			*		*
CO				*	*	*
HCl				**		*
H_2O					*	**
CO_2				**	*	*
SO_2		*			*	
N_2O		*				*
NH_3				*	*	*
CH_4	**		*		*	*
CCl_4	*	*		*		
SF_6	**	**		*		*
C_2H_6	*		*			*

Born Approximation. The Born approximation for cross sections is valid at high incident electron energies $E \geq 100\,\text{eV}$. At these energies, the interaction potential $V(r)$ of the scattered electron with a molecule is small and can be considered as a perturbation. Calculations are usually performed in the first Born approximation which coincides with the first order of perturbation theory. In this approximation, the differential cross section can be written as

$$\frac{\mathrm{d}\sigma_{if}}{\mathrm{d}\Omega}(\boldsymbol{q}) = \frac{k_f}{k_i}\left|\frac{1}{2\pi}\int \mathrm{d}\boldsymbol{r}\langle i|\exp(\mathrm{i}\boldsymbol{q}\boldsymbol{r})V(\boldsymbol{r},X)|f\rangle\right|^2 , \tag{6.1.3}$$

where $\langle i|$ and $|f\rangle$ are the initial and final wave functions of a molecule, $\boldsymbol{q} = \boldsymbol{k}_f - \boldsymbol{k}_i$ is the momentum transfer, and $V(\boldsymbol{r},X)$ is the effective potential, depending on coordinates $\boldsymbol{r}$ of the scattered electron and X of the molecule. Strictly speaking, the Born approximation is valid only at energies much higher than the mean energy of the molecular electrons. However, in some cases, this simple approximation gives quite good results for low energies as well, for example, in the case of excitation of rotational states in the hydrogen molecule [6.5].

Distorted Wave Approximation (DWA). In this method, the influence of the long-range interactions V_L on the scattered electron are taken into account. The differential cross section has the more complicated form:

$$\frac{\mathrm{d}\sigma_{if}}{\mathrm{d}\Omega}(\boldsymbol{q}) = \frac{k_f}{k_i}\left|\frac{1}{2\pi}\int \mathrm{d}\boldsymbol{r}\big\langle \varphi_i(\boldsymbol{k}_i)\big|V_{if}\big|\varphi_f(\boldsymbol{k}_f)\big\rangle\right|^2 . \tag{6.1.4}$$

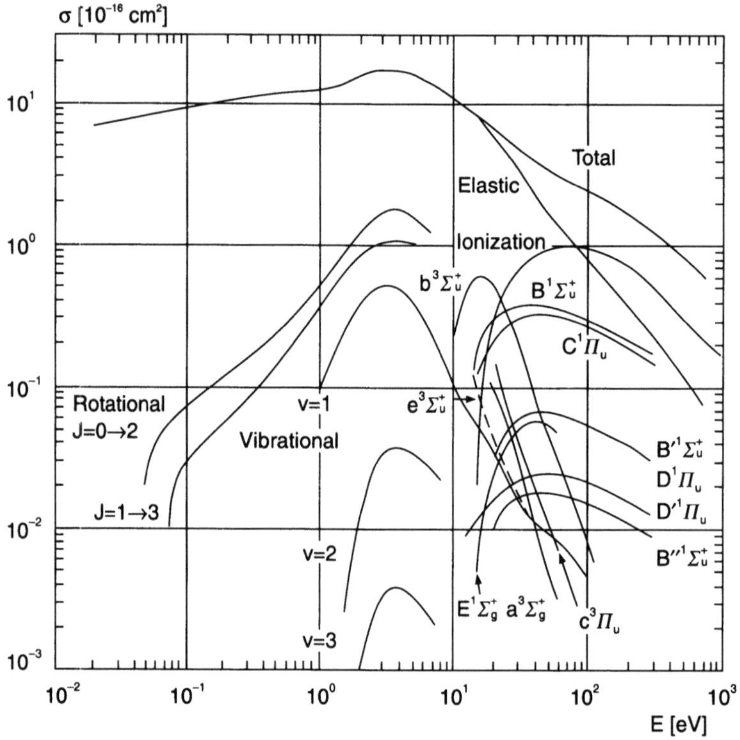

Fig. 6.1. Comparison of cross sections for various collision processes for the neutral H_2 molecule. For comparison, cross sections of ionizations of atomic hydrogen are shown as well (Data at room temperature)

Here, φ_i and φ_f are the plane waves distorted by the long-range interaction in the initial and final states: $V_{if} \equiv \langle i|V|f\rangle$. The DWA method is applied to the calculation of the cross sections in the Coulomb and other long-range fields at high and medium incident electron energies [6.6].

Close-Coupling Approximation (CC). This method is applied to the description of the low and medium energy region (from one hundredth up to hundreds eV). According to the close-coupling approximation, the molecular radial functions $u_\gamma(r)$ are found from the solution of coupled integro- differential equations:

$$\left(\frac{d^2}{dr^2} - \frac{l(l+1)}{r^2} + k_\gamma^2\right) u_\gamma^{(l)}(r) - 2\sum_{\gamma' l'} \langle \gamma l|V|\gamma' l'\rangle u_{\gamma'}^{(l')}(r)$$
$$= 2r \sum_{\gamma'' l''} \left\langle \gamma \middle| \hat{A} r^{-1} u_{\gamma''}^{(l'')}(r) \middle| \gamma'' \right\rangle . \tag{6.1.5}$$

Here l, l' and l'' are the orbital quantum numbers; the indices γ, γ' and γ'' correspond to the different molecular states, r is the radial coordinate, and $\hat{A}$ is the operator of antysymmetrisation of molecular and scattered electrons. In practice, one deals only with a finite number of scattering channels γ, γ' and γ'' and orbital numbers l, l' and l''. The question of the choice of the basis and the effective interaction potential is very important since the efficiency and accuracy of the method depends very strongly on this choice. Neglecting all nondiagonal matrix elements in (6.1.5), one can obtain the equation in the DWA approximation. Usually, solutions of (6.1.5) are obtained as an expansion of the functions u_γ on a certain basis, for example, on the Slater nodeless wave functions or hydrogen orbitals:

$$u_\gamma(r) = \sum_{n=0}^{N} c_{\gamma_n}\varphi_n(r) \ . \tag{6.1.6}$$

Here, φ_n denotes the basic functions, and N their number. Accounting for (6.1.6), the system of (6.1.5) is reduced to a purely algebraic one. The close-coupling method allows one to calculate the cross sections of molecular electronic excitation, resonance processes and excitation of vibrational and rotational states of molecules as well.

R-matrix Method. This method is used for the determination of cross sections of electronic, vibrational and rotational excitation of diatomic molecules and resonant processes [6.8]. In this method, the complete coordinate space of the scattered electron is divided into two regions: internal region (1) which contains the short-range part of the interaction, and the external region (2) which extends to infinity from the molecule. The border of these two areas is a sphere of some radius r_0. This border is called the R-matrix surface. To find the R-matrix, one should determine the bound states in region (1) with additional boundary conditions, for example, with a logarithmic derivative, equalling a certain arbitrary constant on the border [6.8]. Wave functions of these bound states u_n are called eigenfunctions of the R-matrix. They are defined in the case of one channel as

$$R = \frac{1}{r_0}\sum_{n=1}^{\infty}\frac{[u_n(r_0)]^2}{k_n^2 - k^2} \ . \tag{6.1.7}$$

Here, $k_n^2/2$ is the energy of the bound state n.

Usually, the function $u_\gamma^{(l)}$ are found from the solution of the system (6.1.5) by means of an expansion of u_n on a certain basis. In the case of electron–molecule scattering, the R-matrix basis contains a number of molecular bound states with additional pseudo or virtual states to provide the completeness of the basis [6.9]. The characteristics of scattering, for example, elements of the K-matrix, are found from the common condition in the regions (1) and (2) on their border. In region (2), one can neglect the coupling between different scattering channels and solve isolated equations with the effective one-particle potential. The R-matrix equations automatically provide its unitarity and consequently the conservation law for a number of particles. This constitutes one of the advantages of the R-matrix method.

As for the other theoretical methods, one should mention the variational method [6.10], the method of projection operators, algebraic methods and others [6.11–13].

6.2 Types of Collision Processes

Atoms in molecules can be excited or ionized via inelastic collisions with electrons. *Single ionization*

$$e^- + A \rightarrow A^+ + 2e^- ,$$

as well as *multiple ionization*

$$e^- + A \rightarrow A^{z+} + (z+1)\, e^-$$

may take place [6.14]. The total positive current and the total number of secondary electrons are proportional to the cross section

$$\sigma_i^{\mathrm{t}} = \sum_{z \geq 1} z\sigma_{iz} , \tag{6.2.1}$$

which is called the *gross* or *net ionization cross section.*

Ionization processes are important not only as a mechanism of energy loss by the incident electrons but also as a source of free electrons in matter; ionization strongly affects the energy distribution of free electrons. An ionized molecule may dissociate into two or more fragments one of which at least is an ion. We call this process *dissociative ionization.* A molecule may also dissociate into fragments one of which at least is in an excited state. This process is called the *dissociative excitation.* The electron can take up the energy from excited species by deexciting them in collisions called *superelastic collisions.*

Molecules can capture a slow electron and form a negative ion (*electron attachment*). A negative molecular ion may dissociate into neutral and negatively charged fragments; because the electron attachment processes reduce the number of slow electrons, they are important in forming the energy spectrum of free electrons in matter.

The *stopping power*, or the mean rate of energy loss by an electron having kinetic energy E in passing through matter at a distance x, is usually expressed as $-\mathrm{d}E/\mathrm{d}x$. It is proportional to the density N_{t} of atoms or molecules in target:

$$-\frac{\mathrm{d}E}{\mathrm{d}x} = N_{\mathrm{t}} S , \tag{6.2.2}$$

where S is the *stopping cross section* defined as

$$S = \sum_n \Delta E_{0n} \sigma_{0n} \tag{6.2.3}$$

in terms of the transition energy ΔE_{0n} and the excitation cross section σ_{0n}, when a molecule is excited to level n.

Various experimental techniques have been developed to measure the collision cross sections. At relatively high energies (more than a few eV), the *crossed-beam technique* is the most reliable [6.15] and can provide detailed information on collision processes (for example, double differential cross sections). In this method, however, measurements at extremely forward (<10°) and backward (>150°) angles are difficult to be carried out and extrapolations to such angles are necessary with some assumptions to obtain partial cross sections, resulting in additional uncertainties. On the other hand, at energies less than 10 eV, the *swarm technique* is the most often used [6.16]. In this method, the electron drift velocities, diffusion coefficients, ionization coefficients etc. are measured, and the final collision cross sections are calculated using the Boltzmann equations or Monte-Carlo technique incorporating elastic scattering cross sections determined before by other methods such as the crossed-beam technique. Therefore, the final cross sections strongly depend on the determination of elastic scattering cross sections.

6.3 Elastic Scattering

The mean rate of a linear energy loss due to elastic collisions of molecules with electrons is the product of the molecular density N_t and the stopping power cross section S, where

$$S = \left(\frac{2m}{M}\right) E\sigma_m \tag{6.3.1}$$

and σ_m is the *momentum-transfer cross section* related to the differential cross section $d\sigma/d\Omega$ for elastic scattering at an angle θ by

$$\sigma_m = \int \frac{d\sigma}{d\Omega}(1-\cos\theta)\,d\Omega \ . \tag{6.3.2}$$

Here, m and M are the electron and molecular masses. The quantity σ_m is also called the *transport cross section.*

The *viscosity cross section* σ_{vis} is defined by

$$\sigma_{vis} = \int \frac{d\sigma}{d\Omega}(1-\cos^2\theta)\,d\Omega \ , \tag{6.3.3}$$

where $d\sigma/d\Omega$ is the elastic differential cross section at the incident electron energy E.

Figures 6.2, 3 exhibit the total (sum of inelastic cross sections over all channels), elastic, momentum-transfer and viscosity cross sections of N_2 molecules colliding with electrons. We note that the elastic cross section σ_{elas} in Figs. 6.2, 3 include effects of rotational transitions. Therefore, the present elastic cross sections should be called "vibrationally elastic".

Tables 6.2, 3 summarize the data of the total electron scattering cross sections for H_2 molecules [6.15].

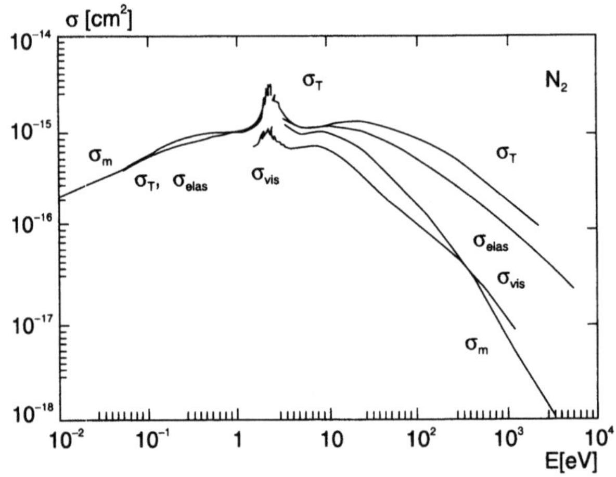

Fig. 6.2. Total scattering (σ_t), elastic (σ_{elas}), momentum-transfer (σ_m), and viscosity (σ_{vis}) cross sections for N_2. Details around 2.3 eV shape resonance are shown in Fig. 6.3

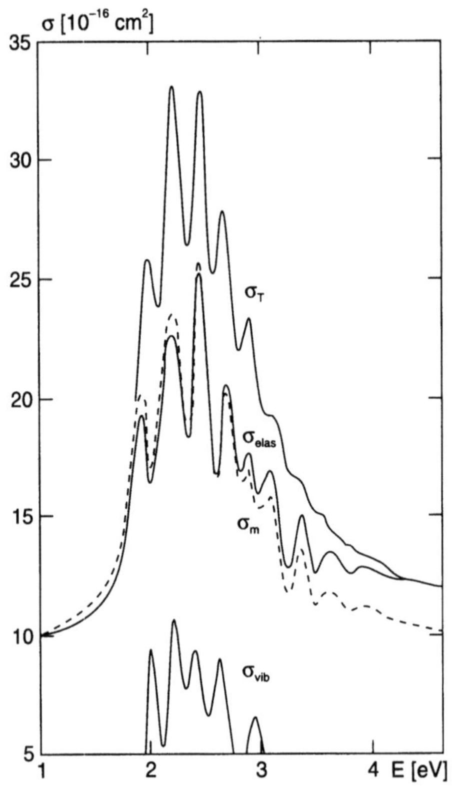

Fig. 6.3. Details of the total scattering (σ_t), elastic (σ_{elas}), and momentum-transfer (σ_m) cross sections for N_2 in the region of the 2.3 eV shape resonance. The sum of the vibrational cross sections $[\sigma_{vib} = \sum_{v'=1}^{g} \sigma_{vib}(0 \rightarrow v')]$ is also presented for comparison

Table 6.2. Total electron scattering cross sections for H_2 (in 10^{-16} cm^2)

	Reference				Reference			Reference		
E [eV]	6.18*	6.19	6.21	*E* [eV]	6.19	6.21	*E* [eV]	6.20 meàs.	6.20 semi-emp.	6.21
0.020	7.41	–	–	3.0	–	16.2	13.0	–	–	8.62
0.030	7.77	–	–	3.2	–	16.2	13.5	–	–	8.39
0.040	8.05	–	–	3.4	–	16.1	14.0	–	–	8.15
0.060	8.52	–	–	3.5	16.55	–	14.5	–	–	7.84
0.080	8.79	–	–	3.6	–	16.2	15.0	–	–	7.75
0.100	9.19	–	–	3.8	–	16.0	16.0	–	–	7.39
0.150	9.75	–	–	4.0	–	15.7	17.0	–	–	6.94
0.200	10.05	–	–	4.2	–	15.8	18.0	–	–	6.71
0.250	10.45	–	–	4.4	–	15.6	19.0	–	–	6.44
0.300	10.71	9.27	–	4.5	15.79	–	20.0	–	6.21	6.23
0.350	10.95	–	–	4.6	–	15.3	21.0	–	–	5.96
0.400	11.21	–	15.1	4.8	–	15.1	22.0	–	–	5.74
0.450	11.36	–	–	5.0	–	14.9	23.0	–	–	5.69
0.500	11.62	–	–	5.5	–	14.4	24.0	–	–	5.49
0.6	(11.9)	11.56	11.9	6.0	14.22	13.9	25.0	5.75	–	5.29
0.8	(12.4)	–	12.4	6.5	–	13.4	26.0	–	–	5.18
1.0	(12.8)	14.00	13.1	7.0	–	13.0	28.0	–	–	4.89
1.2	(13.2)	–	13.6	7.5	–	12.4	30.0	4.89	4.58	4.68
1.4	(13.5)	–	14.0	8.0	12.35	12.0	32.0	–	–	4.43
1.5	(13.7)	15.37	–	8.5	–	11.4	34.0	–	–	4.26
1.6	–	–	14.7	9.0	–	11.1	35.0	4.44	–	–
1.8	–	–	15.0	9.5	–	10.8	36.0	–	–	4.17
2.0	–	16.16	15.4	10.0	10.78	10.3	38.0	–	–	4.16
2.2	–	–	15.7	10.5	–	10.1	40.0	4.00	4.05	3.76
2.4	–	–	15.8	11.0	–	9.68	45.0	3.86	–	3.59
2.5	–	16.16	–	11.5	–	9.40	50.0	3.62	–	3.29
2.6	–	–	15.9	12.0	–	9.11	60.0	–	3.43	–
2.8	–	–	16.1	12.5	–	8.86	70.0	–	–	–
							75.0	–	2.92	–

* numbers in parentheses where obtained from ([6.7], (Fig. 6.9))

Table 6.3. Total electron scattering cross sections for H_2 (in 10^{-16} cm^2) [6.22]

E [eV]	σ	*E* [eV]	σ
2.0	17.0 ± 0.1	35	4.66 ± 0.03
2.9	17.3 ± 0.2	40	4.30 ± 0.01
3.9	17.6 ± 0.13	45	3.98 ± 0.01
4.9	15.6 ± 0.07	50	3.66 ± 0.09
6.8	14.3 ± 0.07	75	2.92 ± 0.02
8.8	12.2 ± 0.09	100	2.56 ± 0.04
10.8	10.56 ± 0.04	150	1.98 ± 0.01
12.9	9.48 ± 0.03	200	1.69 ± 0.03
14.9	8.48 ± 0.02	300	1.27 ± 0.03
20	6.72 ± 0.04	400	1.04 ± 0.03
25	5.79 ± 0.04	500	0.87 ± 0.02
30	4.92 ± 0.05		

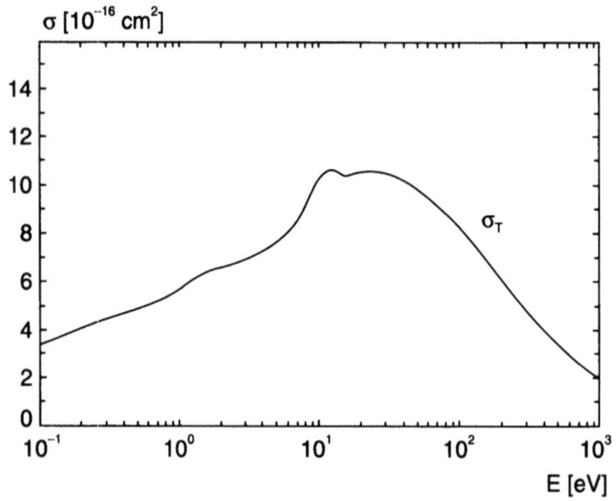

Fig. 6.4. Total cross section for electron scattering from O_2

The total cross section for electron scattering with O_2 is depicted in Fig. 6.4 [6.23, 24].

Total electron scattering cross sections for CO_2 have been reported in [6.22, 25–27]. The results are given in Tables 6.4, 5.

There has been considerable experimental interest in SF_6 molecules because of its commercial use as a gaseous dielectric in high-voltage devices.

Table 6.4. Total electron scattering cross sections for CO_2 (in 10^{-16} cm^2) [6.27]

E [eV]	σ	E [eV]	σ
0.069	68.55 ± 2.50	0.791	10.07 ± 0.30
0.073	66.31 ± 2.50	0.912	9.23 ± 0.25
0.083	60.63 ± 2.20	1.034	8.25 ± 0.20
0.094	54.31 ± 2.05	1.158	7.22 ± 0.20
0.105	49.15 ± 1.67	1.283	6.62 ± 0.20
0.115	45.57 ± 1.55	1.410	6.23 ± 0.20
0.126	42.55 ± 1.49	1.538	6.15 ± 0.18
0.137	39.53 ± 1.20	1.667	6.11 ± 0.18
0.159	35.83 ± 1.20	1.798	6.00 ± 0.18
0.170	33.59 ± 1.16	1.918	6.10 ± 0.22
0.180	31.85 ± 1.15	2.082	6.12 ± 0.35
0.191	30.29 ± 1.08	2.378	7.15 ± 0.42
0.202	29.02 ± 1.05	2.650	7.70 ± 0.45
0.213	27.97 ± 1.01	2.917	9.30 ± 0.60
0.269	25.22 ± 0.90	3.419	12.47 ± 0.80
0.325	22.03 ± 0.73	3.664	13.41 ± 0.80
0.438	17.73 ± 0.50	3.815	13.61 ± 0.80
0.554	14.51 ± 0.35	4.105	12.86 ± 0.80
0.672	12.23 ± 0.30	4.528	8.86 ± 0.55
		4.684	7.27 ± 0.55

Table 6.5. Total electron scattering cross sections for CO_2 (in 10^{-16} cm^2) [6.22]

E [eV]	σ	E [eV]	σ
2.0	6.02 ± 0.08	8.8	11.18 ± 0.04
2.5	6.45 ± 0.04	9.8	11.41 ± 0.05
3.0	8.03 ± 0.20	10.8	13.48 ± 0.08
3.4	12.96 ± 0.16	12.4	15.2 ± 0.10
3.7	14.57 ± 0.04	14.9	16.6 ± 0.10
3.85	17.8 ± 0.10	20	17.4 ± 0.10
3.95	16.9 ± 0.14	25	18.4 ± 0.2
4.15	14.9 ± 0.12	30	19.0 ± 0.2
4.4	13.56 ± 0.08	35	18.0 ± 0.3
4.9	10.12 ± 0.20	40	17.7 ± 0.1
5.9	8.70 ± 0.25	45	17.0 ± 0.1
6.4	8.73 ± 0.04	50	15.7 ± 0.1
7.3	9.97 ± 0.07		

Accurate total cross sections using the time-of-flight method have been reported in [6.28] for 0.5–100 eV electron energy (Fig. 6.5).

The momentum-transfer cross sections for some well-studied molecules are listed in Table 6.6 [6.29].

The results on the integral elastic cross sections for H_2 molecules as a function of the incident electron energy are presented in Table 6.7.

Differential elastic cross sections for N_2 molecules have been measured in [6.38, 47, 48]. The data extrapolated to the 0°–180° range of angles are compiled in Table 6.8. They have been obtained from the measured ratios of the elastic cross sections σ_{N_2}/σ_{He} [6.48] and the revised data on σ_{He} [6.49]. The integral elastic cross sections for N_2 molecules (including pure rotational excitation) are listed in Table 6.9 together with the momentum-transfer cross sections.

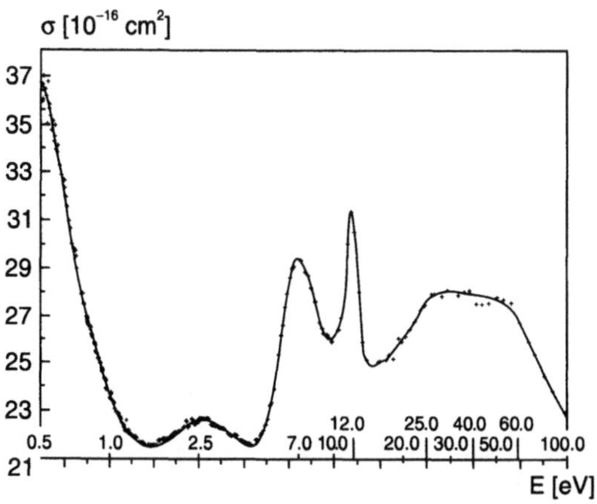

Fig. 6.5. Total electron scattering cross sections for SF_6 [6.28]

Table 6.6. Momentum-transfer cross sections σ_m for scattering of electrons by molecules (in 10^{-16} cm^2) [6.29]

E [eV]	H_2^a	H_2^b	H_2^c	N_2^a	O_2^e	O_2^a	CO^f	CO_2^a	H_2O^g	NH_3^a	N_2O^a
0.000		6.4	6.36	1.10	0.35		60	600	5000	10000	500
0.001		6.4	–	1.36	0.35		40	540	–	5600	170
0.002		6.5	–	1.49	0.36		25	387	–	3850	114
0.005		6.8	–	1.81	0.50		12.3	–	–	2240	69
0.01		7.3	7.26	2.19	0.70		7.8	170	2900	1400	47
0.02		8.0	7.95	2.85	0.99		5.9	119	1380	800	32.4
0.05		9.28	9.22	4.33	1.6		5.4	–	500	322	19.0
0.08		–	10.0	5.35	–		–	58.0	298	189	14.3
0.10		10.5	10.4	5.95	2.5		7.3	52.0	233	144	12.5
0.15		11.4	11.3	7.1	3.1		8.8	40.0	150	83	9.2
0.2		12.0	11.9	7.9	3.6	–	10.0	31.0	110	54	6.3
0.3		13.0	12.9	9.0	4.5	–	12.1	20.3	70	27.5	3.6
0.4		13.9	13.8	9.7	5.2	–	13.4	14.3	51	17	2.75
0.5		14.7	14.6	9.9	5.7	–	13.8	10.9	39.3	12	2.6
0.6		–	15.5	9.95	–	5.70	14.2	–	31.2	9.4	2.8
1.0		17.4	17.3	10.1	7.2	6.38	16.0	5.55	14.6	5.3	4.9
1.2		17.8	–	10.5	7.9	6.71	21.5	5.02	11.2	4.6	6.0
1.3		18.0	–	–	7.9	–	26.5	4.90	–	–	–
1.5		18.25	18.1	11.6	7.6	6.75	38.0	4.83	8.65	4.1	7.3
1.7		18.25	–	13.5	7.3	–	43.0	4.85	7.7	–	–
1.9		18.1	–	19.8	6.9	–	43.0	–	–	–	–
2.0		–	17.9	17.0	–	6.52	40.5	5.10	6.8	3.85	8.8
2.1		17.9	–	19.7	6.6	–	35.0	–	–	–	–
2.2		17.7	–	23.4	6.5	–	31.0	–	6.35	–	–
2.3		–	–	20.5	–	–	27.5	–	–	–	–
2.4		–	–	21.8	–	–	25.0	–	–	–	–
2.5		17.0	17.7	23.1	6.1	–	23.3	5.85	6.07	3.9	9.4
2.7		–	–	20.1	–	–	21.0	–	6.0	–	–
3.0		16.0	16.9	15.2	5.7	6.29	18.5	7.02	6.03	4.15	9.1
4.0		14.0	14.7	11.1	5.5	6.16	13.5	8.37	6.55	4.9	7.7
5		12.2	12.9	9.7	5.6	6.30	12.0	6.46	7.25	5.8	7.3
7		8.9	–	–	6.6	–	10.7	6.60	8.6	–	8.2
8		7.85	–	10.2	7.1	7.21	10.3	7.28	8.95	8.4	8.7
10		6.0	7.60	10.0	8.0	7.63	9.6	9.28	9.3	9.0	9.6
12		5.2	–	9.6	8.5	–	9.1	10.0	9.2	8.8	10.0
15	2.9	4.5	3.25[d]	8.9	8.8	7.10	8.3	10.1	8.7	8.2	9.7
20	1.88	3.9	2.20[d]	7.6	8.6	6.10	7.2	8.88	7.6	7.0	8.5
25	1.35	3.6	1.49[c]	6.7	8.2	–	6.2	7.71	6.47	5.9	7.3
30	1.00	3.4	1.03[d]	5.9	8.0	4.87	5.4	6.91	5.5	5.1	6.3
40	0.62	–	0.640[d]	4.7	–	–	4.2	5.74	3.99	3.9	4.85
50	0.44		–	3.8	7.7	3.49	3.4	5.18	2.94	3.1	3.93
60	0.31		0.295[d]	3.2	–	2.93	2.9	4.56	2.24	2.6	3.25
75	–		–	–	6.8	–	2.3	–	–	–	–
80	0.195		–	2.35	–	2.16	2.2	3.38	1.45	1.9	2.37
100	0.132		0.150[d]	1.80	6.5	1.65	1.72	2.58	1.05	1.5	1.85
120	0.104			1.46		1.35	1.42	2.04	0.84	1.2	–
150	0.073			1.13		1.13	1.07	1.55	0.64	0.89	–
200	0.045			0.80		0.946	0.79	1.08	0.464	0.60	0.88
300	0.022			0.48		0.626	0.48	0.661	0.296	0.33	–

Table 6.6. *Continued*

E [eV]	H_2^a	H_2^b	H_2^c	N_2^a	O_2^e	O_2^a	CO^f	CO_2^a	H_2O^g	NH_3^a	N_2O^a
500	0.0091			0.23		0.311	0.238	0.360	0.152	0.15	0.335
700	–			–		–	0.138	–	0.091	–	–
800	0.0040			0.110		0.164	0.111	0.193	–	0.072	–
1000	0.0027			0.077		0.115	0.076	0.140	0.053	0.050	0.154
2000	–			–		–	0.021	–	–	0.0152	0.0483

[a] Recommended in [6.1]
[b] [6.30]
[c] [6.31]
[d] Average of the results of the crossed-beam experiments in [6.32–34]
[e] [6.35]
[f] Taken from [6.36] for E < 0.4 eV and from [6.37] for 0.4 eV ≤ E ≤ 4.0 eV and calculated from differential cross section in [6.38] for E = 200, 500 and 800 eV. For 5 eV ≤ E ≤ 100 eV the σ_m in [6.39] (as renormalized in [6.15]) are multiplied by an energy dependent factor chosen by comparing the differential cross sections in [6.39] and [6.40]. The total cross sections in [6.14] have also been taken into account. The final σ_m has been determined by smoothing and interpolating these values
[g] [6.42]

The values of elastic and momentum-transfer cross sections for O_2 molecules have been calculated on the basis of differential elastic cross sections (Fig. 6.6).

Elastic scattering cross sections in CO molecule were determined for the 3–100 eV electron energy range [6.39]. Similar to N_2 case, these data have been evaluated using the revised elastic cross sections for He atoms given in [6.49]. Corrected integral elastic and momentum-transfer cross sections are listed in Table 6.10.

The alkali-halide molecules possess very large dipole momenta (≈ 10 Debye), hence, they provide a special class for electron-molecule scattering studies. The dominant interaction is caused by the dipole momenta of a molecule and an electron. This interaction leads to very simple cross-section formulae that depend only on the electron energy, dipole momentum and momentum of inertia [6.50]. The most probable process in these collisions is the rotational excitation corresponding to transitions with $\Delta J = \pm 1$. The modern experimental techniques are not capable to resolve the rotational structure in these excitation processes. Results of measurements have usually been reported as "elastic" cross sections. This is done with the understanding that the cross sections include all transitions with $\Delta J = \pm 1$ as well as $\Delta J = 0$ (elastic) processes, and to a lesser extend $\Delta J \geq 2$ and unresolved vibrational excitation processes. Experimental integral "elastic" cross sections are of the order of $10^{-14}\,\text{cm}^2$ and near-zero-degree differential cross sections are of the order of $10^{-12}\,\text{cm}^2/\text{sr}$. The differential cross sections have extremely narrow peaks and usually a scattering into a few degrees represents about 90% of the integral cross section. Experimental and theoretical results for elastic scattering on LiF molecules at 20 eV are depicted in Fig. 6.7.

Table 6.7. Integral elastic cross sections (in $10^{-16} cm^2$) for H_2 molecules as a function of the incident electron energy. From [6.15]

E (eV)	[6.43], [6.44][a]	[6.45][b]	[6.34]	[6.46]	semi-emp. [6.20]
0.29	9.60				
0.58	11.15				
0.98	12.34				
1.48	13.23				
2.00	–		14.4		
2.50	14.11		–		
3.00	–	14.8	15.8		
3.49	13.61	–	–		
4.49	12.81	–	–		
5.0	–	13.7	15.9		
5.97	11.26	–	–		
7.0	–	11.1	13.58		
8.01	9.68	–	–		
9.98	8.32	–	–		
10.0	–	8.27	11.29		
10.08	8.04	–	–		
15		5.89	7.55		
20		4.15	5.61		
30		2.30	3.36		
40		1.70	2.50		
60		1.00	1.27		
75		0.70	–		
100			0.77	0.89	
150			0.50	0.52	
200			0.39	0.36	
300				0.23	
400				0.15	
500				0.12	
600				–	0.095
700				0.073	–
750				–	0.069
800				–	0.066
900				–	0.059
1000				0.052	
2000				0.027	

[a] Rotation excitation is not included
[b] These values were obtained by renormalizing the original data [6.45] on the basis of the improved data [6.49] for He

Elastic differential scattering cross sections for H_2O molecules at energies 2.14 and 6.0 eV were reported in [6.51]. These data were obtained by unfolding the elastic-scattering ($\Delta J = 0$) contribution from their high-resolution measurements with combined elastic and rotational ($\Delta J = \pm 1$) energy-loss spectra. The results are displayed in Fig. 6.8.

Elastic integral and momentum-transfer cross sections for CO_2 molecules are listed in Table 6.11 [6.52, 53].

Table 6.8. Experimental differential elastic cross sections (in 10^{-16} cm^2/sr) for N_2 molecules as a function of the scattering angle θ and the incident electron energy E_0. The values in parentheses are the ratios of the cross sections σ_{N_2}/σ_{He}. The data in square brackets refer to the extrapolated cross sections. From [6.48]

	E_0[eV]									
θ [deg]	5.0	7.0	10.0	15.0	20.0	30.0	40.0	50.0	60.0	75.0
[0]	1.85	2.15	3.28	5.91	7.35	9.75	11.0	11.4	11.9	7.10
[10]	1.83	2.14	3.05	4.83	6.09	7.21	8.00	8.60	7.45	4.80
20	1.70	2.07	2.76	3.68	4.54	4.83	4.54	4.14	3.56	2.81
	(6.49)	(6.73)	(7.15)	(7.85)	(8.78)	(9.23)	(9.33)	(9.49)	(9.13)	(8.57)
30	1.73	1.92	2.32	2.56	2.77	2.39	2.07	1.60	1.26	0.939
	(6.49)	(6.54)	(6.75)	(6.50)	(6.69)	(6.15)	(6.17)	(5.61)	(5.15)	(4.88)
40	1.67	1.86	1.93	1.67	1.61	1.14	0.843	0.632	0.459	0.333
	(6.00)	(6.49)	(6.18)	(4.98)	(4.81)	(3.97)	(3.58)	(3.30)	(2.91)	(2.81)
50	1.44	1.61	1.54	1.26	0.974	0.593	0.398	0.295	0.216	0.155
	(4.84)	(5.55)	(5.21)	(4.33)	(3.56)	(2.73)	(2.35)	(2.19)	(2.01)	(1.98)
60	1.20	1.31	1.24	0.902	0.623	0.332	0.230	0.177	0.122	0.0952
	(3.69)	(4.34)	(4.31)	(3.43)	(2.69)	(1.93)	(1.83)	(1.79)	(1.57)	(1.71)
70	0.918	1.01	0.867	0.563	0.382	0.180	0.143	0.113	0.0742	0.0619
	(2.58)	(3.14)	(3.00)	(2.28)	(1.87)	(1.28)	(1.45)	(1.48)	(1.26)	(1.48)
80	0.743	0.770	0.602	0.352	0.248	0.132	0.0944	0.0807	0.0576	0.0569
	(1.89)	(2.22)	(2.00)	(1.46)	(1.31)	(1.08)	(1.16)	(1.30)	(1.23)	(1.74)
90	0.628	0.603	0.434	0.276	0.191	0.112	0.0758	0.0597	0.0563	0.0742
	(1.45)	(1.59)	(1.36)	(1.13)	(1.04)	(1.00)	(1.06)	(1.13)	(1.45)	(2.81)
100	0.585	0.556	0.475	0.396	0.265	0.130	0.0791	0.0772	0.0697	0.0685
	(1.23)	(1.34)	(1.39)	(1.57)	(1.44)	(1.23)	(1.20)	(1.64)	(2.08)	(3.10)
115	0.584	0.648	0.607	0.512	0.444	0.213	0.154	0.173	0.149	0.117
	(1.08)	(1.37)	(1.58)	(1.87)	(2.32)	(2.09)	(2.47)	(4.07)	(5.11)	(6.40)
135	0.657	0.830	0.865	0.799	0.672	0.457	0.412	0.363	0.244	0.219
	(1.06)	(1.52)	(1.97)	(2.61)	(3.29)	(4.37)	(6.60)	(8.79)	(9.07)	(13.46)
[140]	0.69	0.89	0.93	0.87	0.75	0.56	0.50	0.43	0.29	0.26
[150]	0.75	1.05	1.12	1.04	0.90	0.71	0.75	0.60	0.35	0.32
[160]	0.83	1.28	1.35	1.21	1.05	0.91	1.08	0.78	0.42	0.41
[170]	0.93	1.46	1.60	1.36	1.18	1.09	1.42	0.95	0.46	0.49
[180]	1.06	1.80	1.90	1.47	1.27	1.23	1.79	1.09	0.49	0.53

Differential elastic scattering cross sections for CH_4 including rotational excitation were reported in [6.54] at 10 eV and lower (Figs. 6.9, 10). Integral elastic (σ_{el}) and momentum-transfer (σ_m) cross sections for CH_4 were collected in [6.55] (Table 6.12).

In [6.56], inelastic scattering by CH_4 were studied in the 7.5–15.0 eV energy-loss region at 20, 30 and 200 eV electron-impact energies. The energy-loss spectrum was divided into five regions, each corresponding to a number of unassigned electronic transitions. The summary of notations is listed in Table 6.13. Inelastic integral and momentum-transfer cross sections obtained from these data are given in Table 6.14.

Table 6.9. Integral (σ) and momentum-transfer (σ_m) elastic cross sections in N_2 (in 10^{-16} cm^2)

	[6.38]	[6.47]		[6.48][a]	
E [eV]	σ	σ	σ_m	σ	σ_m
1.5		9.6	10.0		
1.9		16.7	17.1		
2.1		17.5	18.1		
2.4		17.4	16.8		
3.0		14.8	12.6		
4.0		11.1	9.6		
5.0		11.2	9.2	11.7	9.25
7.0		12.5	10.8	13.3	10.9
10.0		11.7	9.3	13.3	10.5
15.0		11.3	8.2	12.1	8.85
20.0	9.13	12.1	8.3	11.2	7.33
30.0	–	10.3	6.0	8.89	5.01
40.0	–	9.4	4.7	8.12	4.66
50.0	7.22	8.5	4.1	7.03	3.82
60.0	–	–	–	5.45	2.55
70.0	–	7.3	2.8	–	–
75.0	–	–	–	4.31	2.29
100.0	4.96	5.6	1.76		
150.0	–	4.5	1.17		
200.0	3.53	3.7	0.75		
300.0	–	2.6	0.46		
400.0	2.11	2.3	0.35		
500.0	1.65				
800.0	1.10				

[a] Calculated from the renormalized data in Table 6.8

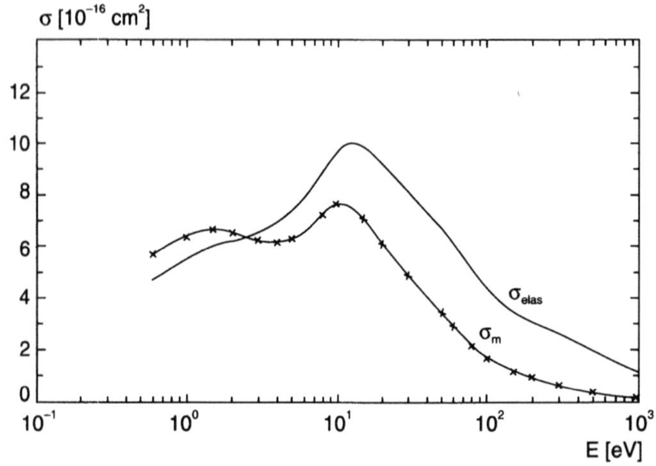

Fig. 6.6. Values of elastic (σ_{elas}) and momentum-transfer (σ_m) cross sections for electron collisions with O_2 [6.24]

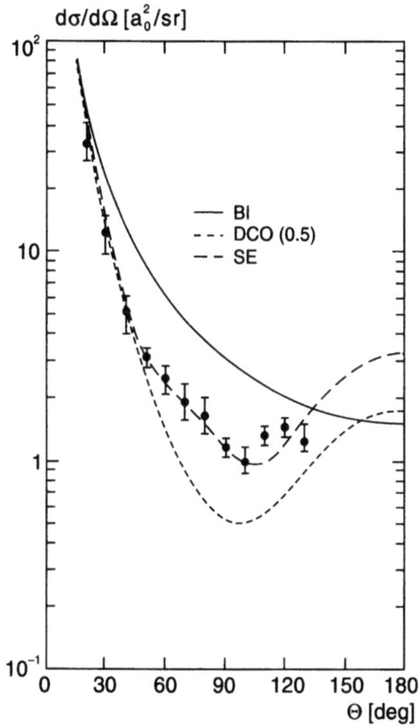

Fig. 6.7. Differential "elastic" cross sections for LiF (in units a_0^2/sr) at 20 eV impact energy. The experimental electron scattering results are normalized to the theory at $\theta = 40°$ are indicated by *dots* with *error bars*. The theoretical results were obtained from the BI, DCO (0.5) and SE models as indicated [6.50]

The elastic electron scattering cross sections for C_2H_2, C_2H_4 and C_2H_6 molecules from 100 to 1000 eV electron-impact energy at scattering angles from 3° to 130° can be found in [6.57] (Tables 6.15, 16).

Elastic integral (σ_{elas}) and momentum-transfer (σ_m) cross sections for SF_6 have been reproduced in Table 6.17 [6.58].

Table 6.10. Elastic integral (σ) and momentum-transfer (σ_m) cross sections in CO (in 10^{-16} cm^2) [6.39]

E_0 [eV]	σ	σ_m
3	18.3	12.0
5	15.1	12.0
7.5	11.2	9.25
9.9	11.5	9.56
15	11.1	8.31
20	9.50	5.93
30	6.34	3.88
50	4.54	2.55
75	3.50	1.42
100	2.61	0.88

Table 6.11. Elastic integral σ and momentum-transfer σ_m cross sections for CO_2 (in 10^{-16} cm^2) [6.52, 53]

	σ		σ_m	
E [eV]	[6.52]	[6.53]	[6.52]	[6.53]
3.0	10.9	–	12.0	–
3.5	14.8	–	15.0	–
3.8	17.3	–	16.6	–
4.0	14.7	8.73	13.7	7.34
5.0	12.4	–	11.2	–
7.0	11.2	–	10.8	–
10.0	14.1	10.61	13.1	9.52
15.0	17.6	–	12.8	–
20.0	24.5	12.32	13.3	8.89
30.0	22.3	–	11.5	–
40.0	20.6	–	9.4	–
50.0	15.6	8.75	8.0	5.55
70.0	12.5	–	4.8	–
90.0	9.5	–	2.4	–

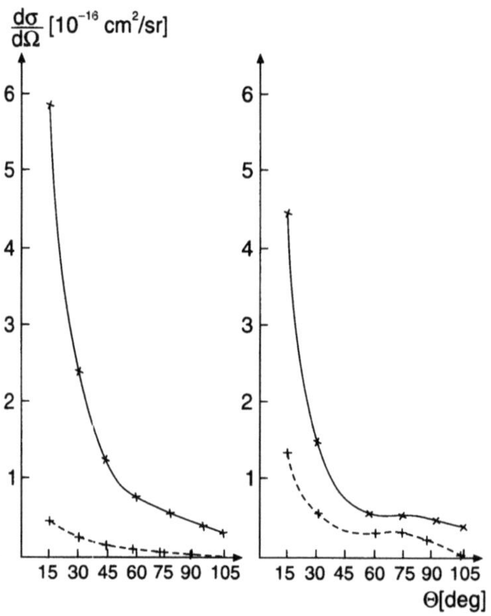

Fig. 6.8. Differential cross sections for elastic scattering and rotational excitation in H_2O at 2.14 and 6.0 eV impact energies. The *full curve* shows the rotational excitation and de-excitation cross section and the *dashed curve* represents pure elastic scattering [6.51] (+ elastic; × rotational excitation)

6.4 Excitation of Molecules by Electron Impact

6.4.1 Rotational Excitation

Rotational inelastic collisions significantly contribute to the electron energy losses only at low electron energies. The cross sections are relatively small for

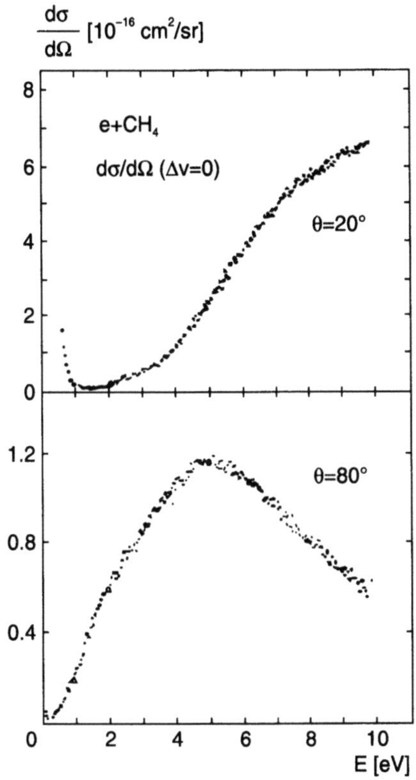

Fig. 6.9. Energy dependence of the differential elastic scattering cross section for CH_4 at 20° and 80° scattering angles [6.54]

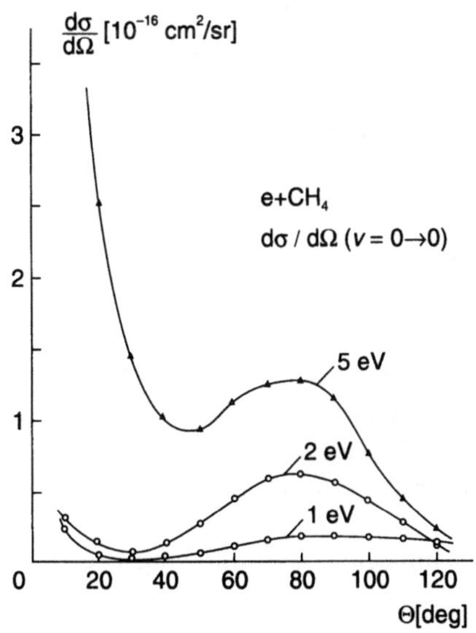

Fig. 6.10. Angular dependence of the elastic scattering cross sections for CH_4 at 1, 2 and 5 eV impact energies [6.54]

Table 6.12. Integral elastic (σ) and momentum-transfer (σ_m) cross sections for CH_4 (in 10^{-16} cm^2) [6.55]

E [eV]	σ	σ_m
3	7.4	6.6
5	13.7	12.4
6	16.1	14.5
7.5	19.6	16.0
9	18.9	14.4
10	18.4	13.1
15	15.7	7.6
20	14.3	6.5

Table 6.13. Summary of notations and assignments for the inelastic cross sections of CH_4 [6.56]

Region number	Energy-loss range [eV]
1	7.5–9.0
2	9.0–10.5
3	10.5–12.0
4	12.0–13.5
5	13.5–15.0

transitions with a change of the rotational quantum number of $\Delta J \geq 6$. Homonuclear diatomic molecules in the $^1\Sigma$ state, such as H_2, N_2 and F_2 in the electronic ground state satisfy the selection rules allowing a change of J with even numbers. Therefore, only transitions $J \to J, J \to (J \pm 2)$ and $J \to (J \pm 4)$ are important in this case. Transitions with odd ΔJ numbers are possible for polar molecules.

Information on cross sections for the transitions $\Delta J < 6$ is sufficient for calculations of nearly all important rotational cross sections. The detailed balance principle yields for direct and inverse transitions

$$k_J^2(2J+1)\frac{\mathrm{d}\sigma(J \to J')}{\mathrm{d}\Omega} = k_{J'}^2(2J'+1)\frac{\mathrm{d}\sigma(J' \to J)}{\mathrm{d}\Omega} . \tag{6.4.1}$$

Due to rotational transitions, the stopping cross section for a non-polar molecule is independent of J, and its Boltzmann-averaged rate is independent of the electron temperature T [6.59], provided the condition $k_B T \ll E$ is satisfied; k_B being the Boltzmann constant. The stopping cross section for a polar molecule has a weakly T-dependent correction term proportional to ln T and to the square of the dipole moment d.

In the rotational excitation of the H_2 molecule, the following processes have been investigated experimentally

$$e + H_2(X^1\Sigma_g\, v = 0,\ J = 0) \Rightarrow e + H_2(X^1\Sigma_g\, v = 0,\ J = 2) \ , \tag{6.4.2}$$

Table 6.14. Inelastic integral (σ) and momentum-transfer (σ_m) cross sections for CH_4 (in 10^{-16} cm^2) [6.56]

	20 eV		30 eV		200 eV	
Region number	σ	σ_m	σ	σ_m	σ	σ_m
1	0.0148	0.0132	0.0107	0.0079	–	–
2	0.242	0.164	0.263	0.104	0.091	0.0044
3	0.442	0.340	0.435	0.214	0.135	0.0097
4	0.568	0.478	0.647	0.357	0.299	0.002
5	0.443	0.393	0.578	0.354	–	–

Table 6.15. Elastic scattering cross sections for C_2H_2 (in 10^{-16} cm^2/sr) [6.57]

	E_0 [eV]				
θ [deg]	100	200	400	600	1000
(0)	136.3	106	81.7	57.8	51.48
3	82.4	57.42	41.4	27.28	20.46
4	70.3	45.30	32.8	21.32	15.45
5	59.3	35.34	25.9	16.98	11.65
6	49.5	29.82	20.1	13.19	6.67
7	41.8	24.30	15.81	10.28	6.5
8	35.2	19.88	12.66	7.950	4.74
9	29.7	16.12	10.24	6.124	3.52
10	25.1	13.03	8.05	4.878	2.62
12	17.7	8.83	5.12	2.891	1.60
14	12.7	6.01	3.22	1.825	0.990
16	9.0	4.42	2.19	1.247	0.745
18	6.6	3.18	1.50	0.827	0.609
20	4.95	2.26	1.06	0.672	0.529
24	2.86	1.29	0.649	0.529	0.297
28	1.70	0.791	0.493	0.386	0.155
32	1.10	0.548	0.395	0.227	0.0935
36	0.74	0.426	0.279	0.147	0.0645
40	0.53	0.365	0.200	0.0975	0.0497
45	0.39	0.309	0.129	0.0676	0.0348
50	0.31	0.243	0.0904	0.0538	0.0226
55	0.27	0.183	0.0712	0.0424	0.0165
60	0.23	0.143	0.0603	0.0321	0.0132
65	0.185	0.115	0.0507	0.0242	0.0105
70	0.16	0.092	0.0411	0.0192	0.00790
80	0.129	0.072	0.0273	0.0136	0.00532
90	0.124	0.066	0.0203	0.01	0.00387
100	0.126	0.061	0.0170	0.0074	0.00300
110	0.132	0.057	0.0145	0.0061	0.00239
120	0.15	0.054	0.0129	0.0054	0.00197
130	0.17	0.051	0.0112	0.0049	0.00168

Table 6.16. Elastic scattering cross sections for C_2H_4 (in 10^{-16} cm^2/sr) [6.57]

	E_0 [eV]				
θ [deg]	100	200	400	600	1000
(0)	222.0	147	101	66.5	68.4
3	130.9	78	44	27.7	23.4
4	108.8	63	33	21.8	17.4
5	91.9	49	25	17.4	12.4
6	79.4	38	19.4	13.2	8.45
7	65.4	31	14.6	9.6	5.7
8	55	25	11.3	7.0	4.0
9	46	19.6	8.6	5.4	3.0
10	38.8	15.8	6.4	4.1	2.2
12	26.5	10.2	3.8	2.4	1.45
14	18.5	6.5	2.4	1.66	1.03
16	12.9	4.1	1.7	1.2	0.75
18	8.75	2.9	1.26	0.92	0.59
20	6.1	2.2	0.99	0.74	0.48
24	3.3	1.31	0.66	0.53	0.28
28	2.0	0.88	0.48	0.36	0.142
32	1.4	0.64	0.37	0.23	0.092
36	1.0	0.50	0.27	0.14	0.068
40	0.765	0.41	0.18	0.099	0.048
45	0.54	0.32	0.12	0.075	0.031
50	0.44	0.24	0.091	0.058	0.022
55	0.37	0.18	0.076	0.040	0.0176
60	0.31	0.135	0.062	0.0295	0.0132
65	0.25	0.114	0.048	0.024	0.0099
70	0.19	0.100	0.037	0.020	0.00795
80	0.143	0.084	0.026	0.0142	0.0055
90	0.13	0.079	0.023	0.0099	0.0039
100	0.15	0.068	0.0174	0.0082	0.0030
110	0.154	0.056	0.0144	0.0068	0.00246
120	0.65	0.051	0.0126	0.0058	0.0021
130	0.18	0.053	0.0115	0.0051	0.00173

Table 6.17. Integral elastic (σ) and momentum-transfer (σ_m) cross sections for SF_6 (in 10^{-16} cm^2) [6.58]

E_0 [eV]	σ	σ_m
5	17.9	14.9
10	19.5	14.1
15	16.4	12.0
20	14.0	11.5
30	15.0	9.46
40	11.7	6.82
50	10.0	5.69
60	8.86	5.51
75	6.58	4.79

$$e + H_2(X^1\Sigma_g\, v = 0,\ J = 1) \Rightarrow e + H_2(X^1\Sigma_g\, v = 0,\ J = 3)\ . \qquad (6.4.3)$$

Experimental rotational excitation cross sections in H_2 are displayed in Fig. 6.11. At low energies, the isotope effects (H_2 and D_2) are clearly seen. The resonance structure of the rotational cross sections of N_2 near 2.3 eV is exhibited in Fig. 6.12.

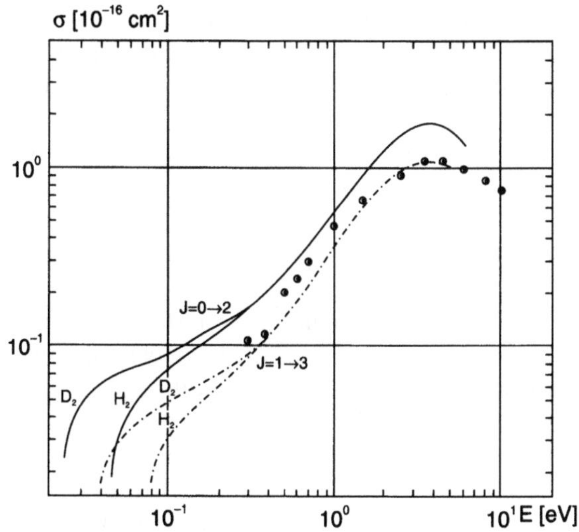

Fig. 6.11. Cross sections of rotational excitation of H_2 and D_2 (— $J = 0$–2 [6.60]; -·-·- $J = 1$–3 [6.61]; ○○○ $J = 1$–3 [6.44])

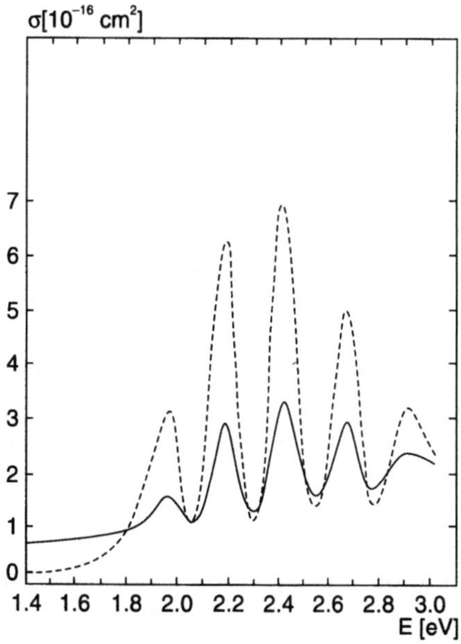

Fig. 6.12. Details of the rotational cross sections in N_2 for $J = 0$–2 (——) and $J = 0$–4 (- - -) calculated by *Onda* [6.62] in the region of the shape resonance

6.4.2 Vibrational Excitation

The cross sections for *vibrational transitions* $v \to v'$ are much smaller as compared to those for rotational transitions. In the absence of resonances they are of the order of $10^{-18}\,\text{cm}^2$ [6.63]. At energies below a few eV the non-resonant vibrational transition cross sections for the *infrared-active modes* are larger than for infrared-inactive modes. Non-resonant excitation usually takes place predominantly by two vibrational quanta.

Sharp peaks are found in the cross sections for each channel of vibrational excitation just above its threshold for many polar molecules (e.g., HF, HCl and H_2O) and for some non-polar molecules (CH_4 and SF_6) [6.64]. Sometimes, the cross-section maximum exceeds $10^{-15}\,\text{cm}^2$. No peaks at threshold have been reported yet for homonuclear diatomic molecules.

Usually, resonances appear in electron–molecule collisions below a few tens of eV and lead to an increase of the vibrational transition cross sections up to $10^{-17}\,\text{cm}^2$ or even up to the geometrical cross section [6.63, 65, 66]. Excitation by up to twenty or thirty vibrational quanta is likely to occur in some resonant processes.

If the lifetime $\hbar/\Gamma$ of the resonant state is much longer than the typical vibrational period (of about 10^{-14} s), many vibrations of the nuclei take place during the lifetime, where Γ is the width of the level. This condition leads to the formation of sharp peaks in the energy dependence of the vibrational excitation cross sections at the positions of the vibrational levels of the resonant state. Most of the so-called Feshbach resonances belong to this category [6.11]. Examples belonging to this category are those of O_2 below 2.5 eV [6.29, 92].

If $\hbar/\Gamma$ is shorter than 10^{-14} s, the nuclei cannot vibrate even one period during the lifetime of the resonant state. Thus, the cross section shows a broad hump without any resonance structure. Only low vibrational excitation takes place, for example, in the H_2 molecule near 3 eV, in N_2O near 2 eV, and in H_2O near 6 eV.

If $\hbar/\Gamma$ is comparable to a typical vibrational period, the nuclei can experience one or a few vibrations before the resonant state decays. In this case, the peaks appear in the energy dependence of the vibrational cross section, but none of them is associated with any vibrational level of the resonant state, e.g., as it was found in the N_2 molecule around 2.3 eV and in CO around 1.8 eV. Figure 6.13 exhibits the cross sections for vibrational excitation $v = 0$–1 in N_2. Figure 6.13 shows that in the region $1.8\,\text{eV} < E < 3\,\text{eV}$ the shape resonance strongly affects the vibrational cross section.

Cross sections for vibrational excitation of N_2 to $v = 2$ and $v = 3$ were reported in [6.72]. The data for $v = 2$ are given in Fig. 6.14. Cross sections for vibrational levels up to $v = 24$ can be found in [6.73] for 9 eV incident electron energy and a scattering angle of 30°.

Cross sections for vibrational excitation in N_2 have a pronounced resonance character showing a maximum at $E = 1$–4 eV. It corresponds to

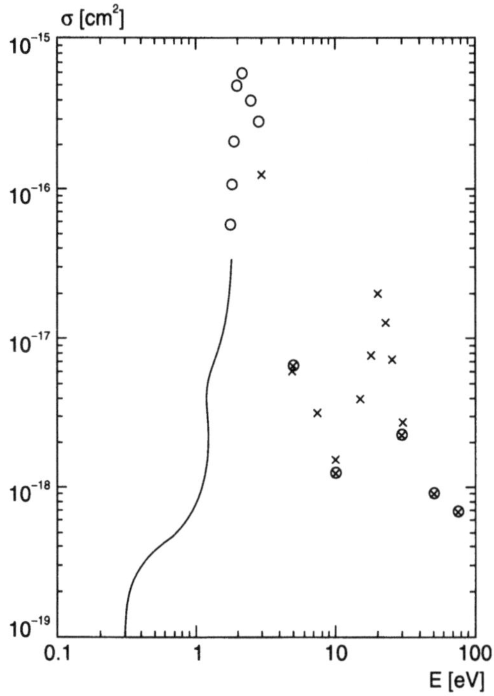

Fig. 6.13. Cross sections for the vibrational excitation = 01 in N_2. Only representative values are shown for the theoretical results of *Onda* [6.67] (○○○). Experimental data are from the crossed beam measurements by *Tanaka* [6.68] (× × ×) and *Truhlar* [6.69, 70] (⊗⊗⊗), and from swarm analysis by *Engelhardt et al.* [6.71] (- -). The results of [6.69, 70] have been renormalized in [6.15]

the $^2\Pi_g$ resonance. The resonance at $E = 20$–$30\,\mathrm{eV}$ has a $^2\Sigma^+$ character. In [6.74], cross sections for excitation of the first 17 vibrational states of N_2 have been reported for 0–5 eV energies. The absolute cross sections at maximum for vibrational excitation of the N_2 molecule with quantum numbers $v = 1$–17 are listed in Table 6.18.

A summary of cross sections of vibrational excitation of the O_2 molecule for transitions from the electronic and vibrational ground states has been given in [6.65] (Fig. 6.15).

The vibrational cross sections of O_2 at energies from 4 to 45 eV have been reported in [6.75, 76] and correspond to the sum (Fig. 6.16)

$$\sigma_{\mathrm{vib}}(\Sigma) = \sum_{v'} \sigma_{\mathrm{vib}}(0\text{–}v') \ . \tag{6.4.4}$$

The sharp resonance structure in the vibrational cross sections at energies 0.3–1.0 eV were discussed in [6.77]. This was interpreted as a temporary capture of the electron to form the O_2^- ($^2\Pi_g$) state. In Fig. 6.16 these values are denoted as σ_{vib} (0–1) under the assumption that $\Delta E = 50\,\mathrm{meV}$.

Vibrational excitation of the CO molecule in the (1–5) eV energy range is enhanced by the $^2\Pi$-type shape resonance. Vibrational excitation in this range has been documented in [6.51, 78, 79]; the previous data have been reviewed in [6.65] (Table 6.19).

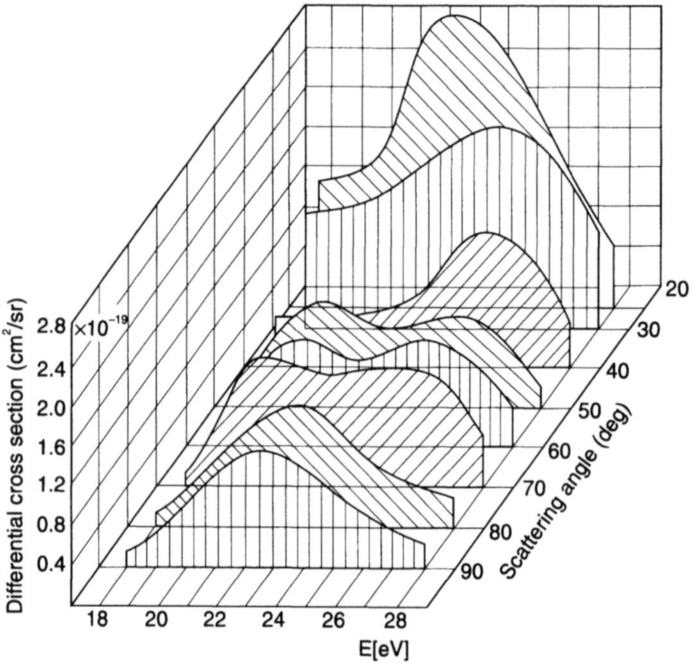

Fig. 6.14. Combined energy and angular dependence of the differential cross section for vibrational excitation to $v = 2$ of N_2 [6.72]

Table 6.18. Absolute values of integral cross sections σ_{0v} of vibrational excitation of N_2 in their maximum (in cm^2)[a] [6.74]

Transition	E_{max} [eV]	S_{rel} ($\pm 20\%$)	σ_{0v} [cm^2]
0–1	1.95	1	5.6(−16)
0–2	2.00	0.66	3.7(−16)
0–3	2.15	0.55	3.1(−16)
0–4	2.22	0.37	2.1(−16)
0–5	2.39	0.23	1.3(−16)
0–6	2.48	0.13	7.1(−17)
0–7	2.64	6.8(−2)	3.8(−17)
0–8	2.82	2.8(−2)	1.6(−17)
0–9	2.95	1.1(−2)	6.1(−18)
0–10	3.09	3.9(−3)	2.2(−18)
0–11	3.30	1.1(−3)	6.3(−19)
0–12	3.87	2.5(−4)	1.4(−19)
0–13	4.02	8.0(−5)	4.5(−20)
0–14	4.16	2.2(−5)	1.3(−20)
0–15	4.32	6.5(−6)	3.6(−21)
0–16	4.49	1.6(−6)	9.1(−22)
0–17	4.66	4.3(−7)	2.4(−22)

[a] S_{rel} is the sum of differential cross sections of transition $0 \to v$ for the angles $\theta = 0°$ and $\theta = 180°$, given in relative units; E_{max} is the energy, at which the cross section has its maximum value

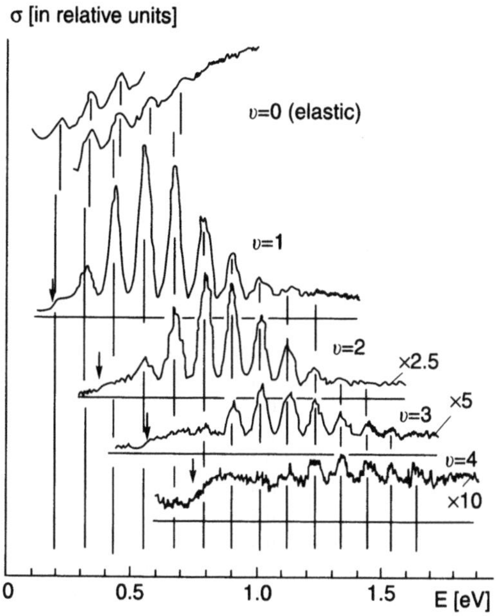

Fig. 6.15. Cross sections of vibrational excitation of O_2 from the ground electronic and vibrational state [6.65]

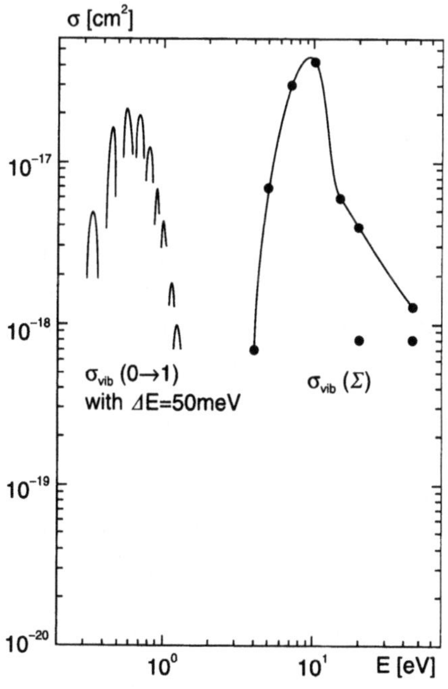

Fig. 6.16. Vibrational excitation cross sections in O_2. In the energy range below 2 eV, resonant cross sections [6.77] for $v = 0 \rightarrow 1$ are shown. In the higher energy region, the cross section summed over the final vibrational states [6.75, 76] are given

Electron interactions with hydrogen halides were extensively investigated because of their important role in various laser applications and partly because of the pronounced low-energy resonances associated with dissociative attach-

Table 6.19. Differential cross sections for $v = 0 \rightarrow 1$ vibrational excitation in CO (in 10^{-19} cm^2/sr) [6.79]. Entries in parentheses refer to extrapolated values[a]

	E_0 [eV]							
θ [deg]	3.0	5.0	9.0	20.0	30.0	50.0	75.0	100.0
(0)	87	51	38	26	6.9	5.8	3.7	4.4
(5)	85	49	28	22	5.6	4.1	2.7	3.2
(10)	81	46	19	18	4.6	3.0	1.9	2.0
15	75.1	42.7	9.15	14.4	3.09	1.30	1.31	1.2
20	72.0	37.0	4.83	11.6	2.03	0.661	0.829	0.956
25	58.6	26.8	3.35	9.37	1.48	0.428	0.767	0.656
30	50.3	17.2	2.11	8.10	1.21	0.364	0.878	1.07
35	43.6	12.7	1.88	8.32	1.14	0.489	0.808	0.986
40	40.5	10.1	1.75	8.10	1.01	0.476	0.799	0.941
50	34.5	6.53	1.59	7.85	1.13	0.499	0.527	0.692
60	30.4	5.40	1.62	7.92	1.23	0.477	0.385	0.420
70	30.9	5.73	1.43	7.38	1.57	0.437	0.327	0.294
80	28.1	5.30	1.29	6.79	1.60	0.412	0.319	0.235
90	25.1	4.67	1.04	6.52	1.57	0.435	0.277	0.281
100	20.6	4.14	0.913	6.03	1.42	0.476	0.256	0.361
110	16.8	3.89	0.886	6.73	1.34	0.496	0.242	0.413
120	17.9	4.37	0.866	7.26	1.62	0.528	0.281	0.497
130	19.9	5.66	1.47	7.80	1.93	0.633	0.323	0.547
(140)	22	8.6	2.1	8.7	2.2	0.78	0.36	0.59
(150)	25	11	2.7	9.0	2.4	0.97	0.38	0.63
(160)	28	14	3.2	9.4	2.6	1.1	0.40	0.66
(170)	30	16	3.6	9.8	2.8	1.3	0.41	0.68
(180)	31	17	4.0	10	2.9	1.4	0.42	0.71

[a] Numbers in this table differ slightly from those of the tabulation in [6.79] due to small changes in the He elastic cross sections [6.49]

ment and vibrational excitation of these molecules. A survey on near-threshold vibrational excitation cross sections may be found in [6.80]. For HF, HCl and HBr molecules, the peak cross sections are associated with a sharp threshold resonance and a broader shape resonance at slightly higher electron-impact energies (Table 6.20).

The vibrational excitation of the symmetric-stretch mode $(v_1, v_2, v_3) = (100)$ of CO_2 due to electron impact is illustrated in Fig. 6.17.

The vibrational mode in the 4 eV resonance region of the SO_2 molecule was studied in [6.82]. The cross sections for the $(n, 0, 0)$ vibrational progression at 3.4 eV impact energy and 90° scattering angle are depicted in Fig. 6.18.

Table 6.20. Peak cross sections for the $v = 0 \rightarrow 1$ vibrational excitation (in 10^{-16} cm^2) [6.80]

Molecule	Near threshold region	Shape resonance region
HF	7	1
HCl	20	3
HBr	40	27

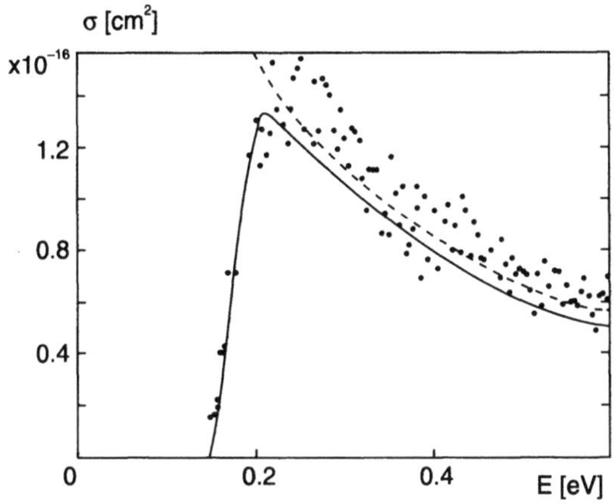

Fig. 6.17. Threshold behavior of the cross section for vibrational excitation of the symmetric-stretch mode $(_{1},_{2},_{3}) = (100)$ of CO_2 by electron impact [6.81]

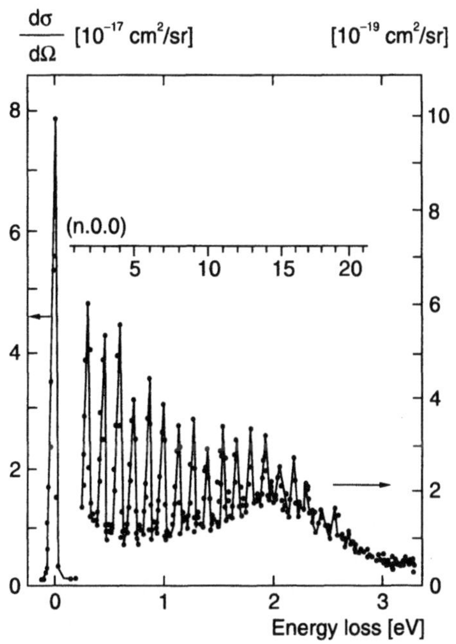

Fig. 6.18. Differential cross sections for the excitation of the $(n, 0, 0)$ vibrational progression in SO_2 at 3.4 eV impact energy and 90° scattering angle [6.82]

6.4.3 Electronic Excitation

At high electron-impact energies the interaction between the incident electron and the target molecule may be considered as a weak perturbation on the free relative motion.

In first-order perturbation theory, the excitation differential cross section for the transition $i \to f$ has the form:

$$\frac{d\sigma_{if}}{d\Omega} = 4\frac{k_f}{k_i}\frac{Ry}{\Delta E_{if}}\frac{F_{if}(K)}{K^2} , \qquad (6.4.5)$$

where k_i and k_f are momenta of the incident and scattered electron, ΔE_{if} is the transition energy, $\mathrm{Ry} = 13.6\,\mathrm{eV}$, K is the momentum transfer $(\boldsymbol{K} = \boldsymbol{k}_i - \boldsymbol{k}_f)$, and $F_{if}(K)$ is the generalized oscillator strength (Sect. 5.3).
The integral cross section is defined as

$$\sigma_{if}\,[\mathrm{cm}^2] = 7.04 \times 10^{-16} \left(\frac{\Delta E_{if}}{\mathrm{Ry}}\right)^{-1} \left(\frac{E}{\mathrm{Ry}}\right)^{-1} \int\limits_{K_{\min}}^{K_{\max}} \left[\frac{F_{if}(K)}{K}\right] \mathrm{d}K \,, \tag{6.4.6}$$

where $K_{\min} = |\boldsymbol{k}_i - \boldsymbol{k}_f|$ corresponds to the angle $\theta = 0$ and $K_{\max} = k_i + k_f$ to $\theta = \pi$.

For optically allowed (dipole) transitions, the Born cross sections comprise the Bethe logarithm (Bethe-Born approximation [6.84, 85]):

$$\sigma^{\mathrm{B}}\,[\mathrm{cm}^2] = 3.52 \times 10^{-16} \left(\frac{E}{\mathrm{Ry}}\right)^{-1} \left(\frac{\Delta E}{\mathrm{Ry}}\right)^{-1} f_{if} \ln\left[4c_{if}\left(\frac{E}{\mathrm{Ry}}\right)\right] \,, \tag{6.4.7}$$

where f_{if} is the *optical oscillator strength* for the transition i–f, and c_{if} is a dimensionless constant. At high electron energies, σ_{if} behaves as

$$\sigma_{if} \sim E^{-1} \ln\left(\frac{E}{\Delta E}\right), \quad E \gg \Delta E \,. \tag{6.4.8}$$

Therefore, a plot of $(E/\mathrm{Ry})\sigma_{if}$ against $\ln(E/\mathrm{Ry})$ approaches a straight line for high E, called a *Fano* (or *Bethe*) *plot* [6.84, 85]. It is used for the interpolation of the cross sections at high energies; and even at low energies where the plot is nonlinear.

For electron energies higher than 10 keV, the relativistic correction to (6.4.7) gives

$$\sigma\,[\mathrm{cm}^2] = 3.52 \times 10^{-16} \left(\frac{mv^2}{2\mathrm{Ry}}\right)^{-1} \left(\frac{\Delta E_{if}}{\mathrm{Ry}}\right)^{-1} \times f_{if} \left\{ \ln\left[\frac{c_{if}\beta^2}{1-\beta^2}\right] - \beta^2 + 11.2 \right\} \,, \tag{6.4.9}$$

where $\beta = v/c$, c being the velocity of light. In this energy range, the Fano plot represents the dependence of $\beta^2\sigma$ as a function of $\ln[\beta^2/(1-\beta^2)] - \beta^2$.

The generalized oscillator strength $F_{if}(K)$ approaches the optical oscillator strength f_{if} for $k \to 0$. This makes it possible to simulate photoabsorption experiments using small-angle electron scattering and to measure extensive photoionization cross sections for atoms and molecules at wavelengths inaccessible with conventional light sources [6.86, 87].

The total cross section σ_{tot} for inelastic scattering is obtained by a summation of the cross section σ_{if} over the final states including the continuum. The Born-Bethe approximation leads to the asymptotic formula

$$\sigma_{\mathrm{tot}}\,[\mathrm{cm}^2] = 3.52 \times 10^{-16} \frac{\mathrm{Ry}}{E} M_{\mathrm{tot}} \ln\left[4c_{\mathrm{tot}} \frac{E}{\mathrm{Ry}}\right] \,, \tag{6.4.10}$$

where M_{tot} and $\ln c_{\mathrm{tot}}$ are constants.

The available data on electron-impact excitation cross sections for electronic states in H_2 obtained by means of the electron-scattering techniques are extremely limited. This is mainly due to the difficulty of resolving the heavily overlapping band structure of electronic transitions in the electron-energy-loss spectra. The situation is further complicated by the large variety of other (non-discrete) excitation processes which lead to dissociation into the ground, excited and ionized fragments. Most of the available data were derived from studying the optical radiation caused by electron impact.

Direct excitation of the $b^3\Sigma_u^+$ state in the hydrogen molecule is illustrated in Fig. 6.19. Direct excitation of this antibonding state is the dominant process leading to the dissociation of H_2 into H atoms in their ground state. However, with increasing impact energy, the cascade contributions from higher electronic states to the $b^3\Sigma_u^+$ state, which also produces atomic hydrogen, become significant.

The excitation of electronic states of N_2 has been most widely studied by the measurement of the emission cross section σ_{emis} which is proportional to the measured radiation intensity. In Fig. 6.20, the cross sections for excitation to the $A^3\Sigma_u^+$, $B^3\Pi_g$ and $W^3\Delta_u$ states of N_2 are shown.

Electron excitation scattering cross sections have been reported in various papers. The most extensive coverage [6.89–91] is based on N_2 elastic scattering cross sections. In Table 6.21, the corrected results of [6.90] are listed, based on the updated elastic N_2 cross sections.

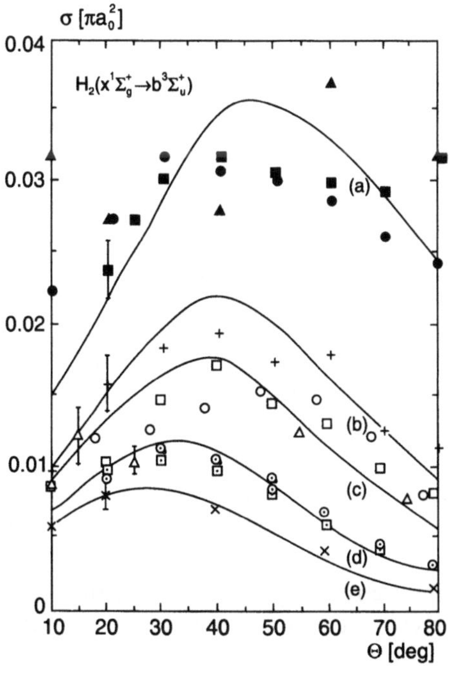

Fig. 6.19. Differential cross section of the $X^1\Sigma_g^+ \rightarrow b^3\Sigma_u^+$ transition in H_2 as a function of scattering angle. *Solid curves* are theoretical and symbols are experimental data. Incident energies are (*a*) 25 eV: ●●●, ■■■, ▲▲▲; (*b*) 35 eV: + + +; (*c*) 40 eV: ○○○, □□□, △△△; (*d*) 50 eV: ⊙ ⊙ ⊙, □□□, △△△; (*e*) 60 eV: × × × [6.88]

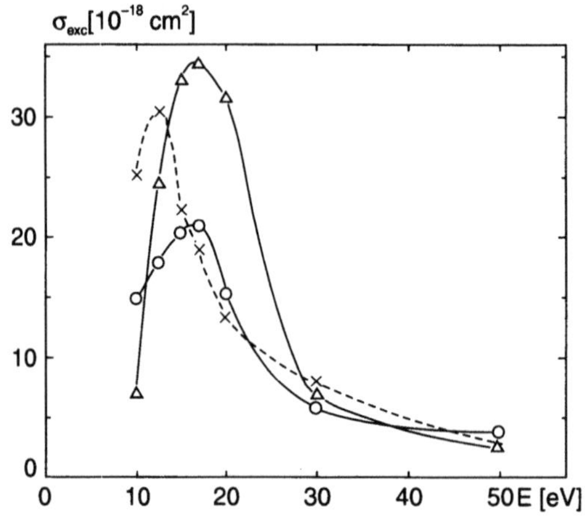

Fig. 6.20. Cross sections for the electronic excitations from the ground state of N_2 to A_u^3+, B_g^3 and W_u^3 states, measured in [6.89] and renormalized in [6.15]. Curves are drawn to guide the eyes (○○○: A_u^3+; × × ×: B_g^3; △△△: W_u^3)

Table 6.21. Differential cross sections for the excitation of the $A^1\Pi_g$ state of N_2. Entries in parentheses beneath the given energy refer to cross section multiplier (cm^2/sr) for that column. Entries in parentheses in the first column refer to extrapolated cross sections[a] [6.90]

	Energy [eV]						
Angle [deg]	10.0 (10^{-19})	12.5 (10^{-18})	15.0 (10^{-18})	17.0 (10^{-18})	20.0 (10^{-18})	30.0 (10^{-18})	50.0 (10^{-18})
(0)	2.13	1.51	15.5	29.1	24.0	23.6	43.8
(10)	3.37	2.39	10.5	23.4	22.8	19.0	30.6
20	4.94	3.69	6.90	16.7	17.9	12.6	8.27
30	6.15	3.54	4.52	9.71	9.61	4.77	1.24
40	7.39	2.99	3.06	5.52	4.05	1.32	0.474
50	6.21	2.52	2.54	3.20	1.43	0.767	0.267
60	3.73	2.12	2.05	1.84	0.812	0.642	0.206
70	3.75	1.76	1.71	1.28	0.793	0.602	0.184
80	3.59	1.51	1.38	1.08	0.846	0.653	0.155
90	3.41	1.32	1.36	1.07	0.942	0.734	0.162
100	2.50	1.13	1.40	1.13	1.09	0.718	0.185
110	2.08	1.09	1.40	1.16	1.29	0.798	0.237
120	1.51	0.950	1.26	1.09	1.22	0.823	0.285
130	1.08	0.736	1.04	0.960	1.04	0.810	0.363
(140)	0.717	0.532	0.788	0.807	0.878	0.798	0.485
(150)	0.533	0.413	0.612	0.652	0.769	0.735	0.676
(160)	0.459	0.344	0.476	0.634	0.663	0.722	0.892
(170)	0.368	0.294	0.348	0.544	0.545	0.735	1.14
(180)	0.310	0.263	0.266	0.456	0.435	0.728	1.41

[a] The cross section values in this Table were obtained from that of [6.90] by correcting their values with a factor representing the ratio of the elastic cross section of Table 6.8 to those used in [6.90]

In contrast to the N_2 molecule, there are relatively few experimental data and even less theoretical calculations on cross sections of the excitation of the electronic states of the O_2 molecule. The corresponding experimental data were

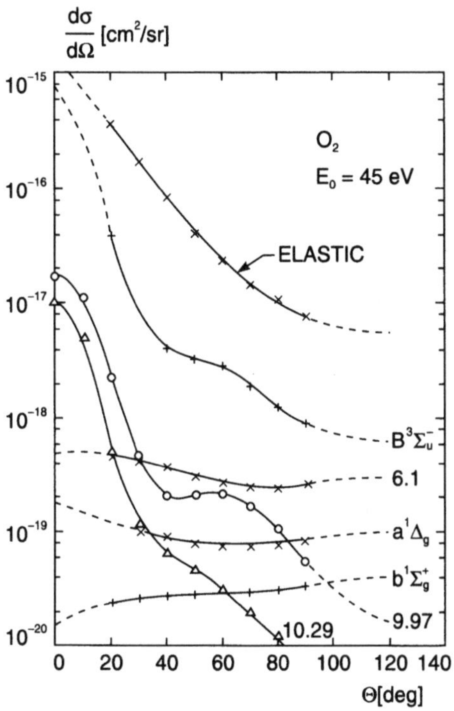

Fig. 6.21. Differential cross sections in O_2 at 45 eV impact energy [6.76]. The *solid curves* are drawn through the experimental data; the *dashed* ones are extrapolations

reviewed in [6.15, 92]. The main peak in the energy-loses spectra is the so-called Schuman-Runge continuum at an energy of about 8 eV, although there are many other absorption lines in the 4–13 eV energy range.

The differential cross sections for electronic excitation of the O_2 molecule with 20 and 45 eV electrons for the 6.1 eV transition ($c^1\Sigma_u^-$, $C^3\Delta_u$, $A^3\Sigma_u^+$ states), for the $B^3\Sigma_u^-$ state at 9.97 eV ("longest" band) and for the 10.29 eV ("second" band) transitions were reported in [6.76]. The results for 45 eV impact energy are presented in Fig. 6.21.

Figure 6.22 depicts the experimental cross section for the excitation of the CO molecule from the ground to the excited $A^1\Pi$ states with $v = 0$–4 [6.93]. Cross sections for other quantum numbers v show a similar energy dependence.

6.4.4 Resonance Processes

Elastic or inelastic scattering of electrons by molecule A often occurs via a temporarily bound intermediate state A^- called a *resonance* or *autoionizing state*. Then, the cross section reveals a sharp peak structure as a function of the energy E,

$$\sigma = \frac{\sigma_a(\tilde{E} + q)^2}{(1 + \tilde{E}^2)} + \sigma_b \ , \tag{6.4.11}$$

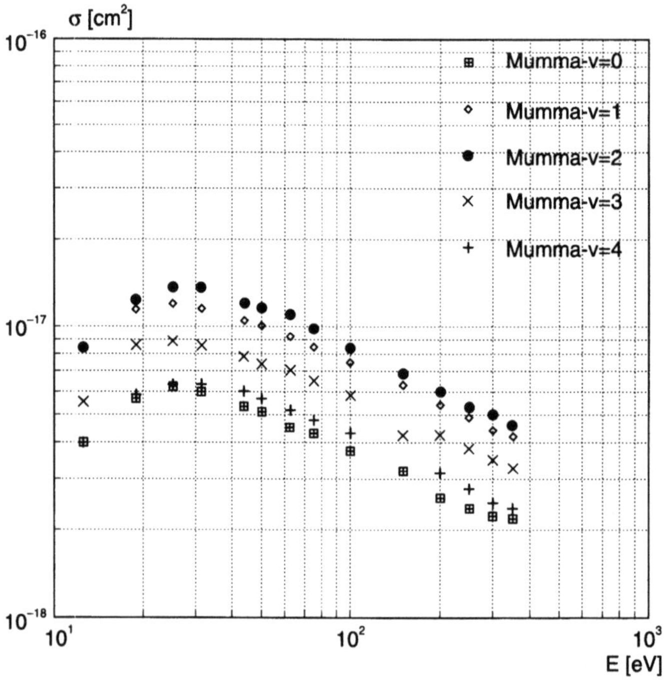

Fig. 6.22. Cross sections for excitation of transitions in CO ($A^1\Pi$, $v = 0 \to 4$) [6.93]

where $\tilde{E} = (E - E_r)/(\Gamma/2)$, σ_a and σ_b are slowly varying functions of E, E_r, Γ, and q being a constant. The constant q determines the shape of the resonance curve, Γ determines its width, the latter is related to the lifetime τ of the resonance state as $\Gamma = \hbar/\tau$. The cross sections averaged over resonances can strongly differ from those without them.

There are several mechanisms leading to the formation of autoionizing states [6.65, 66] such as *shape resonance*, *Feshbach resonance* and so on (see [6.29] for details). The low-energy behavior of the cross sections for electron–molecule collisions was discussed in [6.64].

6.5 Dissociation and Dissociative Attachment

According to the Born-Oppenheimer principle, the effective potential governing the motion of the nuclei is given by the eigenvalues of the electronic Hamiltonian. Thus, a change in the electronic structure of the molecule strongly affects the nuclear motion, and an electronic transition is often accompanied by the vibration, excitation or dissociation of a molecule.

6.5.1 Dissociative Processes

The main dissociative processes in scattering of electrons on molecules include: dissociation and predissociation, dissociative ionization, dissociative attachment, and dissociative recombination. The resulting atoms may be both in the ground and excited states.

(i) Dissociation and predissociation. The dissociative process in a diatomic molecule under electron impact may be represented as

$$e^- + A_2 \to e^- + A + A \ . \qquad (6.5.1)$$

At low incident electron energies, the dissociation may occur through a molecular transition to a repulsive level with subsequent decay. Another possible mechanism is a transition to an attractive level with a subsequent transition to the attractive term and then a transition to a repulsive level with dissociation of the molecule if there is a quasicrossing of levels. The latter mechanism is known as the *predissociation.*

(ii) Dissociative ionization is a process of the type

$$e^- + A_2 \to 2e^- + A + A^+ \ . \qquad (6.5.2)$$

The removal of one of the target electrons from a stable molecule almost always leads to the creation of a weakly bound electronic state. For example, the excited electronic states of H_2^+ have potential curves which are either completely repulsive or have a small region of very weak attractive forces at large interatomic distances R.

(iii) The dissociative attachment consists of the formation of a free atom and a negative atomic ion:

$$e^- + A_2 \to A^- + A \ . \qquad (6.5.3)$$

Usually, molecules taking part in the dissociative attachment consist of atoms which may form a stable negative ion. The incident electron becomes temporally attached to the target molecule, which leads to the formation of an unstable intermediate state which can decay by electronic emission or dissociation. The probability of dissociation (or electronic emission) is large if the lifetime of that state is of the same order or larger than the vibrational period (10^{-14} s). In this case, the nuclear motion can be disturbed significantly by the presence of the extra electron.

(iv) Dissociative recombination is the process

$$e + A_2^+ \to A + A \ . \qquad (6.5.4)$$

Dissociative recombination is a transition of the system from the initial state of the ion A_2^+ to a certain repulsive level of molecule A_2. Usually, it is the level corresponding to the doubly excited electronic configuration. The transition probability is defined by the final molecular levels.

The adiabatic approximation is not applicable for the description of the dissociation processes. Therefore, the corresponding cross sections are usually calculated in the Born approximation or by using resonant-state models [6.94].

The dissociation of the N_2 molecule by electron impact can occur via two main channels: dissociative ionization and dissociation through the electronically excited state of the molecule, for example

$$N_2 + e \rightarrow N^+ + N(^4S, ^2D) + 2e \ , \tag{6.5.5}$$

$$N_2 + e \rightarrow (N_2)^{**} + e \rightarrow N(^4S) + N(^4S, ^2D, ^2P) + e \ . \tag{6.5.6}$$

The total experimental dissociation cross section [6.95–97] and the cross section for dissociative excitation of the N_2 molecule (6.5.6) are depicted in Fig. 6.23 in comparison with theoretical calculations [6.98].

The total dissociation cross section of CH_4 has been obtained in the 10–500 eV energy range [6.99] through measurements of the pressure variation in molecular dissociation (Fig. 6.24). These results, which include the contribution of all inelastic exciting and ionizing channels, indicate that, at relatively high impact energies (above 50 eV), the dissociation probabilities into ionic and neutral fragments are roughly equal; whereas at low energies, the fragments are mostly uncharged molecules in the ground state.

The total cross section for dissociation of ethane (C_6H_6) was reported in [6.100]. The data are shown in Fig. 6.25 combined with those for methane (CH_4) and hydrogen (H_2) [6.101]. In [6.100], the dissociation cross sections for deuterated ethane were also described and a small isotope effect [$\sigma(C_2H_6) = 1.09\,\sigma(C_2D_6)$] was reported.

The dissociative excitation usually results in the emission of photons and is investigated by optical spectroscopy. Recommended cross sections for H_2 are

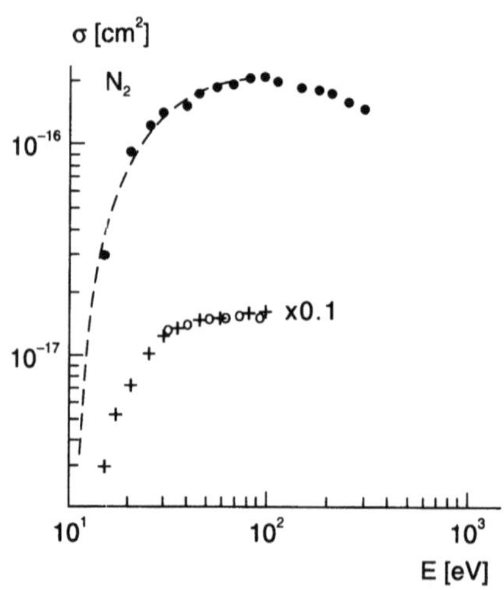

Fig. 6.23. Total cross section of dissociation of N_2 by electron impact: - - - [6.98]; ●●● [6.95]. Cross section of the dissociative excitation of N_2 [process, given by (6.6.6)]: + + + [6.96]; ○○○ [6.95, 97]

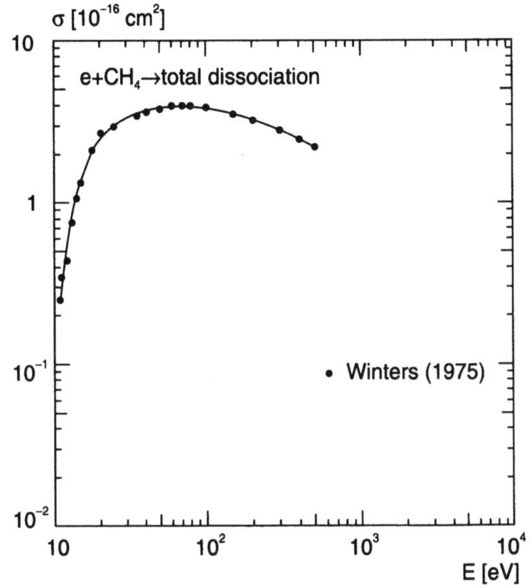

Fig. 6.24. Total dissociation cross section of CH_4 by electron impact [6.99]

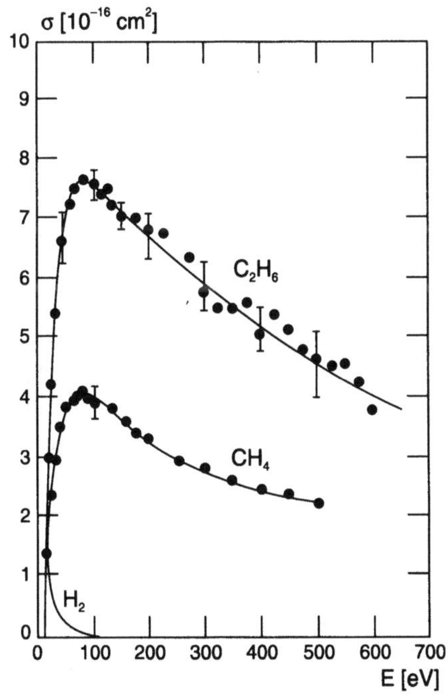

Fig. 6.25. Comparison of the experimental cross sections for dissociation of H_2 [6.101], CH_4 [6.101] and C_2H_6 [6.100] as a function of electron impact energy

shown in Fig. 6.26. We note that these cross sections include the contribution from dissociative ionization where one of the fragmented products is an ionized atom.

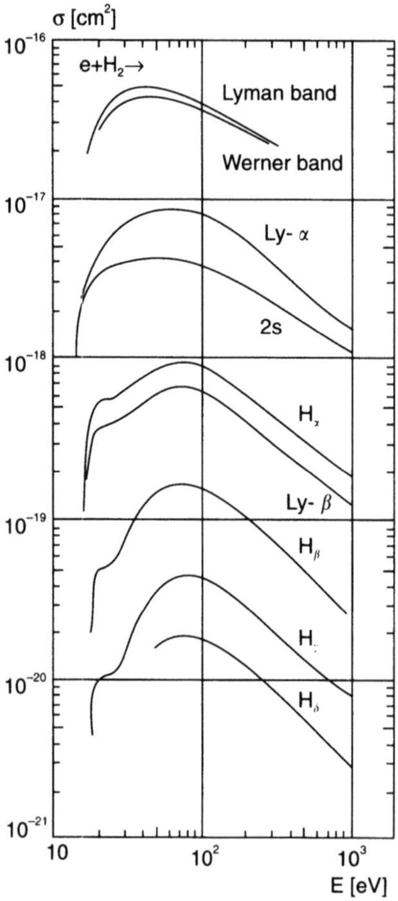

Fig. 6.26. Cross sections of photon emission for Lyman band, Werner band, Lyman-α, Lyman-β, $2S$ state, Balmer-α, -β, -γ and -δ lines [6.102]. The normalization of cross sections is based on the data [6.103] for 100 eV

Cross sections for dissociative excitation in the CO molecule $(A^1\Pi$, $v = 0$–$4)$ were reported in [6.93] for incident electron energies up to 300 eV. The results for the states with $v = 0$ and $v = 3$ are given in Fig. 6.27. The corresponding cross sections for the states $v = 0, 1, 2, 3$ and 4 are practically of the same order.

The energy dependence of Lyman-α emission cross sections in $e + H_2O$ collisions are displayed in Fig. 6.28. They were obtained by renormalization of the original values [6.104].

The cross sections for OI (8447 Å) emission from the H_2O molecule which are due to the transition $3p\,^3P$–$3s\,^3S^\circ$, are reproduced in Fig. 6.29.

In [6.110], the energy distribution of atomic hydrogen in high Rydberg states was measured at electron energies $E > 5$ eV. These measurements were carried out using the time-of-flight technique. The energy range from 1.7 to 3.5 eV at 25 eV electron impact is shown in Fig. 6.30. With increasing impact energy the new high-energy peaks appear. At 100 eV electron energy, the most

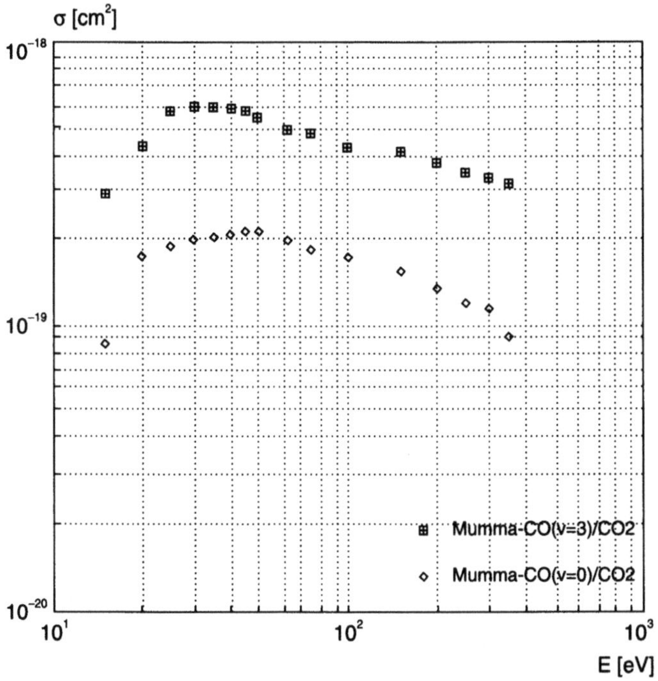

Fig. 6.27. Cross sections for dissociative excitation to CO ($A^1\Pi$, $v = 0, 3$) by electron impact on CO_2 [6.93]

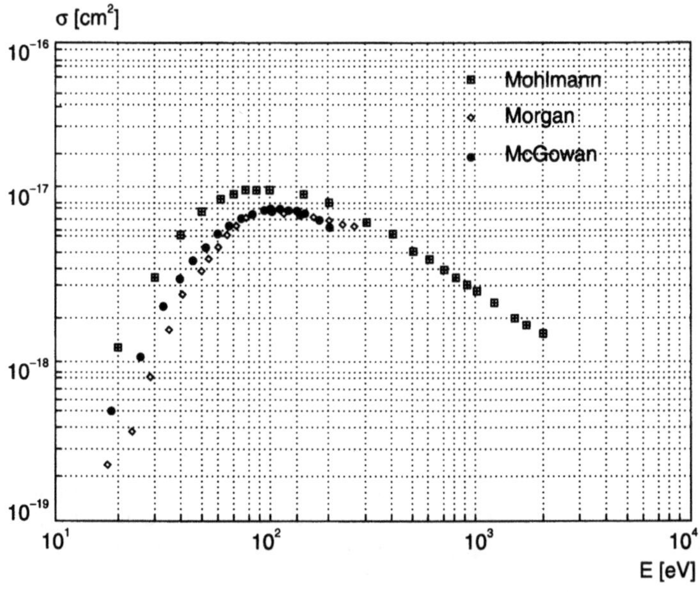

Fig. 6.28. Lyman-α emission cross sections from H_2O by electron impact (□□□ Mohlmann [6.105]; ◇◇◇ Morgan [6.106]; ○○○ McGowan [6.107])

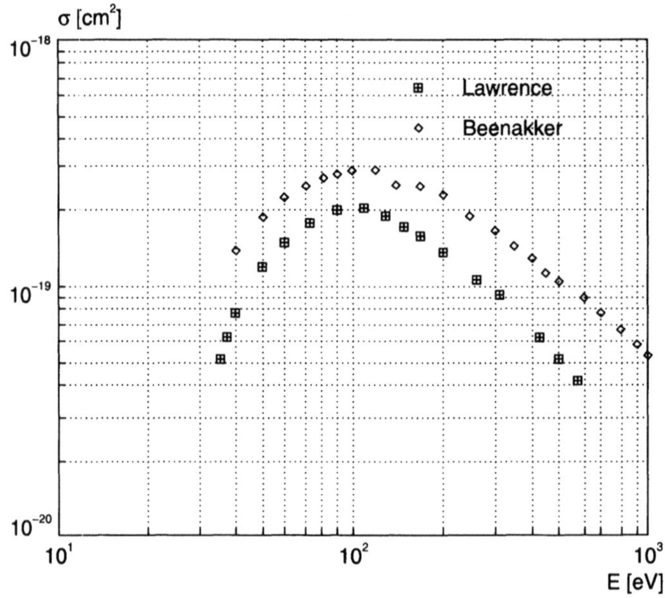

Fig. 6.29. Cross sections for OI (8447 Å) emission from H_2O by electron impact: ⊞⊞⊞ Lawrence [6.108]; ◇◇◇ Beenakker [6.109]

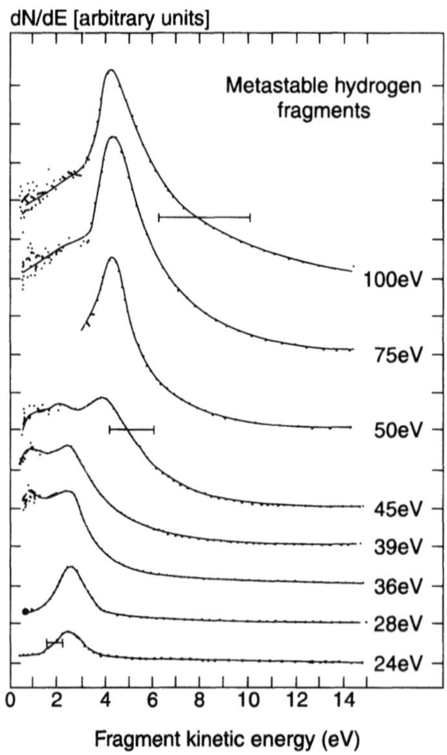

Fig. 6.30. Energy distributions of metastable atomic hydrogens produced in $e + CH_4$ collisions under different electron energy impact [6.110]

intense peak of atomic hydrogen is observed at 4 eV, meanwhile their energy extends up to 14 eV. However, no absolute cross sections have been measured for the production of energetic hydrogen atoms.

6.6 Ionization by Electron Impact

Ionization in an electron–molecule collision results at least in three outgoing particles, two of which are electrons. Therefore, specification of the energies $\varepsilon_i (i = 1, 2)$ and directions Ω_i of the two outgoing electrons is necessary for defining even the simplest type of ionization

$$e^-(E) + A \to A^+ + e^-(\varepsilon_1, \Omega_1) + e^-(\varepsilon_2, \Omega_2) \ . \tag{6.6.1}$$

If the states of the atom A and ion A^+ are known, only two of three energies E, ε_1, and ε_2 are independent because of the energy conservation law.

If both outgoing electrons are detected, one has the (triple) differential cross section $\mathrm{d}^3\sigma/\mathrm{d}\varepsilon\,\mathrm{d}\Omega_1\mathrm{d}\Omega_2$ as a function of E, ε, Ω_1, and Ω_2, where ε is either ε_1 or ε_2. If only one electron is detected, one has the (double) differential cross section $\mathrm{d}^2\sigma/\mathrm{d}\varepsilon\mathrm{d}\Omega$ as a function of E, ε, and Ω. It is also possible to define a single differential cross section $\mathrm{d}\sigma/\mathrm{d}\varepsilon$ as a function of E and ε.

Because two electrons in the continuum are indistinguishable (exchange effect), the integral ionization cross section $\sigma(E)$ is calculated by integrating the single differential cross section over ε from zero to $(E - I)/2$ where I is the binding energy.

For high energy region $E \gg I$ the *Rutherford formula* is used

$$\frac{\mathrm{d}\sigma^{\mathrm{Ruth}}}{\mathrm{d}\varepsilon} = \frac{3.52 \times 10^{-16}\,\mathrm{cm}^2}{\left\{E[(\varepsilon + I)/\mathrm{Ry}]^2\right\}} \tag{6.6.2}$$

for the energy transfer $\varepsilon + I$ to a single free electron. Here $\mathrm{Ry} = 13.6\,\mathrm{eV}$.

Indirect ionization processes can strongly enhance the total ionization cross section. One of the most important physical mechanisms is *excitation-autoionization* when the target is first excited to an autoionizing state and then decays in creating an additional electron.

For the H_2 molecule, the following ionization processes are possible (see also Fig. 6.31):

$$e^- + H_2 \to e^- + \mathrm{H}_2^+ + e^- \tag{6.6.3}$$
$$\to e^- + \mathrm{H} + \mathrm{H}^+ + e^- \tag{6.6.4}$$
$$\to e^- + \mathrm{H}^+ + \mathrm{H}^+ + 2e^- \tag{6.6.5}$$
$$\to \text{total ionization} \ . \tag{6.6.6}$$

Equation (6.6.6) represents the total ion production including H_2^+ and H^+. A compilation of the cross sections (6.6.3, 4, 6) was given in [6.111]. A differentiation between reactions (6.6.4 and 5) can be made using the

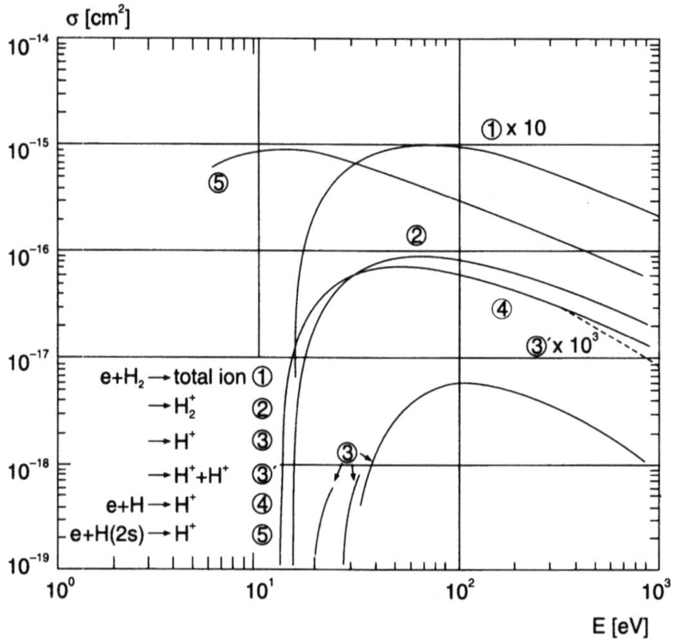

Fig. 6.31. Cross sections of the production for total ionization, molecular hydrogen ions, protons and two protons. Those of proton production from H and H (2*S*) are also shown for comparison (see [6.102]). Note that the *dashed curves* for proton production at lower energies correspond to the processes via $^2\Sigma_g$ and $^2\Sigma_u$ states

coincidence technique. Cross sections for the reaction (6.6.5) have been reported in the 400–1900 eV energy range in [6.112] where the ratios of cross sections of (6.6.5) to (6.6.3) are of the order of 10^{-3} and decrease with increasing impact energy. The cross sections for ionization of atomic hydrogen from the ground and excited 2*s* states are displayed in Fig. 6.31.

The most important cross sections of ionization type are the gross (or net) ionization cross section. For the ionization of N_2 they are written as

$$\sigma_{\text{gross}} = \sum_n n\sigma(N_2^{n+}) + \sum_m m\sigma(N^{m+}) \ . \qquad (6.6.7)$$

Here $\sigma(N_2^{n+})$ is the cross section for the production of n-fold ionized molecular ions and $\sigma(N^{m+})$ is the cross section for dissociative ionization which produces N^{m+}. The dissociative ionization cross sections of N_2

$$\sigma_{\text{diss}} = \sigma_{\text{ion}}(N^+) + 2\sigma_{\text{ion}}(N^{++}) \qquad (6.6.8)$$

have been discussed in [6.97]. In [6.17], the results of [6.97] were increased by 5% to correct for the fragment ions with lower velocities and decreased subsequently by 8.8% to correct the pressure gauge. The obtained values are shown in Fig. 6.32.

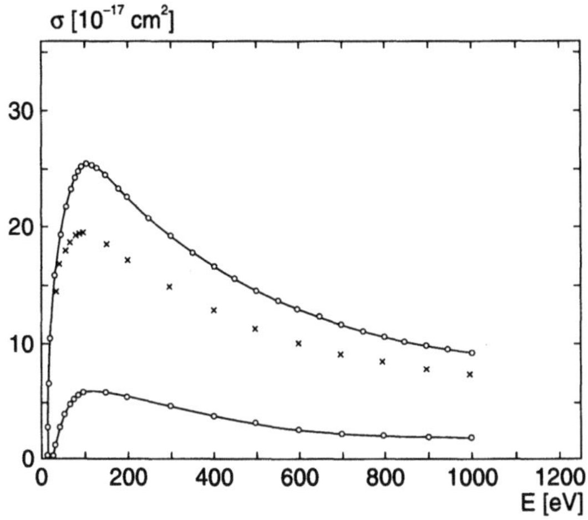

Fig. 6.32. Electron impact ionization cross sections of the N_2 molecule. *1* is the gross ionization cross section [6.113]; *2* is the cross section of production of N_2^+ [6.17]; *3* is the dissociative ionization cross section [6.17]

Several papers report the data on the emission of N_2^+ ions formed in collisions of electrons with N_2 molecules. If Fig. 6.33, the cross section is depicted for the production of the $A^2\Pi_u$ state of N_2^+, which has been deduced from the emission cross section [6.114, 115].

In Fig. 6.34, the cross sections for electron-impact ionization of O_2 are presented. The most reliable data on σ_{gross} for O_2 are plotted in Fig. 6.34.

The total and partial (different final fragments of the target) ionization cross sections for the CO molecule are displayed in Fig. 6.35. In Table 6.22, the

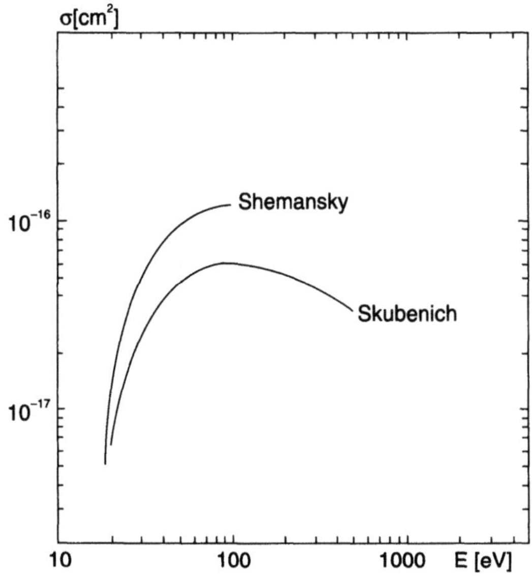

Fig. 6.33. Cross sections for the production of N_2^+ in its $A^2\Pi_u$ state in collision of electrons with N_2. The values derived from the emission cross sections of [6.114] and [6.115] are shown

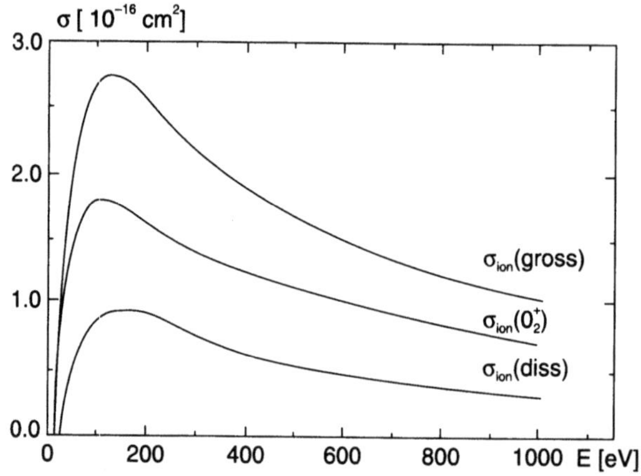

Fig. 6.34. Cross sections for electron impact ionization of O_2. Gross ionization cross section σ_{ion} (gross) [6.113], cross section for production of O_2^+ $\sigma_{\text{ion}}\left(O_2^+\right)$ [6.97], dissociative ionization cross section σ_{ion} (diss) [6.24]

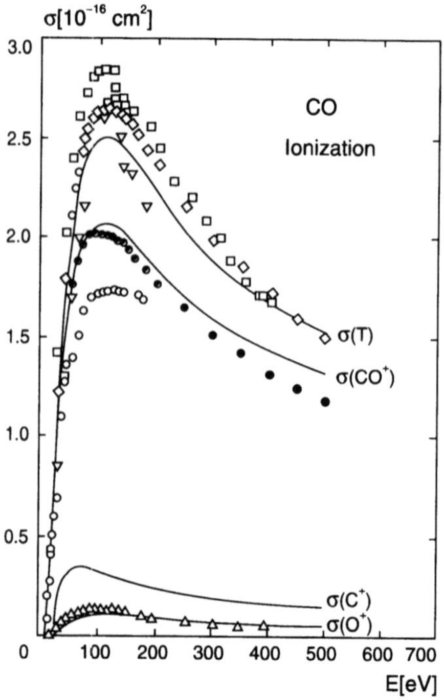

Fig. 6.35. Total ionization cross sections $\sigma(\text{T})$, partial and dissociative ionization cross sections $\sigma(\text{CO}^+)$, $\sigma(\text{C}^+)$ and $\sigma(\text{O}^+)$ for productions of various ions by electron impact on CO (▽ [6.136]; □△ CO^+ [6.113]; ◇ [6.113]; ⊙ [6.97]. The cross sections for production of ions with kinetic energy larger than 0.23 eV; ○ [6.117]; — [6.116])

Table 6.22. Total ionization cross section $\sigma(T)$, partial and dissociative ionization cross sections $\sigma(CO_2^+)$, $\sigma(CO^+)$, $\sigma(O^+)$ and $\sigma(C^+)$ in (10^{-16} cm^2), for production of various ions by electron impact on CO_2

E [eV]	$\sigma(CO_2^+)^a$	$\sigma(CO^+)^a$	$\sigma(O^+)^a$	$\sigma(C^+)^a$	$\sigma(T)^a$	$\sigma(T)^b$	$\sigma(T)^c$
15	0.05	0.002	0.00	0.00	0.05	0.10	
19.5	–	–	–	–	–	0.45	
20	0.40	0.007	0.001	0.00	0.41	–	
25	0.86	0.019	0.022	0.001	0.90	–	
26	–	–	–	–	–	1.12	
30	1.38	0.049	0.062	0.002	1.50	1.51	–
40	1.95	0.089	0.122	0.017	2.18	2.11	–
50	2.37	0.098	0.155	0.037	2.66	2.59	2.43
60	2.60	0.103	0.180	0.050	2.93	2.93	–
80	2.82	0.110	0.200	0.061	3.19	3.35	–
100	2.88	0.111	0.210	0.064	3.27	3.52	3.32
120	2.89	0.109	0.215	0.065	3.27	3.55	–
140	2.85	0.106	0.216	0.064	3.24	3.52	–
200	2.62	0.095	0.205	0.057	2.99	3.26	2.85
300	2.30	0.080	0.177	0.046	2.61	2.81	–
400	2.07	0.072	0.165	0.038	2.34	2.42	2.23
500	1.86	0.067	0.155	0.034	2.11	2.14	
600						1.91	
800						1.57	
1000						1.40	

[a] [6.116], [b] [6.113], [c] [6.118]. See also [6.119] for the production of CO_2^+ and CO_2^{++} and [6.120] for the production of singly charged ions CO_2^+, CO^+, O^+ and C^+

total, partial and dissociative ionization cross sections are listed for production of various ions by electron impact on CO_2.

Total ionization cross sections of CH_4 were published in several papers, and agreement among the observed data has been found (Fig. 6.36). We note that the cross sections for CD_4 practically coincide with those for CH_4 indicating that the isotope effect is of minor importance in ionization processes, which was confirmed experimentally [6.122]. The partial cross sections for ionization of CH_4 molecules have been measured [6.123] in a wide electron-energy range (Fig. 6.37).

Partial ionization cross sections of ethane (C_2H_6) were measured over the energy range of 15–30 eV (Fig. 6.38). The most probable channel of ionization is the formation of $C_2H_4^+$.

Experimental single and double ionization and dissociative ionization cross sections of CO, CO_2, NO and NO_2 by electron impact for $20\,\mathrm{eV} \leq E \leq 1000\,\mathrm{eV}$ were reported in [6.124].

The energy distribution of the secondary (ejected and scattered) electrons from N_2 was determined in the incident energy range 50 to 2000 eV [6.125]. The spectrum of secondary electrons (Fig. 6.39) is well described by the empirical formula [6.125]

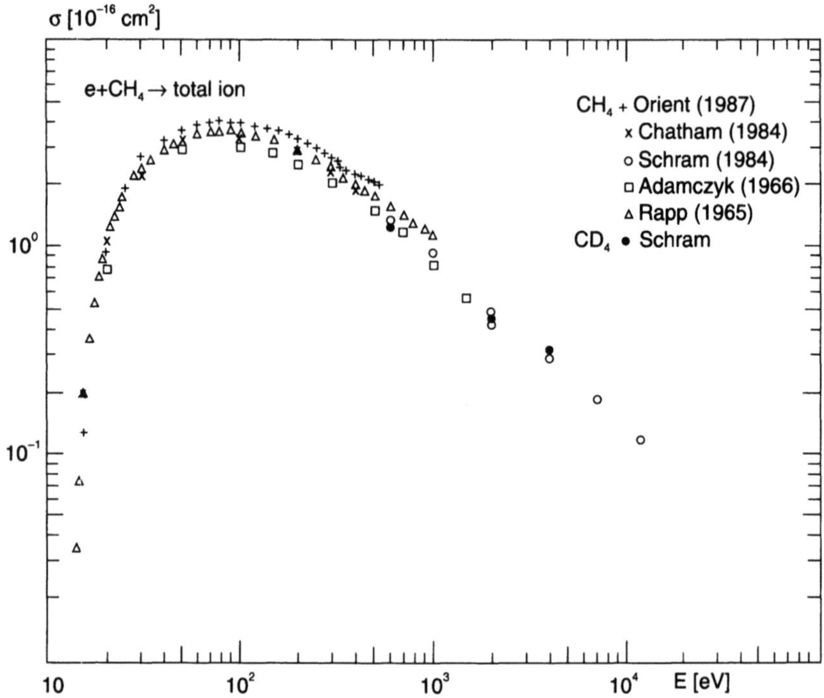

Fig. 6.36. Total ionization cross sections of CH_4 by electrons as a function of electron energy (CH_4: + + + [6.116]; × × × [6.121]; ○○○ [6.122]; □□□ [6.123]; △△△ [6.97]; CD_4 [6.122])

$$\frac{d\sigma}{dE_s} = \frac{F(E)}{E_s^2 + E_0^2} , \tag{6.6.9}$$

where $F(E)$ is a slowly varying function of the incident electron energy E, E_s is the secondary electron energy, and $E_0 = 11.4\,\mathrm{eV}$. The expansion coefficients for the Legendre polynomials of the double differential cross sections for the emission of secondary electrons in the ionization of N_2 by electron impact are listed in Table 6.23.

Figure 6.40 represents the energy distributions of the secondary electrons with energy E_s for the electron impact ionization of O_2. The form of the spectrum is described by the same formula as (6.6.9):

$$\frac{d\sigma}{dE_s} = \frac{F(E)}{E_s^2 + (15.2\,\mathrm{eV})^2} . \tag{6.6.10}$$

The energy distributions of O^+ and CO^+ ions from CO_2 show several peaks ranging from 5 to 10 eV when the incident electron energy increases between 45–150 eV [6.127]. The angular distribution of O^+, C^+ and CO^+ ions have also been measured between 30–300 eV electron energy [6.120]. At low

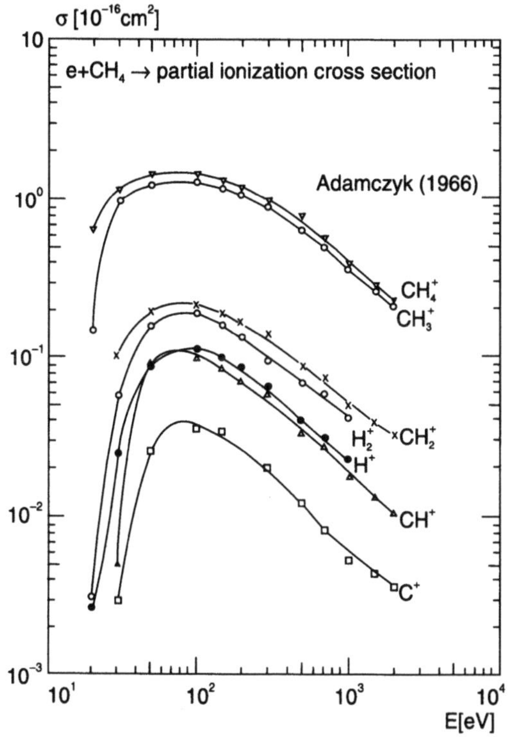

Fig. 6.37. Partial ionization cross sections of CH_4 by electron impact [6.123]

energies, the angular distributions reveal a relatively large asymmetry but become isotropic at higher electron impact energy.

6.7 Electron–Ion Recombination and Electron Attachment

A stable atomic negative ion exists if its energy E^- is less than the energy E^0 of the neutral atom in the ground state. The difference $E^0 - E^-$ is called the *electron affinity* E^A. For an electron having a kinetic energy E to be caught by a molecule into a stable orbit the system must lose the excess energy $E + E^A$ either by emission of radiation or by providing the energy to the third body. The former process, namely,

$$e^- + A \to A^- + \hbar\omega \tag{6.7.1}$$

is called *radiative electron attachment*. The latter process occurs in a *three-body collision*

$$e^- + A + M \to A^- + M^* \ , \tag{6.7.2}$$

M being either an electron or a neutral molecule. The three-body collisions are important only when the density of the particles is quite high.

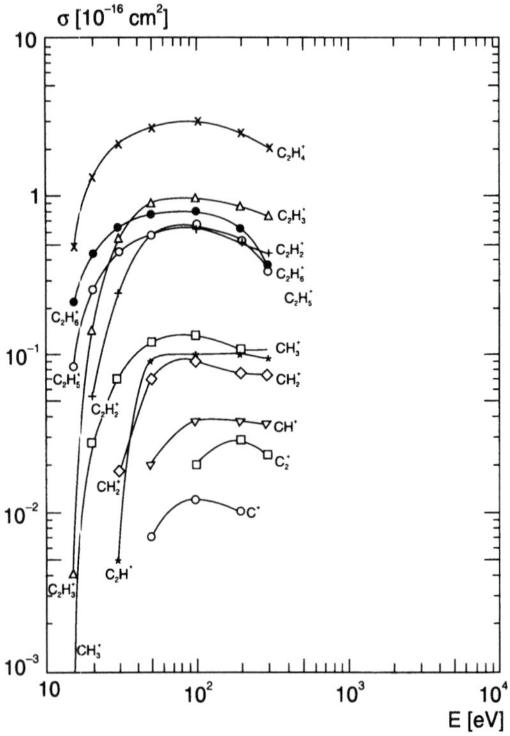

Fig. 6.38. Partial ionization cross sections of C_2H_6 by electron impact as a function of electron energy [6.121]

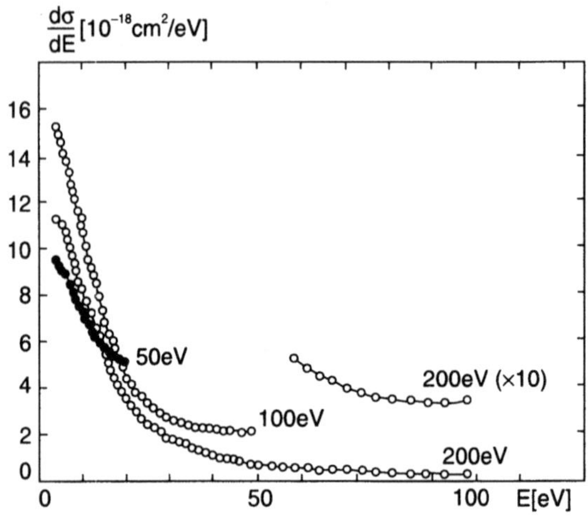

Fig. 6.39. Energy distribution of the secondary electrons at the ionizing collisions of electron with N_2. Energy of the incident electron is indicated along each curve [6.125]

Let N_e be the electron density and N_A the density of neutral molecules. The rate of decrease of N_e per unit time due to two-body electron-attachment processes is proportional to N_eN_A, the coefficient of proportionality being the attachment rate coefficient α:

Table 6.23. Coefficients in (10^{-20} cm^2/eV · sr) for the Legendre-polynomial expansion $d^2\sigma/d\varepsilon\, d\Omega = \Sigma A_n(E,\varepsilon)P_n(\cos\theta)$ of the doubly differential cross section for emission of secondary electrons with an energy ε in ionization of N_2 by the impact of electrons with energy E. The singly differential cross section $d\sigma/d\varepsilon$ is $4\pi A_0$ (in 10^{-20} cm^2/eV) [6.126]. See [6.125] for further information on double differential cross sections for ionization of atoms and molecules

ε [eV]	A_0	A_1	A_2	A_3	A_4	A_5
$E = 200$ eV						
2.0	1.08[2][a]	−2.32[1]	−1.51[1]	−7.95	7.50[−1]	
4.0	9.03[1]	−4.13	−3.80	−1.81	6.08	
6.0	8.15[1]	3.55[−1]	−4.57	1.44[−1]	5.08	
8.0	7.04[1]	3.22	−3.29	−2.89	4.83	
10.0	6.14[1]	3.02	−4.57	−4.20	3.83	
15.9	4.04[1]	5.09	−3.76	−4.66	2.24	
20.0	2.77[1]	4.39	−3.24	−4.98	1.06	
40.0	8.87	2.94	−1.98[−1]	−3.18	−1.27	1.21[−1]
50.0	6.09	2.48	7.88[−1]	−2.19	−1.26	2.88[−1]
$E = 500$ eV						
2.0	4.93[1]	−1.95[1]	−6.56	3.82	4.99	
4.0	4.62[1]	−3.03	−4.60	5.30[−1]	2.71	
6.0	4.50[1]	1.34	−4.64	2.85[−2]	3.36	
8.0	4.03[1]	2.96	−4.75	−1.10	2.83	
10.0	3.57[1]	4.34	−4.84	−1.50	1.71	
15.9	2.49[1]	4.29	−4.46	−2.60	1.66	
20.0	1.73[1]	3.67	−3.99	−2.93	1.00	
40.0	5.57	2.11	−1.86	−1.97	7.97[−2]	9.80[−1]
50.0	3.57	1.57	−1.28	−1.55	−2.78[−1]	9.02[−1]
$E = 1000$ eV						
2.0	3.76[1]	−1.03	−4.70	1.97	2.77	
4.0	3.06[1]	3.32[−1]	−3.84	1.67[−1]	1.19	
6.0	2.82[1]	1.01	−4.43	3.88[−1]	1.28	
8.0	2.46[1]	1.47	−4.74	−1.46[−1]	1.67	
10.0	2.20[1]	1.63	−4.73	−3.49[−1]	1.56	
15.9	1.52[1]	1.76	−3.83	−7.37[−1]	1.37	
20.0	1.03[1]	1.53	−3.19	−1.11	1.23	
40.0	3.33	8.81[−1]	−1.57	−8.38[−1]	5.37[−1]	5.05[−1]
50.0	2.13	6.69[−1]	−1.06	−7.23[−1]	3.54[−1]	5.01[−1]
$E = 2000$ eV						
2.0	2.09[1]	1.14	−2.49	1.66[−1]	3.87[−1]	
4.0	1.73[1]	9.67[−1]	−3.82	−4.87[−1]	1.93[−2]	
6.0	1.61[1]	1.09	−3.92	−3.94[−1]	−3.27[−1]	
8.0	1.43[1]	1.07	−4.05	−4.49[−1]	−2.21[−1]	
10.0	1.28[1]	1.11	−3.94	−5.77[−1]	−2.06[−1]	
15.9	9.03	8.96[−1]	−3.25	−6.47[−1]	1.01[−1]	
20.0	6.15	7.66[−1]	−2.63	−5.93[−1]	1.65[−1]	1.03[−1]
40.0	1.97	4.06[−1]	−1.21	−4.39[−1]	3.14[−1]	2.14[−1]
50.0	1.27	2.84[−1]	−8.38[−1]	−3.42[−1]	2.90[−1]	2.33[−1]

[a] $1.08[2] = 1.08 \times 10^2$

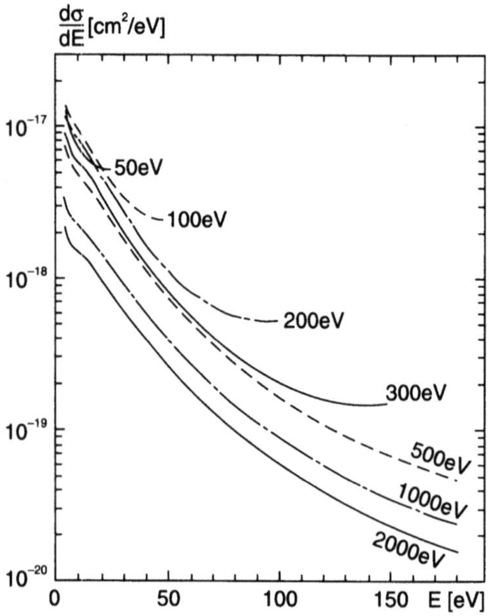

Fig. 6.40. Energy distribution of secondary electrons at the ionizing collisions of electrons with O_2. The energy of the incident electron is indicated [6.125]

$$-\partial N_e/\partial t = \alpha N_e N_A \ . \tag{6.7.3}$$

The rate coefficient α is normally given in units of cm^3/s. If a third body M is involved, the rate coefficient α as defined by (6.7.3) depends on the density N_M of M. The three-body attachment rate coefficient K may be conveniently defined by

$$-\partial N_e/\partial t = K N_e N_A N_M \ . \tag{6.7.4}$$

If the third body is an electron, the right-hand side of (6.7.4) is replaced by $KN_e^2 N_A$, where K is normally given in units of cm^6/s.

If A is a molecule, say, a diatomic molecule BC, the excess energy $E + E^A$ may be transferred to the nuclear motion, and the molecule may dissociate without emitting a captured electron:

$$\begin{aligned} e^- + BC &\rightarrow B + C^- \\ &\rightarrow B^- + C \ . \end{aligned} \tag{6.7.5}$$

This is *dissociative attachment*. This additional mechanism of excess energy redistribution in the ion BC^- increases the cross section for electron attachment to a large extent. The neutral dissociation fragments B and C are often formed in excited states. Thus, the dissociative attachment processes often play the role of a source of excited species [6.128]. Dissociative attachment proceeds efficiently through a resonant state:

$$e^- + BC \rightarrow (BC^-)^* \rightarrow B + C^-, \text{ or } B^- + C \ . \tag{6.7.6}$$

Furthermore, the cross section for dissociative attachment is enhanced by many orders of magnitude if the molecule is initially in an excited vibrational state. This is because the vibrational motion of the nuclei facilitate the dissociative motion [6.128]. A similar but less remarkable enhancement of the cross section occurs for a molecule being initially in an excited rotational state.

Dissociative recombination of CO^+ is described as

$$e^- + CO^+ \rightarrow CO^* \rightarrow C + O \ . \tag{6.7.7}$$

This process plays a role in low-temperature plasmas. The absolute cross section (6.7.7) decreases as $1/E$ [6.129]. The sum of the cross sections for the processes is expressed by (Fig. 6.41)

$$\begin{aligned} e^- + CO^+ \rightarrow e^- + \left(CO^+\right)^* &\rightarrow e^- + C^+ + O^* \\ &\rightarrow e^- + C^* + O^+ \ . \end{aligned} \tag{6.7.8}$$

Dissociative recombination processes are dominant at low electron energies. The cross sections for H_3^+O, and D_3O^+ processes have been measured with the help of the merged electron–ion-beam technique [6.130] from 0.01 to

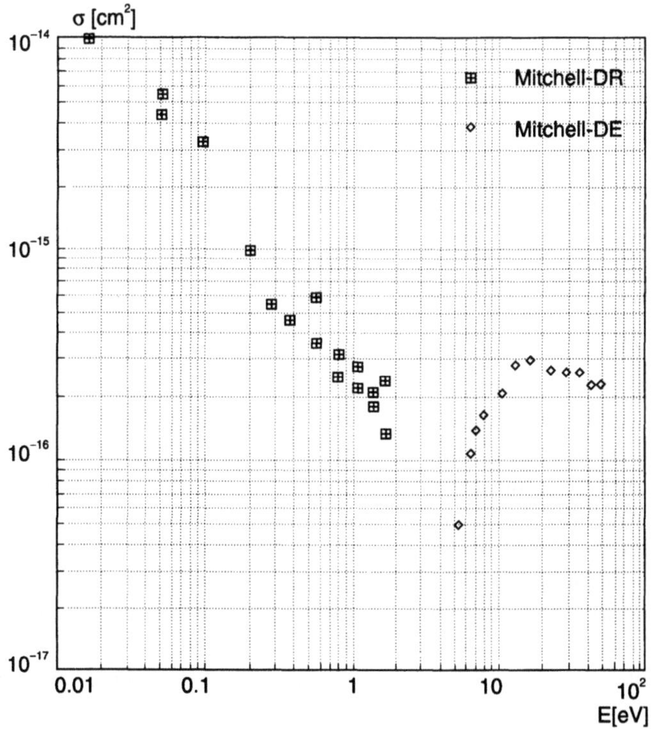

Fig. 6.41. Cross sections for Dissociative Recombination (DR) and Dissociative Excitation (DE) of CO^+ by electron impact [6.129]

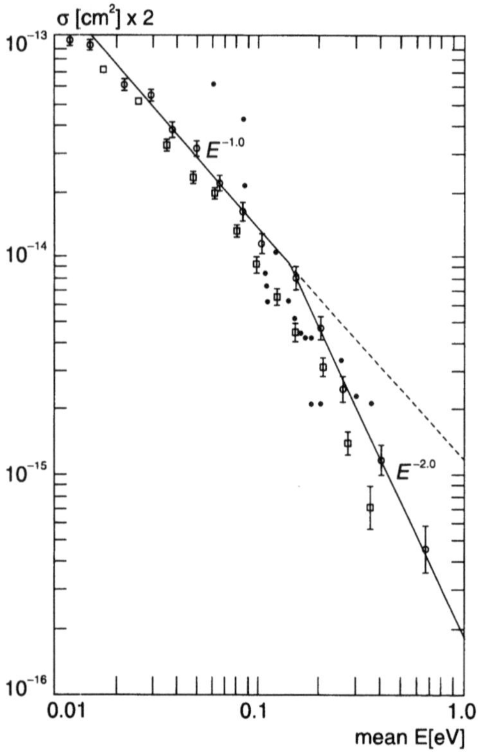

Fig. 6.42. Cross sections for dissociative recombination of H_3O^+ and D_3O^+ by electron impact. The original cross sections of [6.130] are divided by a factor of two (*solid curve*: 0.010.1 eV [6.130]; 0.10.6 eV [6.131]. Other measurements shown have been discussed in [6.130])

1 eV. A trapped-ion technique was used [6.131] to measure the same cross sections at lower energies (0.06–0.6 eV). Both results are in a good agreement (Fig. 6.42).

The cross sections for dissociative electron attachment

$$e^- + H_2 \rightarrow H + H^- \tag{6.7.9}$$

are depicted in Fig. 6.43. It has been shown both experimentally and theoretically that a small peak near the impact energy of 4 eV is strongly enhanced if H_2 molecules are in either vibrationally or rotationally excited states [6.132]. For example, the cross sections for H_2^* in the vibrationally excited state $v = 4$, are four orders of magnitude larger than those for the vibrationally ground state $v = 0$.

The dissociative attachment of electrons at O_2 molecule has been investigated and general agreement in the cross section has been obtained [6.133]. A typical example is shown in Fig. 6.44. The cross section has a broad peak with a maximum at about 6.7 eV. This peak is interpreted as the resonance process

$$e^- + O_2\left(X^3\Sigma_g^-\right) \rightarrow O_2^-\left({}^2\Pi_u\right) \rightarrow O^-\left({}^2P\right) + O\left({}^3P\right) \ . \tag{6.7.10}$$

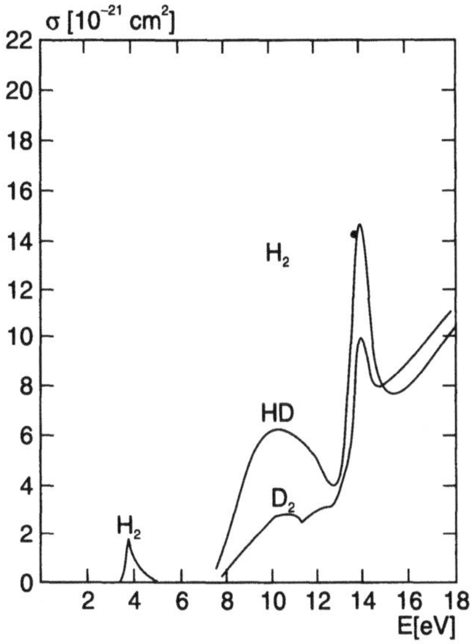

Fig. 6.43. Cross sections of dissociative attachment for HD and D_2 as well as H_2 [6.2]

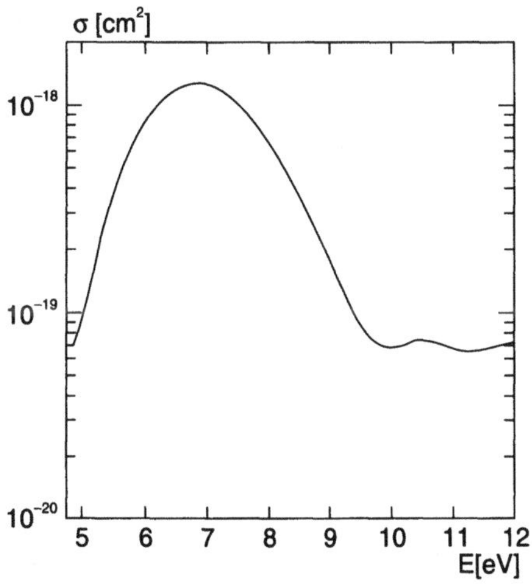

Fig. 6.44. Cross section for the production of O^- in electron collision with O_2 [6.134]

For collision energies above 17 eV, negative ions O^- can be produced through the process of ion pair formation

$$e^- + O_2 \to O^+ + O^- + e^- \ . \tag{6.7.11}$$

For many applications it is important to know the electron energy distribution for different types of inelastic collision processes. Figure 6.45 exhibits such a distribution for the case of electron collision with O_2 molecules.

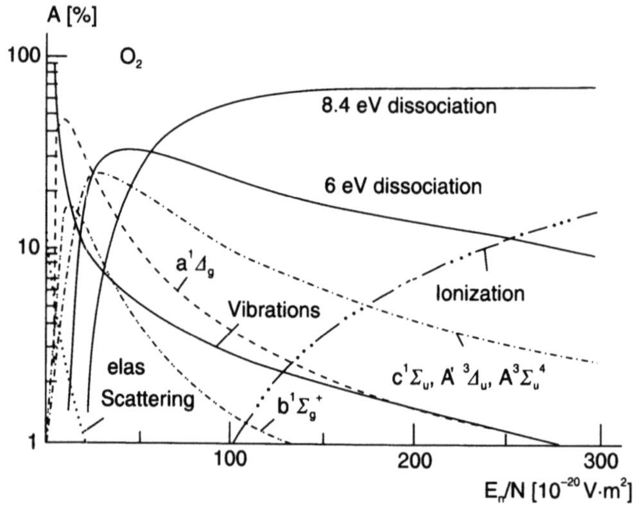

Fig. 6.45. Balance of energy-loses of scattered electrons on O_2 molecule (in %) [6.135]. A is the part of the electron energy lost in the given process

The abscissa represents the ratio of strength of electric field E_n in which the electrons are accelerated to the density of the target N. The quantity E_n/N is approximately proportional to the mean kinetic energy of the scattered electrons.

7 Interatomic Potentials

Particle interactions within molecules are considered. The problems related with potential-energy surfaces and particle-interaction constants were discussed in many reviews and monographs [7.1–8]. Bibliographies on molecular particle surfaces and collisions with photons, electrons and heavy particles can be found in [7.9, 10].

Tables and figures for energy potentials of simple molecules and molecular ions such as H_2, H_2^+, H_2^-, and H_3^+ are given in this chapter.

7.1 Interaction Constants

Let us consider an electrostatic interaction $U(R)$ between two atomic particles A and B. If the distance R between the nuclei is larger than the particle sizes, then $U(R)$ can be written in the form [7.7]:

$$\begin{aligned} U(R) &= e^2\left(-\sum_i \frac{Z_A}{|\mathbf{R}-\mathbf{r}_i|} - \sum_k \frac{Z_B}{|\mathbf{R}-\mathbf{r}_k|} + \sum_{i,k} \frac{1}{|\mathbf{R}+\mathbf{r}_k-\mathbf{r}_i|} + \frac{Z_A Z_B}{R}\right) \\ &= \sum_{l=0}^{\infty} \frac{e^2}{R^{l+1}} \left\{ -\sum_i Z_A r_i^l \mathrm{P}_l(\cos\theta_i) + \sum_k (-1)^l Z_B r_k^l \mathrm{P}_l(\cos\theta_k) \right. \\ &\quad \left. + \sum_{i,k} (-1)^l |\mathbf{r}_i - \mathbf{r}_k|^l \mathrm{P}_l[\cos(\mathbf{r}_i - \mathbf{r}_k, \mathbf{R})] + Z_A Z_B \delta_{lo} \right\}, \quad R \gg r_i,\ r_k\ , \end{aligned} \tag{7.1.1}$$

where r_i and r_k are the electron coordinates of the particles B and A, respectively, Z_A and Z_B are the nuclear charges, and $\mathrm{P}_l(\cos\theta)$ is the Legendre polynomial.

In the case of an interaction between an ion B and an atom A one can deduce from (7.1.1):

$$U(R) = \frac{e\,Z(\mathbf{D}_A\mathbf{n})}{R^2} + \frac{1}{R^3}\left(\frac{e\,Z\,Q_A}{2} - 3(\mathbf{D}_A\mathbf{n})(\mathbf{D}_B\mathbf{n}) + \mathbf{D}_A\mathbf{D}_B\right) + O(1/R^4) + \cdots\ , \tag{7.1.2}$$

where Z is the charge of ion B, $\mathbf{D}$ is the dipole moment, and $Q_A = \sum_k e^2 r_k^2 (1 - 3\cos^2\theta_k)$ is the quadrupole moment of the atom A.

In first-order perturbation theory, the multipole moments of the particle averaged over all possible orientations of the electron momenta J_A and J_B are zero. In second-order perturbation theory, the average long-range interaction potential $V(R)$ has the form

$$V(R) = -\frac{C_4}{R^4} - \frac{C_6}{R^6} , \tag{7.1.3}$$

where $C_4 = \beta Z e^2/2$ and β is the polarizability of atom A; the constant C_6 describes the interaction of the charge Z with the atomic quadrupole moment and the interaction between dipole moments of the particles.

In the case of interactions between two atoms A and B one derives from (7.1.1):

$$\begin{aligned} V(R) = \frac{1}{R^3}\Bigg((\mathbf{D}_A\,\mathbf{D}_B) - 3(\mathbf{D}_A\mathbf{n})\,(\mathbf{D}_B\mathbf{n}) + \frac{3e}{2R^4}\Bigg\{ (\mathbf{D}_A\mathbf{n}) \sum_k r_k^2 \\ \times (5\cos^2\theta_k + 1) - (\mathbf{D}_B\mathbf{n}) \sum_i r_i^2 (5\cos^2\theta_i - 1) \\ -2\left[\sum_i (\mathbf{r}_i\mathbf{D}_k) r_i \cos\,\theta_i + \sum_k (\mathbf{r}_k\mathbf{D}_i) r_k \cos\,\theta_k\right]\Bigg\}\Bigg) + O(1/R^5) + \cdots . \end{aligned} \tag{7.1.4}$$

For the average interaction one obtains

$$V(R) = -C_6 R^{-6} - C_8 R^{-8} - C_{10} R^{-10} + \cdots , \tag{7.1.5}$$

where C_n denotes the van-der-Waals interaction constant (Sect. 7.1.1).

7.1.1 Van-der-Waals Interaction

Van-der-Waalş interaction between atoms, molecules and microscopic particles is of electro-magnetic nature and dominates at large interparticle distances when the overlap of the wave functions of interacting particles can be neglected.

In second-order perturbation theory, the interaction potential of two spherically symmetric atoms A and B (in the S-state) at a large interparticle distance R is expressed in the form:

$$V_{AB}(R) = -C_6 R^{-6} - C_8 R^{-8} - C_{10} R^{-10} + \cdots . \tag{7.1.6}$$

The coefficients C_n are termed Van-der-Waals coefficients. The first term in (7.1.6) describes the dipole–dipole interaction; the second the dipole–quadrupole interaction and the third one the sum of the quadrupole–quadrupole and dipole–octupole interactions, respectively. These constants are defined as a sum of integrals of the type [7.3]

Table 7.1. Lower and upper limits of C_6 coefficients (in atomic units) [7.12]*

	C_6	
	Lower	Upper
(A) Noble-gas pairs		
He–He	1.44(0)	1.47(0)
He–Ne	3.03(0)	3.20(0)
He–Ar	9.43(0)	1.01(1)
He–Kr	1.30(1)	1.42(1)
He–Xe	1.88(1)	2.14(1)
Ne–Ne	6.48(0)	7.27(0)
Ne–Ar	1.95(1)	2.20(1)
Ne–Kr	2.65(1)	3.09(1)
Ne–Xe	3.81(1)	4.65(1)
Ar–Ar	6.36(1)	7.08(1)
Ar-Kr	8.86(1)	1.00(2)
Ar–Xe	1.30(2)	1.51(2)
Kr–Kr	1.24(2)	1.42(2)
Kr–Xe	1.83(2)	2.15(2)
Xe–Xe	2.72(2)	3.25(2)
(B) Alkali atom pairs		
H–H	6.47(0)	6.51(0)
H–Li	6.58(1)	6.68(1)
H–Na	7.02(1)	1.75(2)
H–K	1.11(2)	1.16(2)
H–Rb	1.18(2)	1.26(2)
H–Cs	1.34(2)	1.39(2)
Li–Li	1.38(3)	1.39(3)
Li–Na	1.43(3)	2.16(3)
Li–K	2.34(3)	2.36(3)
Li–Rb	2.53(3)	2.55(3)
Li–Cs	3.01(3)	3.03(3)
Na–Na	1.47(3)	5.00(3)
Na–K	2.41(3)	3.67(3)
Na–Rb	2.60(3)	3.97(3)
Na–Cs	3.08(3)	4.50(3)
K–K	3.97(3)	4.03(3)
K–Rb	4.29(3)	4.37(3)
K–Ca	5.12(3)	5.18(3)
Rb–Rb	4.64(3)	4.74(3)
Rb–Cs	5.53(3)	5.62(3)
Cs–Cs	6.63(3)	6.70(3)
(C) Rare-earth atom pairs		
Be–Be	2.19(2)	2.21(2)
Be–Mg	3.69(2)	3.74(2)
Be–Ca	7.62(2)	7.77(2)
Mg–Mg	6.30(2)	6.38(2)
Mg–Ca	1.30(3)	1.33(3)
Ca–Ca	2.74(3)	2.83(3)

	C_6	
	Lower	Upper
(D) Noble-gas-alkali atom pairs		
He–H	2.80(0)	2.83(0)
He–Li	2.21(1)	2.28(1)
He–Na	2.37(1)	6.87(1)
He–K	3.78(1)	4.18(1)
He–Rb	4.01(1)	4.63(1)
He–Cs	4.43(1)	4.92(1)
Ne–H	5.64(0)	5.78(0)
Ne–Li	4.29(1)	4.48(1)
Ne–Na	4.60(1)	1.37(2)
Ne–K	7.38(1)	8.39(1)
Ne–Rb	7.80(1)	9.42(1)
Ne–Cs	8.60(1)	9.92(1)
Ar–H	1.97(1)	2.03(1)
Ar–Li	1.71(2)	1.77(2)
Ar–Na	1.84(2)	5.08(2)
Ar–K	2.92(2)	3.18(2)
Ar-Rb	3.10(2)	3.52(2)
Ar–Cs	3.45(2)	3.79(2)
Kr–H	2.80(1)	2.91(1)
Kr–Li	2.55(2)	2.62(2)
Kr–Na	2.73(2)	7.37(2)
Kr–K	4.32(2)	4.69(2)
Kr–Rb	4.60(2)	5.18(2)
Kr–Cs	5.15(2)	5.61(2)
Xe–H	4.18(1)	4.41(1)
Xe–Li	4.02(2)	4.14(2)
Xe–Na	4.30(2)	1.13(3)
Xe–K	6.80(2)	7.37(2)
Xe–Rb	7.25(2)	8.13(2)
Xe–Cs	8.14(2)	8.86(2)
(E) Noble-gas–alkaline-earth pairs		
He–Be	1.36(1)	1.39(1)
He–Mg	2.11(1)	2.21(1)
He–Ca	4.51(1)	4.85(1)
Ne–Be	2.70(1)	2.80(1)
Ne–Mg	4.15(1)	4.43(1)
Ne–Ca	8.96(1)	9.85(1)
Ar–Be	1.00(2)	1.04(2)
Ar–Mg	1.58(2)	1.66(2)
Ar–Ca	3.33(2)	3.59(2)
Kr–Be	1.46(2)	1.51(2)
Kr–Mg	2.32(2)	2.43(2)
Kr–Ca	4.84(2)	5.21(2)
Xe–Be	2.23(2)	2.33(2)

* 1.30 (1) means 1.30×10^1

Table 7.1. *Continued*

	C_6			C_6	
	Lower	Upper		Lower	Upper
Xe–Mg	3.58(2)	3.76(2)	Na–Mg	8.97(2)	1.75(3)
Xe–Ca	7.43(2)	8.06(2)	Na–Ca	1.92(3)	3.61(3)
(F) Alkali–alkaline-earth atom pairs			K–Be	8.01(2)	8.18(2)
			K–Mg	1.43(3)	1.45(3)
H–Be	3.55(1)	3.58(1)	K–Ca	3.08(3)	3.16(3)
H–Mg	5.74(1)	5.85(1)	Rb–Be	8.60(2)	8.86(2)
H–Ca	1.19(2)	1.23(2)	Rb–Mg	1.53(3)	1.57(3)
Li–Be	4.78(2)	4.82(2)	Rb–Ca	3.32(3)	3.43(3)
Li–Mg	8.52(2)	8.57(2)	Cs–Be	9.91(2)	1.01(3)
Li–Ca	1.83(3)	1.85(3)	Cs–Mg	1.78(3)	1.81(3)
Na–Be	5.06(2)	1.05(3)	Cs–Ca	3.86(3)	3.95(3)

$$\hbar \int_0^\infty \beta^A_{æ_1}(i\omega)\, \beta^B_{æ_2}(i\omega)\, d\omega \ , \tag{7.1.7}$$

where $\beta_{æ}(i\omega)$ are the $2^{æ}$-pole dynamic polarizabilities of atoms A and B at the frequency ω. For example, the dipole van-der-Waals constant C_6 is given by [7.11]

$$C_6 = \frac{3\hbar}{\pi} \int_1^\infty \beta^A_1(i\omega)\, \beta^B_1(i\omega)\, d\omega \ , \tag{7.1.8}$$

where $\beta^A_1(i\omega)$ denotes the dynamic dipole polarizability.

All values in (7.1.6) are expressed in atomic units. To find $V(R)$ in eV one can use the relation.

$$V(R)[\mathrm{eV}] = -0.5975\left[C_6 R^{-6} + 0.280\, C_8 R^{-8} + 0.0784\, C_{10} R^{-10}\right] \ ,$$

where R is in units of 10^{-10} m, and C_n is in atomic units.

For two interacting hydrogen atoms in their ground state one has (Tables 7.1, 2): $C_6^{\mathrm{H-H}} = 6.49$ a.u. $= 6.49\ e^2 a_0^5 \simeq 6.2 \times 10^{-60}$ erg · cm^6, $C_6^{\mathrm{H-H}} = 124.4$ a.u. $\simeq 3.3 \times 10^{-75}$ erg · cm^8, and $C_{10}^{\mathrm{H-H}} = 3286$ a.u. $\simeq 2.5 \times 10^{-90}$ erg · cm^{10}.

To account for multipole-multipole van-der-Waals forces between atoms is necessary for the quantitative description of the interatomic energy $V(R)$ at distances $R \leq 12 a_0 \simeq 6.4$ Å. This region is of special importance because the majority of interatomic potentials has a minima at $R_m \simeq (3–5)$ Å (Table 2.1). The interaction energy $V^{\mathrm{H-H}}(R)$ for H–H interaction has a minimum at $R_m \simeq 4.2$ Å, where the contributions from the three terms in (7.1.6) yield: 70.45, 21.10 and 8.45 % . At larger $R \simeq 5.3$ Å, the contributions from the higher multipoles are considerably 81.30, 14.64 and 4.06 % , respectively.

Tables 7.1, 2 give the lower and upper limits of the van-der-Waals coefficients C_6, C_8 and C_{10} for hydrogen, noble-gas, alkali and rare-earth atoms

Table 7.2 Lower and upper limits of C_8 and C_{10} coefficients (in atomic units) [7.12]

	C_8		C_{10}	
	Lower	Upper	Lower	Upper
(A) Noble-gas pairs				
He–He	1.39(1)	1.42(1)	1.81(2)	1.84(2)
He–Ne	2.83(1)	3.70(1)	3.86(2)	5.34(2)
He–Ar	1.29(2)	1.85(2)	2.86(3)	4.35(3)
He–Kr	2.08(2)	3.01(2)	5.55(3)	7.75(3)
He–Xe	3.77(2)	5.70(2)	1.26(4)	1.77(4)
Ne–Ne	5.55(1)	9.65(1)	8.26(2)	1.52(3)
Ne–Ar	2.62(2)	4.41(2)	6.08(3)	1.10(4)
Ne–Kr	4.22(2)	6.98(2)	1.17(4)	1.88(4)
Ne–Xe	7.65(2)	1.29(3)	2.63(4)	4.13(4)
Ar–Ar	1.18(3)	1.88(3)	3.49(4)	6.09(4)
Ar–Kr	1.87(3)	2.93(3)	6.24(4)	1.02(5)
Ar–Xe	3.31(3)	5.25(3)	1.30(5)	2.10(5)
Kr–Kr	2.94(3)	4.54(3)	1.09(5)	1.70(5)
Kr–Xe	5.15(3)	8.03(3)	2.20(5)	3.45(5)
Xe–Xe	8.90(3)	1.39(4)	4.28(5)	6.75(5)
(B) Alkali-atom pairs				
H–H	1.24(2)	1.25(2)	3.27(3)	3.29(3)
H–Li	3.06(3)	3.27(3)	1.89(5)	2.17(5)
H–Na	3.68(3)	4.90(3)	2.46(5)	2.96(5)
H–K	7.14(3)	7.67(3)	5.82(5)	6.60(5)
H–Rb	8.38(3)	9.14(3)	7.71(5)	8.54(5)
H–Cs	1.03(4)	1.24(4)	1.11(6)	1.24(6)
Li–Li	7.89(4)	8.19(4)	6.50(6)	7.05(6)
Li–Na	9.16(4)	1.33(5)	7.98(6)	1.11(7)
Li–K	1.80(5)	1.88(5)	1.76(7)	1.91(7)
Li–Rb	2.13(5)	2.23(5)	2.23(7)	2.41(7)
Li–Cs	2.80(5)	3.06(5)	3.13(7)	3.46(7)
Na–Na	1.05(5)	2.01(5)	9.68(6)	1.68(7)
Na–K	2.03(5)	2.97(5)	2.10(7)	3.03(7)
Na–Rb	2.38(5)	3.49(5)	2.63(7)	3.81(7)
Na–Cs	3.08(5)	4.68(5)	3.64(7)	5.44(7)
K–K	3.84(5)	4.00(5)	4.40(7)	4.76(7)
K–Rb	4.46(5)	4.67(5)	5.43(7)	5.87(7)
K–Cs	5.77(5)	6.25(5)	7.44(7)	8.25(7)
Rb–Rb	5.16(5)	5.43(5)	6.64(7)	7.18(7)
Rb–Cs	6.65(5)	7.21(5)	9.03(7)	1.00(8)
Cs–Cs	8.58(5)	9.50(5)	1.23(8)	1.40(8)
(C) Alkaline-earth atom pairs				
Be–Be	1.04(4)	1.09(4)	5.08(5)	5.63(5)
Be–Mg	2.08(4)	2.19(4)	1.20(6)	1.34(6)
Be–Ca	4.33(4)	5.35(4)	3.15(6)	3.84(6)
Mg–Mg	4.11(4)	4.35(4)	2.73(6)	3.04(6)
Mg–Ca	8.74(4)	1.05(5)	6.90(6)	8.44(6)
Ca–Ca	1.90(5)	2.49(5)	1.77(7)	2.28(7)

Table 7.2. *Continued*

	C_8		C_{10}	
	Lower	Upper	Lower	Upper
(D) Noble-gas–alkali-atom pairs				
He–H	4.15(1)	4.20(1)	8.63(2)	8.74(2)
He–Li	9.91(2)	1.08(3)	5.92(4)	7.06(4)
He–Na	1.20(3)	1.50(3)	7.73(4)	9.28(4)
He–K	2.30(3)	2.50(3)	1.84(5)	2.14(5)
He–Rb	2.69(3)	2.98(3)	2.46(5)	2.77(5)
He–Cs	3.26(3)	4.01(3)	3.57(5)	4.02(5)
Ne–H	8.65(1)	9.74(1)	1.85(3)	2.17(3)
Ne–Li	2.00(3)	2.20(3)	1.21(5)	1.47(5)
Ne–Na	2.40(3)	3.19(3)	1.57(5)	1.94(5)
Ne–K	4.57(3)	5.04(3)	3.71(5)	4.38(5)
Ne–Rb	5.33(3)	5.99(3)	4.94(5)	5.63(5)
Ne–Cs	6.45(3)	7.99(3)	7.12(5)	8.13(5)
Ar–H	3.87(2)	4.65(2)	1.08(4)	1.38(4)
Ar–Li	8.63(3)	9.49(3)	5.53(5)	6.69(5)
Ar–Na	1.03(4)	1.53(4)	7.11(5)	9.30(5)
Ar–K	1.95(4)	2.16(4)	1.65(6)	1.94(6)
Ar–Rb	2.27(4)	2.56(4)	2.16(6)	2.48(6)
Ar–Cs	2.76(4)	3.58(4)	3.08(6)	3.55(6)
Kr–H	6.15(2)	7.55(2)	1.95(4)	2.45(4)
Kr–Li	1.34(4)	1.48(4)	8.92(5)	1.08(6)
Kr–Na	1.59(4)	2.48(4)	1.14(6)	1.53(6)
Kr–K	3.01(4)	3.33(4)	2.60(6)	3.06(6)
Kr–Rb	3.50(4)	3.93(4)	3.40(6)	3.89(6)
Kr–Cs	4.26(4)	5.19(4)	4.81(6)	5.56(6)
Xe–H	1.10(3)	1.41(3)	4.17(4)	5.35(4)
Xe–Li	2.30(4)	2.54(4)	1.63(6)	1.97(6)
Xe–Na	2.71(4)	4.52(4)	2.05(6)	2.90(6)
Xe–K	5.07(4)	5.64(4)	4.60(6)	5.42(6)
Xe–Rb	5.87(4)	6.63(4)	5.92(6)	6.84(6)
Xe–Cs	7.15(4)	8.62(4)	8.28(6)	9.69(6)
(E) Noble-gas–alkaline-earth atom pairs				
He–Be	4.17(2)	4.47(2)	1.32(4)	1.56(4)
He–Mg	8.43(2)	9.24(2)	3.55(4)	4.19(4)
He–Ca	1.48(3)	2.19(3)	1.08(5)	1.29(5)
Ne–Be	8.51(2)	9.48(2)	2.76(4)	3.48(4)
Ne–Mg	1.70(3)	1.91(3)	7.35(4)	9.05(4)
Ne–Ca	2.96(3)	4.50(3)	2.20(5)	2.70(5)
Ar–Be	3.61(3)	4.09(3)	1.39(5)	1.76(5)
Ar–Mg	7.20(3)	8.15(3)	3.48(5)	4.31(5)
Ar–Ca	1.32(4)	1.94(4)	9.82(5)	1.25(6)
Kr–Be	5.59(3)	6.38(3)	2.34(5)	2.95(5)
Kr–Mg	1.11(4)	1.26(4)	5.72(5)	7.06(5)
Kr–Ca	2.08(4)	2.99(4)	1.57(6)	2.01(6)
Xe–Be	9.56(3)	1.12(4)	4.58(5)	5.84(5)
Xe–Mg	1.88(4)	2.17(4)	1.07(6)	1.34(6)
Xe–Ca	3.59(4)	5.11(4)	2.82(6)	3.69(6)

Table 7.2 *Continued*

	C_8		C_{10}	
	Lower	Upper	Lower	Upper
(F)Alkali-alkaline-earth atom pairs				
H–Be	1.21(3)	1.27(3)	4.45(4)	4.94(4)
H–Mg	2.49(3)	2.66(3)	1.15(5)	1.29(5)
H–Ca	4.77(3)	6.51(3)	3.35(5)	3.96(5)
Li–Be	2.68(4)	2.82(4)	1.87(6)	2.07(6)
Li–Mg	5.48(4)	5.74(4)	4.15(6)	4.58(6)
Li–Ca	1.27(5)	1.44(5)	1.06(7)	1.26(7)
Na–Be	3.16(4)	4.79(4)	2.36(6)	3.05(6)
Na–Mg	6.34(4)	9.90(4)	5.16(6)	6.99(6)
Na–Ca	1.44(5)	2.41(5)	1.29(7)	1.97(7)
K–Be	6.06(4)	6.41(4)	5.37(6)	5.95(6)
K–Mg	1.20(5)	1.27(5)	1.14(7)	1.26(7)
K–Ca	2.74(5)	3.06(5)	2.81(7)	3.30(7)
Rb–Be	7.08(4)	7.59(4)	6.92(6)	7.59(6)
Rb–Mg	1.39(5)	1.49(5)	1.45(7)	1.59(7)
Rb–Ca	3.18(5)	3.56(5)	3.53(7)	4.11(7)
Cs–Be	8.86(4)	1.02(5)	9.75(6)	1.09(7)
Cs–Mg	1.75(5)	1.98(5)	2.01(7)	2.27(7)
Cs–Ca	4.04(5)	4.71(5)	4.91(7)	5.82(7)

that have been calculated in [7.12] using dynamic multipole polarizabilities of atoms.

In the case of three particles, the long-range interaction potential between atoms A, B and C in the S-state can be written as

$$V_{ABC}(R) = V_{AB}(R) + V_{BC}(R) + V_{AC}(R) + V_{AC}(R) + E_{ABC} \,, \tag{7.1.9}$$

where the potentials with two subscripts correspond to the two-body interaction potential (7.1.1), and E_{ABC} can be expressed by the expansion:

$$E_{ABC} = \sum_{æ_1 æ_2 æ_3} X_{ABC}(æ_1 æ_2 æ_3) \cdot Y_{ABC}(æ_1 æ_2 æ_3) \ . \tag{7.1.10}$$

Here, Y_{ABC} is a geometrical factor and the three-body interaction coefficients X_{ABC} are given by

$$X_{ABC}(æ_1 æ_2 æ_3) = \frac{\hbar}{\pi} \int_0^\infty \beta^A_{æ_1}(i\omega)\, \beta^B_{æ_2}(i\omega)\, \beta^C_{æ_3}(i\omega)\, d\omega \ . \tag{7.1.11}$$

The lower and upper limits of the three-body long-range interaction coefficients X_{ABC} involving atoms of the same kind are compiled in Table 7.3.

Table 7.3. Limits of three-body interaction coefficients X_{ABC} (7.1.5) for atoms of the same kind (in atomic units) [7.12]; D means dipole, Q quadrupole and O octupole interactions, respectively

	DDD		DDQ		DQQ		DDO		QQQ	
Atom	Lower	Upper	Lower	Upper	Lower	Upper	Lower	Upper	Lower	Upper
He	4.91(−1)	4.94(−1)	9.19(−1)	9.26(−1)	1.73(0)	1.74(0)	4.10(0)	4.13(0)	3.27(0)	3.29(0)
Ne	4.11(0)	4.38(0)	7.88(0)	1.16(1)	1.60(1)	3.09(1)	4.36(1)	6.02(1)	3.39(1)	8.29(1)
Ar	1.72(2)	1.80(2)	6.84(2)	9.25(2)	2.77(3)	4.82(3)	7.45(3)	9.82(3)	1.14(4)	2.55(4)
Kr	5.11(2)	5.36(2)	2.57(3)	3.40(3)	1.32(4)	2.19(4)	3.48(4)	4.22(4)	6.88(4)	1.43(5)
Xe	1.79(3)	1.91(3)	1.25(4)	1.66(4)	8.89(4)	1.47(5)	2.15(5)	2.61(5)	6.41(5)	1.32(6)
H	7.20(0)	7.22(0)	2.62(1)	2.63(1)	9.62(1)	9.66(1)	2.37(2)	2.38(2)	3.58(2)	3.60(2)
Li	5.66(4)	5.66(4)	5.70(5)	5.79(5)	6.08(6)	6.37(6)	1.55(7)	1.59(7)	7.07(7)	7.84(7)
Na	5.86(4)	1.99(5)	7.63(5)	1.76(6)	1.03(7)	1.64(7)	2.24(7)	5.70(7)	1.45(8)	1.59(8)
K	2.87(5)	2.87(5)	5.13(6)	5.23(6)	9.42(7)	9.89(7)	1.76(8)	1.81(8)	1.79(9)	1.96(9)
Rb	3.65(5)	3.65(5)	7.64(6)	7.83(6)	1.63(8)	1.72(8)	2.90(8)	2.97(8)	3.53(9)	3.93(9)
Cs	6.62(5)	6.62(5)	1.66(7)	1.76(7)	4.19(8)	4.78(8)	6.67(8)	6.81(8)	1.06(10)	1.33(10)
Be	2.04(3)	2.05(3)	1.82(4)	1.87(4)	1.65(5)	1.74(5)	2.53(5)	2.67(5)	1.53(6)	1.68(6)
Mg	1.12(4)	1.12(4)	1.40(5)	1.44(5)	1.77(6)	1.89(6)	2.59(6)	2.73(6)	2.25(7)	2.53(7)
Ca	1.08(5)	1.09(5)	1.62(6)	1.92(6)	2.45(7)	3.41(7)	4.52(7)	4.71(7)	3.74(8)	6.12(8)

7.1.2 The van-der-Waals Equation of State

There are other constants also called van-der-Waals constants which are related to the van-der-Waals equation of state for N moles of a real gas

$$\left(P+\frac{aN^2}{V^2}\right)(V-bN)=NRT \ , \tag{7.1.12}$$

where P, V and T are the gas pressure, volume and temperature, respectively, a and b are van-der-Waals constants describing a deviation of a real form an ideal gas for which $a=b=0$. The constant R depends on the units employed. If P is atmospheric pressure, V is in liters and T in Kelvin, then

$$R=0.08206\left[\frac{1\cdot \mathrm{atm}}{\mathrm{mol}\cdot \mathrm{degree}}\right] \ .$$

The constants a and b are related to the potential energy of the intermolecular interaction. The constant a is a measure of attractive forces between atoms, and b is related to the repulsion of atoms at distances r smaller than the atomic radius ($r\le a_0$). The constants a and b for different molecules are listed in Table 7.4.

7.1.3 The Lennard-Jones Potential

In some cases, the interaction between atoms and small molecules is well described by the Lennard-Jones (12, 6)-potential:

Table 7.4. Van-der-Walls constants *a* and *b*, (7.1.12) for molecules [7.13]

Name	Formula	*a* l^2 atm mole^2	*b* l mole
Acetic acid	CH_3CO_2H	17.59	0.1068
Acetic anhydride	$(CH_3CO)_2O$	19.90	0.1263
Acetonitrile	CH_3CN	17.58	0.1168
Acetylene	C_2H_2	4.390	0.05136
Ammonia	NH_3	4.170	0.03707
Amyl formate	$HCO_2C_6H_{11}$	27.58	0.1730
Amylene	C_5H_{10}	15.90	0.1207
*Iso*amylene	C_5H_{10}	18.08	0.1405
Aniline	$C_6H_5NH_2$	26.50	0.1369
Argon	A	1.345	0.03219
Benzene	C_6H_6	18.00	0.1154
Benzonitrile	C_6H_5CN	33.39	0.1724
Bromobenzene	C_6H_5Br	28.56	0.1539
n-Butane	C_4H_{10}	14.47	0.1226
iso-Butane	C_4H_{10}	12.87	0.1142
iso-Butyl acetate	$CH_3CO_2C_4H_9$	28.50	0.1833
iso-Butyl alcohol	C_4H_9OH	17.03	0.1143
iso-Butyl benezene	$C_6H_5C_4H_9$	38.59	0.2144
iso-Butyl formate	$HCO_2C_4H_9$	22.54	0.1476
Butyronitrile	C_3H_7CN	25.72	0.1596
Capronitrile	$C_5H_{11}CN$	34.16	0.1984
Carbon dioxide	CO_2	3.592	0.04267
Carbon disulfide	CS_2	11.62	0.07685
Carbon monoxide	CO	1.485	0.03985
Carbon oxysulfide	COS	3.933	0.05817
Carbon tetrachloride	CCl_4	20.39	0.1383
Chlorine	Cl_2	6.493	0.05622
Chlorobenzene	C_6H_5Cl	25.43	0.1453
Chloroform	$CHCl_3$	15.17	0.1022
m-Cresol	$CH_3C_6H_4OH$	31.38	0.1607
Cyanogen	C_2N_2	7.667	0.06901
Cyclohexane	C_6H_{12}	22.81	0.1424
Cymene	$C_{10}H_{14}$	42.16	0.2336
Decane	$C_{10}H_{12}$	48.55	0.2905
Di-isobutyl	C_8H_{18}	34.97	0.2296
Diethylamine	$(C_2H_5)_2NH$	19.15	0.1392
Dimethylamine	$(CH_3)_2NH$	10.38	0.08570
Dimethylaniline	$C_6H_5N(CH_3)_2$	37.49	0.1970
Diphenyl	$(C_6H_5)_2$	52.79	0.2480
Diphenyl methane	$(C_6H_5)_2CH_2$	38.20	0.2240
Dipropylamine	$(C_2H_7)_2NH$	27.72	0.1820
Di-isopropyl	$(C_3H_7)_2$	23.13	0.1669
Durene	$C_{10}H_{14}$	45.32	0.2424
Ethane	C_2H_6	5.489	0.06380
Ethyl acetate	$CH_3CO_2C_2H_5$	20.45	0.1412
Ethyl alcohol	C_2H_5OH	12.02	0.08407
Ethylamine	$C_2H_5NH_2$	10.60	0.08409
Ethyl benzene	$C_2H_5C_6H_5$	28.60	0.1667

Table 7.4. *Continued*

Name	Formula	*a* l^2 atm mole^2	*b* l mole
Ethyl butyrate	$C_3H_7CO_2C_2H_5$	30.07	0.1919
Ethyl isobutyrate	$C_3H_7CO_2C_2H_5$	28.87	0.1994
Ethyl chloride	C_2H_5Cl	10.91	0.08651
Ethyl ether	$(C_2H_5)_2O$	17.38	0.1344
Ethyl formate	$HCO_2C_2H_5$	14.80	0.1056
Ethyl mercaptan	C_2H_5SH	11.24	0.08098
Ethyl propionate	$C_2H_5CO_2C_2H_5$	24.39	0.1615
Ethyl sulfide	$(C_2H_5)_2S$	18.75	0.1214
Ethylene	C_2H_4	4.471	0.05714
Ethylene bromide	$(CH_2Br)_2$	13.98	0.08664
Ethylene chloride	$(CH_2Cl)_2$	16.91	0.1086
Ethylidene chloride	CH_3CHCl_2	15.50	0.1073
Fluorobenzene	C_6H_5F	19.93	0.1286
Germanium tetrachloride	$GeCl_4$	22.60	0.1485
Helium	He	0.03412	0.02370
n-Heptane	C_7H_{16}	31.51	0.2065
n-Hexane	C_6H_{14}	24.39	0.1735
Hydrogen	H_2	0.2444	0.02661
Hydrogen bromide	HBr	4.451	0.04431
Hydrogen chloride	HCl	3.667	0.04081
Hydrogen selenide	H_2Se	5.268	0.04637
Hydrogen sulfide	H_2S	4.431	0.04287
Iodobenzene	C_6H_5I	33.08	0.1658
Krypton	Kr	2.318	0.03978
Mercury	Hg	8.093	0.01696
Mesitylene	$(CH_3)_3C_6H_3$	34.32	0.1979
Methane	CH_4	2.253	0.04278
Methyl acetate	$CH_3CO_2CH_3$	15.29	0.1091
Methyl alcohol	CH_3OH	9.523	0.06702
Methylamine	CH_3NH_2	7.130	0.05992
Methyl butyrate	$C_3H_7CO_2CH_2$	23.94	0.1569
Methyl isobutyrate	$C_3H_7CO_2CH_2$	24.50	0.1637
Methyl chloride	CH_3Cl	7.471	0.06483
Methyl ether	$(CH_3)_2O$	8.073	0.07246
Methyl ethyl ether	$CH_3OC_2H_5$	11.95	0.09775
Methyl ethyl sulfide	$CH_3SC_2H_5$	19.23	0.1304
Methyl fluoride	CH_3F	4.631	0.05264
Methyl formate	HCO_2CH_3	10.84	0.08068
Methyl propionate	$C_2H_5CO_2CH_3$	19.91	0.1360
Methyl sulfide	$(CH_3)_2S$	12.87	0.09213
Methyl valerate	$C_4H_9CO_2CH_3$	28.96	0.1845
Naphthalene	$C_{10}H_8$	39.74	0.1937
Neon	Ne	0.2107	0.01709
Nitric oxide	NO	1.340	0.02789
Nitrogen	N_2	1.390	0.03913
Nitrogen dioxide	NO_2	5.284	0.04424
Nitrous oxide	N_2O	3.782	0.04415

Table 7.4. *Continued*

Name	Formula	*a*	*b*
		l^2 atm mole^2	l mole
n-Octane	C_8H_{18}	37.32	0.2368
Oxygen	O_2	1.360	0.03183
n-Pentane	C_5H_{12}	19.01	0.1460
iso-Pentane	C_5H_{12}	18.05	0.1417
Phenetole	$C_6H_5OC_2H_5$	35.16	0.1963
Phosphine	PH_3	4.631	0.05156
Phosphonium chloride	PH_4Cl	4.054	0.04545
Phosphorus	P	52.94	0.1566
Propane	C_2H_3	8.664	0.08445
Propionic acid	$C_2H_5CO_2H$	20.11	0.1187
Propionitrile	C_2H_5CN	16.44	0.1064
Propyl acetate	$CH_3CO_2C_3H_7$	24.63	0.1619
Propyl alcohol	C_3H_7OH	14.92	0.1019
Propylamine	C_3H_7OH	14.99	0.1090
Propyl benzene	$C_6H_5C_2H_7$	35.85	0.2028
iso-Propyl benzene	$C_6H_5C_2H_7$	35.64	0.2025
Propyl chloride	C_3H_7Cl	15.91	0.1141
Propyl formate	$HCO_2C_3H_7$	18.95	0.1280
Propylene	C_2H_6	8.379	0.08272
Pseudo-cumene	$C_6H_3(CH_3)_3$	36.61	0.2021
Silicon fluoride	SiF	4.195	0.05571
Silicon tetrahydride	SiH_4	4.320	0.05786
Stannic chloride	$SnCl_4$	26.91	0.1642
Sulfur dioxide	SO_2	6.714	0.05636
Thiophene	C_4H_4S	20.72	0.1270
Toluene	$C_6H_5CH_3$	24.06	0.1463
Triethylamine	$(C_2H_5)_3N$	27.17	0.1831
Trimethylamine	$(CH_3)_3N$	13.02	0.1084
Xenon	Xe	4.194	0.05105
m-Xylene	$C_6H_4(CH_3)_2$	30.36	0.1772
e-Xylene	$C_6H_4(CH_3)_2$	29.98	0.1755
p-Xylene	$C_6H_4(CH_3)_2$	30.93	0.1809
Water	H_2O	5.464	0.03049

$$V(R) = 4\varepsilon\left[\left(\frac{\sigma}{R}\right)^{12} - \left(\frac{\sigma}{R}\right)^{6}\right] , \qquad (7.1.13)$$

where σ and ε are semiempirical parameters which have dimensions of length and energy, respectively.

At large distances $r \gg \sigma$, the attractive component to the sixth power is dominant, and the atoms (molecules) are attracted to one another with the force $F(R) = \mathrm{d}V/\mathrm{d}R \propto R^{-7}$, which is typical for dipole–dipole interactions between non-polar molecules. At small distances $r \ll \sigma$, the repulsive component of the 12th power plays a main role, and the molecule, are repelled from each other with the force $F(R) \propto R^{-13}$.

The potential (7.1.13) has a minimum given by

$$V_m(R) = -\varepsilon, \quad R_m = (2)^{1/6}\sigma \simeq 1.122\sigma \ . \tag{7.1.14}$$

For some atoms and molecules the parameters ε and σ are listed in Table 7.5. The forms of the Lennard-Jones potential (7.1.13) as well as the van-der-Waals ones (7.1.5) are useful representations of the interactions between atoms and molecules in their ground state.

The Lennard-Jones potential is often used as a probe potential for the calculation of the *cohesive energy* in the solid state, namely, the energy of all paires of atoms in a crystal as a function of R. The cohesive energy minimum corresponds to the internuclear equilibrium distance $R_m \simeq 1.09\sigma$ for all rare gases.

Table 7.5. Lennard-Jones (12, 6)-potential parameters [7.1]

Substance	ε[k]	σ[Å]
Light gases		
He	10.22	2.576
H_2	33.3	2.968
D_2	39.3	2.948
Noble gases		
Ne	35.7	2.789
Ar	124.0	3.418
Kr	190.0	3.61
Xe	229.0	4.055
Simple polyatomic gases		
Air	97.0	3.617
N_2	91.5	3.681
O_2	113.0	3.433
CO	110.0	3.590
CO_2	190.0	3.996
NO	119.0	3.470
N_2O	220.0	3.879
CH_4	137.0	3.882
CF_4	152.0	4.70
CCl_4	327.0	5.881
SO_2	252.0	4.290
SF_6	201.0	5.51
F_2	112.0	3.653
Cl_2	357.0	4.115
Br_2	520.0	4.268
I_2	550.0	4.982
Inorganic vapors		
HCl	360.0	3.305
HI	324.0	4.123
AsH_3	281.0	4.06
HgI_2	698.0	5.625
$HgBr_2$	530.0	5.414
$SnBr_4$	465.0	6.666
$SnCl_4$	1550.0	4.540
Hg	851.0	2.898
Hydrocarbons		
CH≡CH	185.0	4.221
CH_2=CH_2	205.0	4.232
C_2H_6	230.0	4.418
C_3H_8	254.0	5.061
n-C_4H_{10}	410.0	4.997
i-C_4H_{10}	313.0	5.341
n-C_5H_{12}	345.0	5.769
n-C_6H_{14}	413.0	5.909
n-C_7H_{16}		
n-C_8H_{18}	320.0	7.451
n-C_9H_{20}	240.0	8.448
Cyclohexane	324.0	6.093
C_6H_6	440.0	5.270
Organic vapors		
CH_3OH	507.0	3.585
C_2H_5OH	391.0	4.455
CH_3Cl	855.0	3.375
CH_2Cl_2	406.0	4.759
$CHCl_3$	327.0	5.430
C_2N_2	339.0	4.38
COS	335.0	4.13
CS_2	488.0	4.438

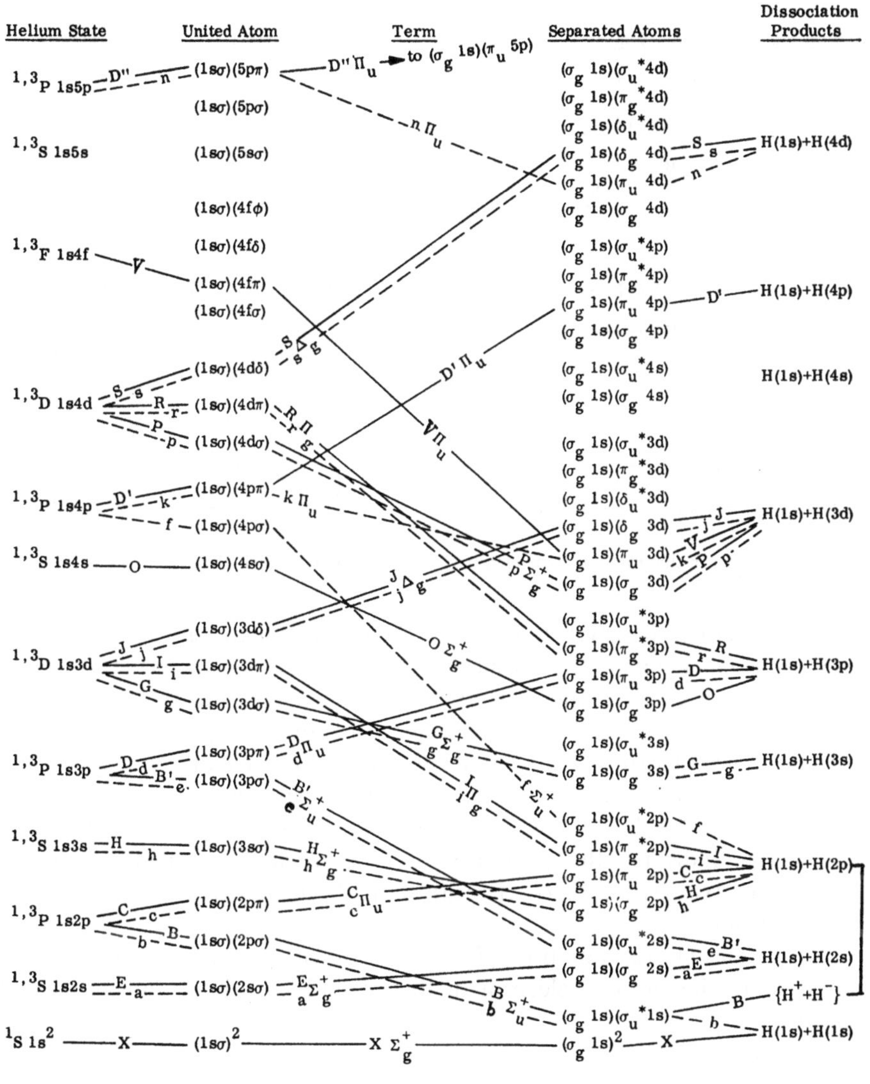

Fig. 7.1. Correlation diagram for hydrogen molecular states with one excited electron (*solid lines*: singlet states; *dashed lines*: triplet states; and *asterisks*: antibonding orbitals [7.15]

7.2 Energy Potentials of Molecular Hydrogen and Its Ions

In this section, we consider the energy potentials of H_2 and its ions, H_2^+ and H_2^- and partiallly H_3 and H_3^+. These results are most often used for molecular potentials of H_2 [7. 15] as one of the simplest but important cases .

In the H_2 molecule, the ground state is denoted by $X\ ^1\sum_g^+(1s\sigma^2)$ labelled historically by the letter X. Excited single states (as the ground state) are

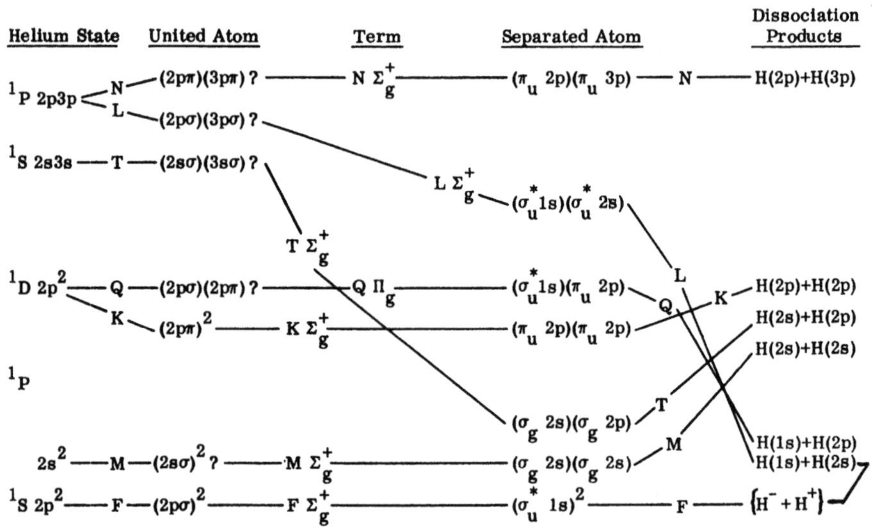

Fig. 7.2. Correlation diagram for hydrogen molecular states with both electron excited [7.15]

denoted by capital letters A, B, C, etc. in the order of increasing energy, triplet states by small letters a, b, c respectively. The hydrogen molecule has no A states. The subscripts g and u denote even and odd symmetry of the state, respectively, under inversion of the wave function through the center of a molecule. The subscripts $+$ or $-$ indicate even or odd symmetry, respectively, under reflection of the wave function at any plane passing through both nuclei. The left superscript 1, 2, 3 etc. denotes the spin multiplicity $2S+1$. A capital greek letter indicates the projection of the angular momentum on the internuclear axis. Thus, Σ, Π, Δ mean 0, 1, 2, etc. Small letter s, p, d, are used for atoms.

Sometimes Dieke's notations [7.16] are adopted as simplified notations for the H_2 states having one electron in the ground 1 s orbital and an excited electron:

Notation	
Dieke	Spectroscopic
A	$s\sigma_g\Sigma_g^+$
B	$p\sigma_u\Sigma_u^+$
C	$p\pi_u\Pi_u$
D	$d\sigma_g\Sigma_g^+$
E	$d\pi_g\Pi_g$
F	$d\delta_g\Delta_g$
G	$f\sigma_u\Sigma_u^+$
H	$f\pi_u\Pi_u$
I	$f\delta_u\Delta_u$
J	$f\phi_u\Phi_u$

Figures 7.1, 2 quantitatively represent the correlation diagrams of the electronic wave functions with internuclear distances increasing from zero (He atom) through small distances (united atom) and larger distances (separated atom) to infinity (dissociation products).

Table 7.6 lists the most important values of molecular hydrogen and its ions. Figures 7.3, 4 shows the potential-energy curves for H_2, H_2^- and H_2^+ molecules. Calculated potential energies for the ground state and the nearest excited states of the H_2 molecule performed by *Kolos* and *Wolniewicz* [7.17–19] are compiled in Tables 7.7–12. Calculated potential energies for the H_2 ion were given in [7.20, 21] and for H_2^+ in [7.22], respectively.

Table 7.6. Energy values E for H_2 [7.15]

Quantity	E[cm^{-1}]	E[eV]	E[a.u.]
Bottom of potential well of the ground state of H_2	−2179.3	−0.270(19)	−1.173 949
Zero-point level of ground the state of H_2	0	0	−1.164 019
Dissociation limit of H_2^- into $H(1s) + H^-$	30 034.5	3.723 (71)	−1.027 172
Dissociation limit of the ground state $H(1s) + H(1s) : D_0^0$	36 117.4	4.477 (87)	−0.999 456
Dissociation limit of H_2^- into $H(2s) + H^-$	112 292.7	13.922(1)	−0.652 374
Dissociation limit into $H(2s) + H(1s)$	118 375.6	14.676(3)	−0.624 660
Bottom of potential well of the ground state of H_2^+	123 275.2	15.283(8)	−0.602 336
Zero-point level of ground the state of $H_2^+(v = 0, J = 0)$	124 418.4	15.425(5)	−0.597 127
Dissociation limit into $H(3s) + H(1s)$	133 608.6	16.564(9)	−0.555 253
Dissociation limit into $H^+ + H^-$	139 712.1	17.321(6)	−0.527 441
Dissociation limit into $H(nl) + H(1s)$	$145\,795.0 - \frac{109\,677.6}{n^2}$	$18.075(8) - \frac{13.597(9)}{n^2}$	$-0.499\,728 \times (1 + n^{-2})$
Dissociation limit of the ion ground state $H(1s) + H^+$	145 795.0	18.075(8)	−0.499 728
Dissociation limit of the ion into $H(2s) + H^+$	228 053.2	28.274(3)	−0.124 932
Dissociation limit of the ion into $H(3s) + H^+$	243 286.2	30.162(9)	−0.055 253
Dissociation limit of the ion into $H(nl) + H^+$	$255\,472.6 - \frac{109\,677.6}{n^2}$	$31.673(7) - \frac{13.597(9)}{n^2}$	$-0.499\,728 \times n^{-2}$
Ionization limit of the ion $H^+ + H^+$	255 472.6	13.673(7)	0

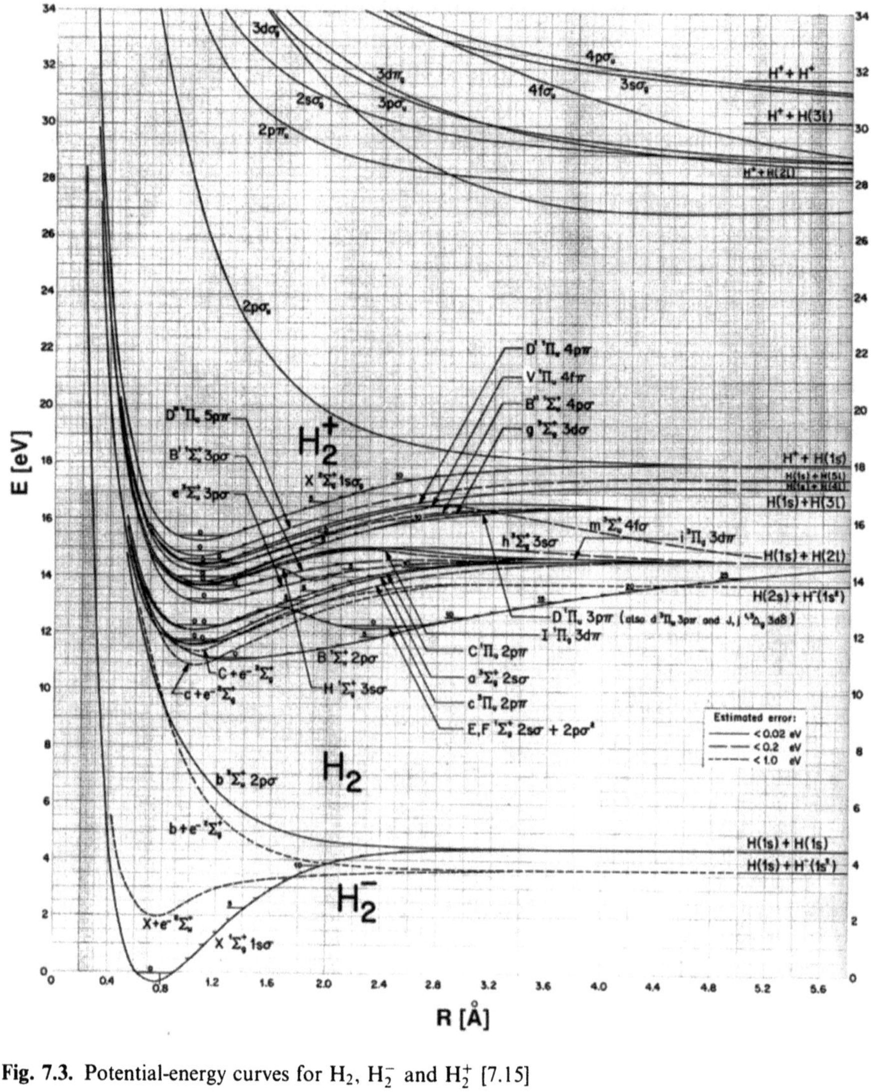

Fig. 7.3. Potential-energy curves for H_2, H_2^- and H_2^+ [7.15]

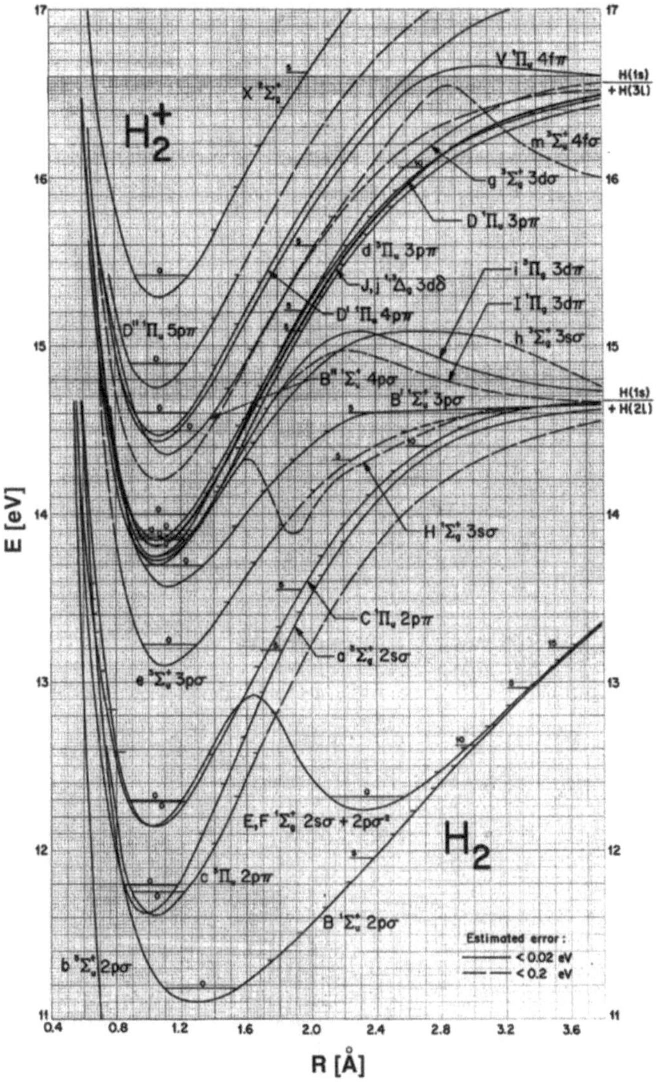

Fig. 7.4. Potential-energy curves for H_2 and H_2^+ [7.15] in expanded scale

Table 7.7. Potential energy E of the ground state $X\,^1\Sigma_g^+\ (1s\sigma)^2$ in H_2 [7.17]

R[Å]	E[eV]	E[cm^{-1}]	R[a_0]	E[a.u.]
0.2117	28.4030	229091.1	0.4000	−0.1202028
0.2381	22.1248	178452.7	0.4500	−0.3509282
0.2646	17.3439	139891.3	0.5000	−0.5266270
0.2910	13.6393	110011.2	0.5500	−0.6627707
0.3440	8.4260	67961.9	0.6500	−0.8543614
0.3704	6.5848	53111.2	0.7000	−0.9220261
0.3969	5.1070	41191.7	0.7500	−0.9763357
0.4233	3.9173	31596.2	0.8000	−1.0200556
0.4763	2.1871	17640.6	0.9000	−1.0836422
0.5292	1.0743	8664.9	1.0000	−1.1245385
0.5821	0.3799	3064.4	1.1000	−1.1500562
0.6350	−.0249	−200.9	1.2000	−1.1649342
0.6879	−.02266	−1827.6	1.3000	−1.1723459
0.7144	−.2706	−2182.5	1.3500	−1.1739627
0.7355	−.2839	−2289.8	1.3900	−1.1744517
0.7408	−.2845	−2294.8	1.4000	−1.1744744
0.7414	−.2845	−2294.8	1.4010	−1.1744746
0.7414	−.2845	−2294.8	1.4011	−1.1744746
0.7461	−.2841	−2291.6	1.4100	−1.1744599
0.7673	−.2731	−2202.9	1.4500	−1.1740558
0.7938	−.2404	−1939.1	1.5000	−1.1728537
0.8467	−.1241	−1001.1	1.6000	−1.1685799
0.8996	0.0425	342.8	1.7000	−1.1624570
0.9525	0.2436	1964.7	1.8000	−1.1550670
1.0054	0.4672	3768.2	1.9000	−1.1468496
1.0583	0.7044	5681.6	2.0000	−1.1381312
1.1113	0.9486	7651.4	2.1000	−1.1291562
1.1642	1.1944	9633.9	2.2000	−1.1201233
1.2171	1.4380	11598.4	2.3000	−1.1111725
1.2700	1.6763	13521.0	2.4000	−1.1024127
1.3229	1.9072	15383.3	2.5000	−1.0939273
1.3758	2.1289	17171.2	2.6000	−1.0857810
1.4288	2.3402	18875.3	2.7000	−1.0780164
1.4817	2.5401	20487.7	2.8000	−1.0706700
1.5346	2.7280	22003.3	2.9000	−1.0637641
1.5875	2.9036	23419.5	3.0000	−1.0573118
1.6404	3.0667	24734.8	3.1000	−1.0513185
1.6933	3.2173	25949.7	3.2000	−1.0457832
1.7463	3.3556	27065.3	3.3000	−1.0407003
1.7992	3.4819	28084.2	3.4000	−1.0360578
1.8521	3.5967	29009.8	3.5000	−1.0318402
1.9050	3.7004	29846.7	3.6000	−1.0280272
1.9579	3.7937	30599.3	3.7000	−1.0245978
2.0108	3.8772	31272.7	3.8000	−1.0215297
2.0638	3.9516	31872.5	3.9000	−1.0187967
2.1167	4.0177	32405.4	4.0000	−1.0163689
2.1696	4.0760	32876.0	4.1000	−1.0142247
2.2225	4.1274	33290.3	4.2000	−1.0123371

Table 7.7. *Continued*

R[Å]	E[eV]	E[cm^{-1}]	R[a$_0$]	E[a.u.]
2.2754	4.1724	33653.7	4.3000	−1.0106810
2.3283	4.2119	33972.1	4.4000	−1.0092303
2.3813	4.2463	34249.1	4.5000	−1.0079682
2.4342	4.2761	34490.1	4.6000	−1.0068703
2.4871	4.3020	34699.1	4.7000	−1.0059178
2.5400	4.3245	34880.3	4.8000	−1.0050923
2.5929	4.3439	35037.0	4.9000	−1.0043782
2.6458	4.3607	35172.1	5.0000	−1.0037626
2.6988	4.3752	35288.8	5.1000	−1.0032309
2.7517	4.3876	35389.1	5.2000	−1.0027740
2.8046	4.3983	35475.6	5.3000	−1.0023800
2.8575	4.4075	35549.7	5.4000	−1.0020423
2.9104	4.4154	35613.4	5.5000	−1.0017521
2.9633	4.4222	35668.1	5.6000	−1.0015030
3.0163	4.4280	35714.8	5.7000	−1.0012899
3.0692	4.4330	35755.0	5.8000	−1.0011069
3.1221	4.4372	35789.5	5.9000	−1.0009498
3.1750	4.4049	35819.1	6.0000	−1.0008150
3.2279	4.4440	35844.3	6.1000	−1.0007002
3.2808	4.4467	35865.6	6.2000	−1.0006030
3.3338	4.4490	35884.6	6.3000	−1.0005162
3.3867	4.4509	35899.9	6.4000	−1.0004466
3.4396	4.4526	35913.1	6.5000	−1.0003869
3.4925	4.4540	35924.9	6.6000	−1.0003328
3.5454	4.4552	3593.4.2	6.7000	−1.0002906
3.5983	4.4564	35943.8	6.8000	−1.0002466
3.6513	4.4572	35950.7	6.9000	−1.0002154
3.7042	4.4579	35956.5	7.000	−1.0001889
3.8100	4.4592	35996.5	7.2000	−1.0001434
3.9158	4.4601	35974.1	7.4000	−1.0001086
4.0217	4.4607	35978.9	7.6000	−1.0000968
4.1275	4.4612	35983.0	7.8000	−1.0000682
4.2333	4.4616	35986.4	8.0000	−1.0000528
4.3656	4.4620	35989.1	8.2500	−1.0000404
4.4979	4.4622	35991.0	8.5000	−1.0000314
4.7625	4.4626	35993.9	9.0000	−1.0000185
5.0271	4.4627	35995.3	9.5000	−1.0000121
5.2917	4.4628	35995.9	10.0000	−1.0000091

Table 7.8. Potential energy E of the singlet state $E, F\ {}^1\Sigma_g^+ 1s\sigma 2s\sigma + 2p\sigma^2$ in H_2 [7.18]

R[Å]	E[eV]	E [cm^{-1}]	R[a$_0$]	E[a.u]
0.5292	15.8909	128171.7	1.0000	−0.5800252
0.6615	13.5525	109310.8	1.2500	−0.6659620
0.7938	12.5461	101193.4	1.5000	−0.7029475
0.8996	12.2222	98580.6	1.7000	−0.7148524
0.9525	12.1557	98044.8	1.8000	−0.7172936
1.0054	12.1335	97865.6	1.9000	−0.7181103
1.0102	12.1333	97864.1	1.9090	−0.7181170
1.0107	12.1333	97864.1	1.9100	−0.7181170
1.0110	12.1333	97864.1	1.9105	−0.7181170
1.0112	12.1333	97864.1	1.9110	−0.7181172
1.0118	12.1333	97864.1	1.9120	−0.7181168
1.0123	12.1333	97864.2	1.9130	−0.7181166
1.0583	12.1452	97959.9	2.000	−0.7176804
1.1642	12.2394	98719.6	2.2000	−0.7142192
1.2700	12.3911	99943.1	2.4000	−0.7086446
1.3758	12.5682	101371.7	2.6000	−0.7021353
1.4817	12.7436	102786.6	2.8000	−0.6956885
1.5875	12.8788	103877.3	3.0000	−0.6907190
1.6404	12.9076	104109.6	3.1000	−0.6896605
1.6457	12.9083	104114.9	3.1100	−0.6896363
1.6510	12.9085	104116.6	3.1200	−0.6896287
1.6563	12.9083	104114.5	3.1300	−0.6896381
1.7463	12.8344	103518.7	3.3000	−0.6923527
1.8521	12.6522	102049.3	3.5000	−0.6990477
1.9844	12.4406	100342.3	3.7500	−0.7068255
2.1167	12.3035	99236.6	4.0000	−0.7118634
2.2754	12.2351	98685.3	4.3000	−0.7143755
2.3220	12.2317	98657.5	4.3880	−0.7145018
2.3225	12.2317	98657.5	4.3890	−0.7145019
2.3230	12.2317	98657.5	4.3900	−0.7145019
2.3236	12.2317	98657.5	4.3910	−0.7145019
2.3241	12.2317	98657.5	4.3920	−0.7145018
2.3246	12.2317	98657.5	4.3930	−0.7145019
2.3283	12.2317	98657.8	4.4000	−0.7145005
2.3813	12.2364	98695.2	4.5000	−0.7143302
2.1535	12.2771	99023.5	4.7500	−0.7128343
2.6458	12.3493	99606.3	5.0000	−0.7101792
2.9104	12.5498	101223.6	5.5000	−0.7028100
3.1750	12.7832	103105.9	6.0000	−0.6942337
3.4396	13.0208	105022.3	6.5000	−0.6855019
3.7042	13.2496	106867.9	7.0000	−0.6770927
3.9688	13.4620	108581.2	7.5000	−0.6692861
4.2333	13.6568	110151.9	8.0000	−0.6621297
4.4979	13.8334	111576.1	8.5000	−0.6556404
4.7625	13.9919	112855.3	9.0000	−0.6498121
5.2917	14.2570	114992.8	10.0000	−0.6400727
6.3500	14.5716	117531.0	12.0000	−0.6285082

Table 7.9. Potential energy E of the singlet state $B\ ^1\Sigma_u^+ 1s\sigma 2p\sigma$ in H_2 [7.19]

R[Å]	E[eV]	E[cm^{-1}]	R[a$_0$]	E[a.u]
0.5292	15.8563	127892.8	1.0000	−0.5812963
0.6615	13.3087	107344.7	1.2500	−0.6749201
0.7408	12.4700	100579.8	1.4000	−0.7057434
0.7938	12.0812	97443.7	1.5000	−0.7200327
0.8467	11.7898	95093.0	1.6000	−0.7307432
0.8996	11.5723	93338.8	1.7000	−0.7387356
0.9260	11.4856	92640.0	1.7500	−0.7419198
0.9525	11.4113	92040.6	1.8000	−0.7446509
1.0054	11.2938	91092.8	1.9000	−0.7489695
1.0583	11.2099	90415.8	2.0000	−0.7520541
1.1113	11.1520	89949.4	2.1000	−0.7541790
1.1800	11.1067	89583.9	2.2300	−0.7558445
1.2171	11.0934	89476.1	2.3000	−0.7563356
1.2700	11.0849	89408.0	2.4000	−0.7566457
1.2753	11.0847	89406.1	2.4100	−0.7566547
1.2806	11.0845	89404.9	2.4200	−0.7566599
1.2859	11.0845	89404.7	2.4300	−0.7566611
1.2912	11.0846	89405.1	2.4400	−0.7566589
1.2965	11.0847	89406.4	2.4500	−0.7566532
1.3018	11.0850	89408.4	2.4600	−0.7566440
1.3229	11.0868	89423.3	2.5000	−0.7565762
1.3917	11.1016	89542.8	2.6300	−0.7560317
1.4288	11.1144	89645.8	2.7000	−0.7555622
1.4817	11.1375	89832.0	2.8000	−0.7547139
1.5875	11.1977	90318.0	3.0000	−0.7524996
1.7198	11.2933	91088.8	3.2500	−0.7489874
1.8521	11.4053	91992.1	3.5000	−0.7448718
1.9844	11.5291	92990.6	3.7500	−0.7403224
2.1167	11.6612	94056.1	4.0000	−0.7354677
2.3813	11.9401	96306.1	4.5000	−0.7252160
2.6458	12.2237	98593.1	5.0000	−0.7147955
2.9104	12.5015	100833.9	5.5000	−0.7045856
3.1750	12.7668	102973.6	6.0000	−0.6948364
3.4396	13.0157	104981.4	6.5000	−0.6856884
3.7042	13.2465	106842.7	7.0000	−0.6772076
3.9688	13.4585	108552.8	7.5000	−0.6694156
4.2333	13.6520	110113.3	8.0000	−0.6623056
4.7625	13.9855	112803.6	9.0000	−0.6500475
5.2917	14.2485	114924.4	10.0000	−0.6403846
6.3500	14.5608	117443.9	12.0000	−0.6289051

Table 7.10. Potential energy E of the singlet state $C\,^1\Pi_u^+ 1s\sigma 2p\pi$ in H_2 [7.17]

R[Å]	E[eV]	E[cm^{-1}]	R[a$_0$]	E[a.u.]
0.5292	16.0959	129825.3	1.0000	−0.5724911
0.5953	14.6525	118183.0	1.1250	−0.6255371
0.6615	13.6809	110346.8	1.2500	−0.6612415
0.7276	13.0383	105163.2	1.3750	−0.6848598
0.7938	12.6137	101738.5	1.5000	−0.7004638
0.8599	12.3514	99622.9	1.6250	−0.7101032
0.9260	12.2030	98426.1	1.7500	−0.7155565
0.9790	12.1447	97955.9	1.8500	−0.7176989
1.0054	12.1314	97848.7	1.9000	−0.7181873
1.0319	12.1270	97813.2	1.9500	−0.7183489
1.0324	12.1270	97813.1	1.9510	−0.7183492
1.0329	12.1270	97813.1	1.9520	−0.7183492
1.0335	12.1270	97813.2	1.9530	−0.7183491
1.0340	12.1270	97813.2	1.9540	−0.7183490
1.0372	12.1271	97813.8	1.9600	−0.7183460
1.0425	12.1275	97817.1	1.9700	−0.7183310
1.0583	12.1304	97840.8	2.0000	−0.7182230
1.1113	12.1571	98055.7	2.1000	−0.1772439
1.1906	12.2353	98686.4	2.2500	−0.7143703
1.3229	12.4310	100264.9	2.5000	−0.7071783
1.4552	12.6683	102179.1	2.7500	−0.6984564
1.5875	12.9192	104202.5	3.0000	−0.6892372
1.8521	13.4005	108084.7	3.5000	−0.6715486
2.1167	13.8070	111363.2	4.0000	−0.6566106
2.3813	14.1193	113882.5	4.5000	−0.6451316
2.6458	14.3409	115669.7	5.0000	−0.6369889
2.9104	14.4866	116844.8	5.5000	−0.6316345
3.1750	14.5762	117567.4	6.0000	−0.6283423
3.4396	14.6292	117995.3	6.5000	−0.6263925
3.7042	14.6562	118212.9	7.0000	−0.6254011
3.9688	14.6701	118325.4	7.5000	−0.6248886
4.1010	14.6741	118357.3	7.7500	−0.6247433
4.2333	14.6769	118379.6	8.0000	−0.6246415
4.3656	14.6786	118393.6	8.2500	−0.6245775
4.4979	14.6796	118401.9	8.5000	−0.6245398
4.6302	14.6801	118405.4	8.7500	−0.6245240
4.7625	14.6802	118406.4	9.0000	−0.6245195
4.8948	14.6801	118405.8	9.2500	−0.6245223
5.0271	14.6798	118403.4	9.5000	−0.6245331
5.2917	14.6789	118396.2	10.0000	−0.6245660

Table 7.11. Potential energy E of the triplet state $a\,{}^3\Sigma_g^+ 1s\sigma 2s\sigma$ in H_2 [7.19]

R[Å]	E[eV]	E[cm^{-1}]	R[a$_0$]	E[a.u]
0.5292	15.2409	122928.9	1.0000	−0.6039132
0.5821	14.0881	113630.8	1.1000	−0.6462787
0.6350	13.2609	106958.8	1.2000	−0.6766786
0.6879	12.6705	102197.1	1.3000	−0.6983742
0.7408	12.2553	98847.6	1.4000	−0.7136358
0.7938	11.9710	96554.9	1.5000	−0.7240820
0.8467	11.7859	95061.5	1.6000	−0.7308864
0.8996	11.6762	94177.5	1.7000	−0.7349145
0.9525	11.6245	93760.0	1.8000	−0.7368166
0.9790	11.6160	93691.4	1.8500	−0.7371293
0.9843	11.6155	93687.4	1.8600	−0.7371476
0.9869	11.6154	93686.5	1.8650	−0.7371515
0.9880	11.6154	93686.4	1.8670	−0.7371521
0.9885	11.6154	93686.3	1.8680	−0.7371524
0.9887	11.6154	93686.4	1.8685	−0.7371522
0.9890	11.6154	93686.3	1.8690	−0.7371523
0.9895	11.6154	93686.4	1.8900	−0.7371520
0.9922	11.6154	93687.0	1.8750	−0.7371492
0.9948	11.6156	93688.4	1.8800	−0.7371430
1.0001	11.6162	93693.2	1.8900	−0.7371209
1.0054	11.6172	93700.9	1.9000	−0.7370861
1.0583	11.6439	93917.0	2.0000	−0.7361014
1.1113	11.6969	94344.1	2.1000	−0.7341555
1.1642	11.7697	94931.6	2.2000	−0.7314786
1.2171	11.8576	95639.9	2.3000	−0.7282511
1.2700	11.9565	96438.2	2.4000	−0.7246140
1.3229	12.0716	97366.6	2.5000	−0.7203839
1.3758	12.1759	98207.9	2.6000	−0.7165507
1.4817	12.4098	100094.2	2.8000	−0.7079559
1.5875	12.6463	102001.9	3.0000	−0.6992638
1.8521	13.2044	106503.3	3.5000	−0.6787537
2.1167	13.6723	110277.1	4.0000	−0.6615590
2.3813	14.0296	113158.8	4.5000	−0.6484291
2.6458	14.2777	115160.3	5.0000	−0.6393097
2.9104	14.4340	116421.0	5.5000	−0.6335655
3.1750	14.5247	117152.7	6.0000	−0.6302316
3.4396	14.5753	117560.7	6.5000	−0.6283729
3.7042	14.6036	117789.1	7.0000	−0.6273322
3.9688	14.6201	117921.9	7.5000	−0.6267268
4.2333	14.6303	118003.8	8.0000	−0.6263539
4.7625	14.6413	118092.5	9.0000	−0.6259496
5.2917	14.6467	118136.1	10.0000	−0.6257510

Table 7.12. Potential energy E of the triplet state $b\ {}^3\sum_u^+ 1s\sigma 2p\sigma$ in H_2 [7.17]

R[Å]	E[eV]	E[cm−1]	R[a0]	E[a.u]	R[Å]	E[eV]	E[cm−1]	R[a0]	E[a.u]
0.5292	14.7617	119064.1	1.0000	−0.6215227	2.7517	4.4882	36200.7	5.2000	−0.9990763
0.5821	13.2868	107167.7	1.1000	−0.6757268	2.8046	4.4840	36167.1	5.3000	−0.9992292
0.6350	12.1103	97678.2	1.2000	−0.7189640	2.8575	4.4805	36138.5	5.4000	−0.9993597
0.6879	11.1459	89900.2	1.3000	−0.7544033	2.9104	4.4775	36114.2	5.5000	−0.9994703
0.7408	10.3365	83371.5	1.4000	−0.7841501	2.9633	4.4749	36093.7	5.6000	−0.9995635
0.7938	9.6437	77783.8	1.5000	−0.8096095	3.0163	4.4728	36076.2	5.7000	−0.9996432
0.8467	9.0420	72930.3	1.6000	−0.8317238	3.0692	4.4710	36061.7	5.8000	−0.9997094
0.8996	8.5138	68670.1	1.7000	−0.8511347	3.1221	4.4695	36049.4	5.9000	−0.9997655
0.9525	8.0470	64904.7	1.8000	−0.8682913	3.1750	4.4682	36039.1	6.0000	−0.9998125
1.0054	7.6326	61562.7	1.9000	−0.8835186	3.2279	4.4671	36030.5	6.1000	−0.9998515
1.0583	7.2640	58589.9	2.0000	−0.8970636	3.2808	4.4662	36023.4	6.2000	−0.9998842
1.1113	6.9359	55943.1	2.1000	−0.9091230	3.3338	4.4655	36017.4	6.3000	−0.9999112
1.1642	6.6438	53586.8	2.2000	−0.9198593	3.3867	4.4649	36012.5	6.4000	−0.9999337
1.2171	6.3838	51490.1	2.3000	−0.9294123	3.4396	4.4644	36008.5	6.5000	−0.9999521
1.2700	6.1527	49626.2	2.4000	−0.9379051	3.4925	4.4640	36005.2	6.6000	−0.9999670
1.3229	5.9475	47971.1	2.5000	−0.9454463	3.5454	4.4636	36002.4	6.7000	−0.9999795
1.3758	5.7655	46503.2	2.6000	−0.9521346	3.5983	4.4634	36000.4	6.8000	−0.9999887
1.4288	5.6043	45203.0	2.7000	−0.9580585	3.6513	4.4632	35998.7	6.9000	−0.9999966
1.4817	5.4617	44053.0	2.8000	−0.9632985	3.7042	4.4630	35997.3	7.0000	−1.0000030
1.5346	5.3360	43038.4	2.9000	−0.9679211	3.7571	4.4629	35996.3	7.1000	−1.0000076
1.5875	5.2247	42140.9	3.0000	−0.9720104	3.8100	4.4628	35995.4	7.2000	−1.0000117
1.6404	5.1268	41351.6	3.1000	−0.9756071	3.8629	4.4627	35994.8	7.3000	−1.0000143
1.6933	5.0407	40657.0	3.2000	−0.9787717	3.9158	4.4626	35994.3	7.4000	−1.0000167
1.7463	4.9651	40046.9	3.3000	−0.9815517	3.9688	4.4626	35994.0	7.5000	−1.0000180
1.7992	4.8987	39511.5	3.4000	−0.9839910	4.0217	4.4626	35993.7	7.6000	−1.0000191
1.8521	4.8405	39042.5	3.5000	−0.9861279	4.0746	4.4625	35993.7	7.7000	−1.0000195
1.9050	4.7897	38632.1	3.6000	−0.9879977	4.1275	4.4625	35993.6	7.8000	−1.0000197
1.9579	4.7452	38273.6	3.7000	−0.9896312	4.1804	4.4625	38993.7	7.9000	−1.0000194
2.0108	4.7064	37960.8	3.8000	−0.9910564	4.2333	4.4625	35993.6	8.0000	−1.0000196
2.0638	4.6727	37688.4	3.9000	−0.9922979	4.2863	4.4625	35993.7	8.1000	−1.0000192

Table 7.12. *Continued*

R[Å]	E[eV]	E[cm−1]	R[a0]	E[a.u]	R[Å]	E[eV]	E[cm−1]	R[a0]	E[a.u]
2.1167	4.6433	37451.3	4.0000	−0.9933781	4.3392	4.4626	35993.7	8.2000	−1.0000191
2.1696	4.6177	37245.3	4.1000	−0.9943164	4.3921	4.4626	35994.0	8.3000	−1.0000180
2.2225	4.5956	37066.7	4.2000	−0.9951304	4.4450	4.4626	35994.1	8.4000	−1.0000173
2.2754	4.5764	36911.9	4.3000	−0.9958355	4.4979	4.4626	35994.3	8.5000	−1.0000164
2.3283	4.5598	36778.0	4.4000	−0.9964456	4.5508	4.4626	35994.5	8.6000	−1.0000158
2.3813	4.5455	36662.6	4.5000	−0.9969715	4.6038	4.4627	35994.6	8.7000	−1.0000152
2.4342	4.5331	36563.0	4.6000	−0.9974252	4.6567	4.4627	35994.8	8.8000	−1.0000143
2.4871	4.5225	36477.3	4.7000	−0.9978159	4.7625	4.4627	35995.2	9.0000	−1.0000127
2.5400	4.5134	36404.0	4.8000	−0.9981500	4.8948	4.4628	35995.5	9.2500	−1.0000109
2.5929	4.5056	36340.7	4.9000	−0.9984382	5.0271	4.4628	35995.9	9.5000	−1.0000095
2.6458	4.4989	36286.6	5.0000	−0.9986849	5.1594	4.4629	35996.3	9.7500	−1.0000076
2.6988	4.4931	36240.2	5.1000	−0.9988960	5.2917	4.4629	35996.5	10.0000	−1.0000067

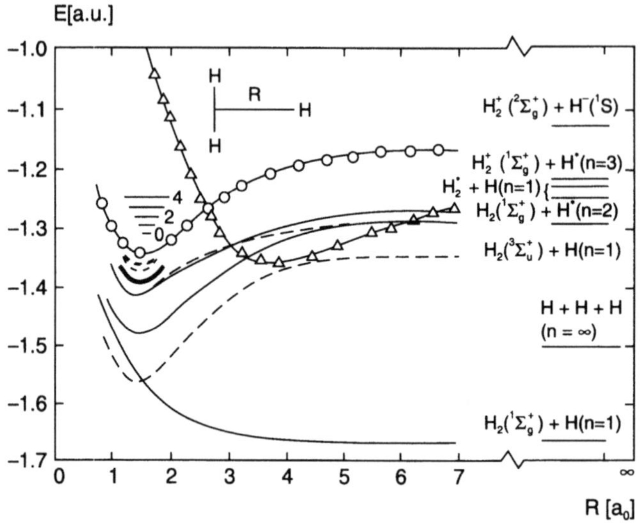

Fig. 7.5. Potential-energy curves for H_3 and H_3^+ [7.23] (*solid curves*: $^2A_1$ states; *dashed curves*: $^2B_2$ states; Δ resonance state of H_3; ooo: $^1A_1$ state of H_3^+

Figure 7.5 exhibits the molecular terms of the H_2 and H_2^+ ion [7.23]. Collisions of electrons with hydrogen species H_2^+, D_2^+, HD^+, H_3^+, H_2D^+ and D_3^+ were considered in [7.24].

References

Chapter 1

1.1 B.H. Bransden, C.J. Joachain: *Physics of Atoms and Molecules* (Longman, London 1981)
1.2 G. Herzberg: *Molecular Spectra and Molecular Structure*, Vol. III *Electronic Spectra and Electronic Structure of Polyatomic Molecules* (Van Nostrand, Princeton 1966)
1.3 J.D. Graibeal: *Molecular Spectroscopy* (McGraw-Hill, New York 1988)
1.4 R. McWeeny: *Methods of Molecular Quantum Mechanics* (2nd edn.) (Univ. of Pisa, Pisa 1992)
1.5 R. Doudel, R. Lefebvre, C. Moser: *Quantum Chemistry* (Wiley Interscience, New York 1959)
1.6 J. Goodisman: *Diatomic Interaction Potential Theory*, Vols. 1, 2 (Academic, New York 1973)
1.7 H.H. Michels: Adv. Chem. Phys. **45**, 225 (1981)
1.8 P.J. Bruna, S.D. Peyerimhoff: Excited-State Potentials, in: *Ab Initio Methods in Quantum Chemistry*, Vol. 1, ed by K.P.Lawleyed (Wiley, Chichester 1987), pp. 1–97
1.9 W. Miller, Jr.: *Symmetry Groups and Their Applications* (Academic, New York 1972)
1.10 R.G. Wooley: Mol. Phys. **30**, 649 (1975)
1.11 E.A. Gastilovich: Uspekhi Fizich. Nauk. **181**, 83 (1991) (in Russian)
1.12 H.W. Kroto: *Molecular Rotation Spectra* (Wiley, London 1975)
1.13 M. Mizushima: *The Theory of Rotating Diatomic Molecules* (Wiley, New York 1975)
1.14 B.R. Judd: *Angular Momentum Theory for Diatomic Molecules* (Academic, New York 1975)
1.15 N.F. Stepanov, B.I. Zhilinskii: J. Mol. Spectra **52**, 277 (1974)
1.16 W. Heitler, F. London: Z. Phys. **44**, 455 (1927)
1.17 L. Pauling: *The Nature of Chemical Bond* (Cornell Univ. Press, Ithaka, NY 1960)
1.18 K. Fukui: *Molecular Orbitals in Physics, Chemistry and Biology* (Academic, New York 1964)
1.19 C.C.J. Roothaan: Rev. Mod. Phys. **23**, 69 (1951)
1.20 A.D. Walsh: Nature (London) **159**, 167, 712 (1947)
1.21 R.B. Woodward, R. Hoffmann: J. Am. Chem. Soc. **87**, 395 (1965)
1.22 H.F. Schaefer III (ed.): *Applications of Electronic Structure Theory* (Plenum, New York 1977)
1.23 D.R. Lide, G.W.A. Milne (eds.): *Handbook of Data on Organic Compounds*, 3rd edn. Vol. 1–7 (CRC, Boca Raton 1994)
1.24 K.P. Huber, G. Herzberg: *Molecular Spectra and Molecular Structure*, Vol. 4, *Constants of Diatomic Molecules* (Van Noostrand-Reinhold, New York 1979)
1.25 W. Demtröder: *Laser Spectroscopy*, 2nd edn. (Springer, Berlin, Heidelberg 1996)
1.26 E. Hirota: *High-Resolution Spectroscopy of Transient Molecules* (Springer, Berlin, Heidelberg 1985)
1.27 R.D. Levine, R.B. Bernstein: *Molecular Reaction Dynamics and Chemical Reactivity* (Oxford Univ. Press, Oxford 1987)
1.28 C.E. Dykstra: *Ab Initio Calculation of the Structures and Properties of Molecules* (Elsevier, Amsterdam 1988)
1.29 W.J. Orville-Thomas (ed.): *Internal Rotation in Molecules* (Wiley, London 1974)
1.30 Y. Ohshima, S. Yamamoto, K. Kuchitsu: Acta. Chem. Scand. A **42**, 307 (1988)

1.31 N.D. Sokolov (ed.): *Hydrogen Bond* (Nauka, Moscow 1981) (in Russian)
1.32 H.B.G. Kasimir: *On the Interaction between Atomic Nuclei and Electrons* (Freeman, San Francisco 1963)
1.33 W.H. Flygare: Chem. Revs. **74**, 653 (1974)
1.34 W.G. Richards: *Spin-Orbit Coupling in Molecules* (Clarendon, Oxford 1981)

Chapter 2

2.1 G. Herzberg: *Molecular Spectra and Molecular Structure*, Vol. 1, *Spectra of Diatomic Molecules*, 2nd edn. (Van Nostrand, Toronto, London 1950)
2.2 K.P. Huber, G. Herzberg: *Moleular Spectra and Molecular Structure*, Vol. 4, *Constants of Diatomic Molecules* (Van Nostrand-Reinhold, New York 1979)
2.3 K.S. Krasnov (ed.): *Molecular Constants of Inorganic Compounds* (Chimia, Moscow 1979) pp. 10–72 (in Russian)
2.4 V.P. Glushko (ed.) *Thermodynamical Properties of Individual Substances, Reference Edition*, Vols. I-IV, 3rd. edn. (Nauka, Moscow 1978, 1979, 1981, 1983) (in Russian)
2.5 B. Starck: Constants of Diamagnetic Molecules. Diatomic Molecules in *Molecular Constants*, Landolt-Börnstein, New Series, Group II, Vol. 4, ed. by K.-H. Hellwege (Springer, Berlin, Heidelberg 1967) pp. 4–14
2.6 R. Tischer: Constants of Diamagnetic Molecules. Diatomic Molecules, in *Molecular Constants*, Landolt-Börnstein, New Series, Group II , Vol. 6, ed. by K.-H. Hellwege (Springer, Berlin, Heidelberg 1974) pp. 2-1–2-37
2.7 R. Tischer: Constants of Radicals. Diatomic Radicals in *Molecular Constants*, Landolt-Börnstein, New Series, Group II, Vol. 6, ed. by K.-H. Hellwege (Springer, Berlin, Heidelberg 1974) pp. 5-1–5-90
2.8 E. Tiemann: Constants of Diamagnetic Molecules. Diatomic Molecules, in *Molecular Constants*, Landolt-Börnstein, New Series, Group II, Vol. 14a, ed. by K.-H. Hellwege (Springer, Berlin, Heidelberg 1982) pp. 5–36
2.9 E. Tiemann: Constants of Radicals. Diatomic Radicals, in *Molecular Constants*, Landolt-Börnstein, New Series, Group II, Vol. 14b, ed. by K.-H. Hellwege (Springer, Berlin, Heidelberg 1983) pp. 24–170
2.10 A.A. Radzig: Spectra of Diatomic Molecules, in *Physical Quantities, Reference Edition*, ed. by I.S. Grigor'ev, E.Z. Mejlichov (Energoatomizdat, Moscow 1991) pp. 849–859 (in Russian)
2.11. E. Tiemann: Constants of Diamagnetic Molecules. Diatomic Molecules, in *Molecular Constants*, Landolt-Börnstein, New Series, Group II, Vol. 19a, ed. by O. Madelung (Springer, Berlin, Heidelberg 1992) pp. 5–40
2.12 R.S. Mulliken: J. Chem. Phys. **23**, 1997 (1955)
2.13 G. Herzbeg, L. Herzberg: Constants of Polyatomic Molecules, in *American Institute of Physics Handbook*, ed. by D.E. Gray, 3rd. edn. (McGraw-Hill, New York 1972) pp. 7-185–7-199
2.14 K.S. Krasnov (ed.): *Molecular Constants of Inorganic Compounds* (Chimia, Moscow 1979) pp. 73–443 (in Russian)
2.15 D.R. Lide (ed.): Fundamental Vibrational Frequencies of Small Molecules in *CRC Handbook of Chemistry and Physics*, 76th edn. (CRC Press, Boca Raton 1995–1996) pp. 9-75–9-78
2.16 K.Kuchitsu (ed.): Structure Data of Free Polyatomic Molecules, Landolt-Börnstein, New Series, Group II, Vol. 21, ed. by O. Madelung (Springer, Berlin, Heidelberg 1992)
2.17 K.-H. Hellwege, A.M. Hellwege (eds.): Structure Data of Free Polyatomic Molecules, Landolt-Börnstein, New Series, Group II, Vol. 7, (Springer, Berlin, Heidelberg 1976)
2.18 K.-H. Hellwege, A.M. Hellwege (eds.): Structure Data of Free Polyatomic Molecules, Landolt-Börnstein, New Series, Group II, Vol. 15 (Springer, Berlin, Heidelberg 1987)

2.19 J. Demaison: Constants of Diamagnetic Molecules in *Molecular Constants*, Landolt-Börnstein, New Series, Group II, Vol. 14a, ed. by K.-H. Hellwege (Springer, Berlin, Heidelberg 1982) pp. 37–583
2.20 J.M. Brown: Constants of Radicals, in *Molecular Constants*, Landolt-Börnstein, New Series, Group II, Vol. 14b, ed. by K.-H. Hellwege (Springer, Berlin, Heidelberg 1983) pp. 171–290
2.21 J. Demaison, G. Wlodarczak: Constants of Diamagnetic Molecules in *Molecular Constants*, Landolt-Börnstein, New Series, Group II, Vol. 19a, ed. by O. Madelung (Springer, Berlin, Heidelberg 1992) pp. 41–143
2.22 J. Demaison, J. Vogt, G. Wlodarczak: Constants of Diamagnetic Molecules, in *Molecular Constants*, Landolt-Börnstein, New Series, Group II, Vol. 19b, ed. by O. Madelung (Springer, Berlin, Heidelberg 1992) pp. 5–488
2.23 D.R. Lide (ed.): Bond Length and Angles in Gas-Phase Molecules, in *CRC Handbook of Chemistry and Physics*, 76th edn. (CRC, Boca Raton, 1995, 1996) pp. 9-15–9-41

Chapter 3

3.1 L.V. Gurvich, G.V. Karachevtsev, V.N. Kondratjev, Y.A. Lebedev, V.A. Medvedev, V.K. Potapov, Y.S. Hodeev: *Chemical Bond Dissociation Energies, Ionization Potentials and Electron Affinity* (Nauka, Moscow 1974) (in Russian)
3.2 K.P. Huber, G. Herzberg: *Molecular Spectra and Molecular Structure*, Vol. 4, *Constants of Diatomic Molecules* (Van Nostrand-Reinhold, New York 1979)
3.3 K.S. Krasnov (ed.): *Molecular Constants of Inorganic Compounds* (Chimia, Moscow 1979) (in Russian)
3.4 V.P. Glushko (ed.): *Thermodynamical Properties of Individual Substances, Reference Edition*, Vols. I-IV, 3rd. edn. (Nauka, Moscow 1978, 1979, 1981, 1983) (in Russian)
3.5 J.A. Kerr: Strength of Chemical Bonds, in *CRC Handbook of Chemistry and Physics*, ed. by D.R. Lide, 76th edn. (CRC, Boca Raton 1995, 1996) pp. 9-51–9-69
3.6 S.G. Lias, J.E. Bartmess, J.F. Liebman, J.L. Holmes, R.D. Levine, W.G. Mallard: Gas-Phase Ion and Neutral Thermochemistry, J. Phys. Chem. Ref. Data **17**, Suppl. 1 (1988)
3.7 S.G. Lias: Ionization Potentials of Gas-Phase Molecules, in *CRC Handbook of Chemistry and Physics*, ed. by D.R. Lide, 76th edn. (CRC, Boca Raton, 1995, 1996) pp. 10-210–10-228
3.8 E.A. Hill: J. Amer. Chem. Soc. **22**, 478 (1900)
3.9 A.A. Christodoulides, D.L. McCorkee, L.G. Christophorou: Electron Affinities of Atoms, Molecules and Radicals, in *Electron–Molecule Interaction and Their Applications*, ed. by L.G. Christophorou, Vol. 3 (Academic, San Diego 1984) pp. 424–641
3.10 H. Hotop, W.C. Lineberger: J. Phys. Chem. Ref. Data **14**, 731–750 (1985)
3.11 A.A. Radzig, V.M. Shustrjakov: Electron Affinity Energy of Atoms and Molecules, in *Physical Quantities, Reference Edition*. ed. by I.S. Grigor'ev, E.Z. Mejlichov (Energoatomizdat, Moscow 1991) pp. 420–422 (in Russian)
3.12 T.M. Miller: Electron Affinities, in *CRC Handbook of Chemistry and Physics*, ed. by D.R. Lide, 76th edn. (CRC, Boca Raton, 1995, 1996) pp. 10-180–10-191
3.13 S.G. Lias, J.F. Liebman, R.D. Levin: J. Phys. Chem. Ref. Data **13**, 695–805 (1984)
3.14 A.A. Radzig, V.M. Shustrjakov: Proton Affinity Energy of Atoms and Molecules, in *Physical Quantities, Reference Edition*. ed. by I.S. Grigor'ev, E.Z. Mejlichov (Energoatomizdat, Moscow 1991) pp. 420–422 (in Russian)

Chapter 4

4.1 A.L. McClellan: *Tables of Experimental Dipole Moments*, Vol. 2 (Rahara Enterpr., El Cerrito 1974)
4.2 R. Tischer, J. Demaison, B. Starck: Constants of Diamagnetic Molecules, Dipole Moments, in *Molecular Constants*, Landolt-Börnstein, New Series, Group II, Vol. 6, ed. by K.-H. Hellwege (Springer, Berlin, Heidelberg 1974) pp. 2-260–2-304
4.3 E. Tiemann, J. Demaison: Constants of Diamagnetic Molecules, Dipole Moments, in *Molecular Constants*, Landolt-Börnstein, New Series, Group II, Vol. 14a, ed. by K.-H. Hellwege (Springer, Berlin, Heidelberg 1982) pp. 584–643
4.4 E. Tiemann, J.M. Brown: Constants of Radicals, in *Molecular Constants*, Landolt-Börnstein, New Series, Group II, Vol. 14b, ed. by K.-H. Hellwege (Springer, Berlin, Heidelberg 1983) pp. 24–290
4.5 E. Tiemann, G. Wlodarczak, J. Demaison, J. Vogt: Constants of Diamagnetic Molecules, Dipole Moments, in *Molecular Constants*, Landolt-Börnstein, New Series, Group II, Vol. 19c, ed. by O. Madelung (Springer, Berlin, Heidelberg 1992) pp. 5–98
4.6 D.R. Lide: Dipole Moments of Molecules in the Gas Phase, in *CRC Handbook of Chemistry and Physics*, ed. by D.R. Lide, 76th edn. (CRC, Boca Raton, 1995-1996) pp. 9-42–9-50
4.7 A.D. Buckingham: Quart. Rev. Chem. Soc. (London) **13**, 183 (1959)
4.8 S. Kielich: Physica **31**, 444 (1965)
4.9 D.E. Stogryn, A.P. Stogryn: Molec. Phys. **11**, 371 (1966)
4.10 A.D. Buckingham: Phys. Chem. **4**, 349 (1970)
4.11 S. Kielich: *Molecular Nonlinear Optics* (PWN, Warsw, 1977) (in Polish)
4.12 R. Tischer, W. Huttner: Magnetic Constants, in *Molecular Constants*, Landolt-Börnstein, New Series, Group II, Vol. 6, ed. by K.-H. Hellwege (Springer, Berlin, Heidelberg 1974) pp. 2-383–2-448
4.13 E. Tiemann, W. Huttner: Magnetic Constants, in *Molecular Constants*, Landolt-Börnstein, New Series, Group II, Vol. 14a, ed. by K.-H. Hellwege (Springer, Berlin, Heidelberg 1982) pp. 745–785
4.14 E. Tiemann, W. Huttner: Magnetic Constants, in *Molecular Constants*, Landolt-Börnstein, New Series, Group II, Vol. 19c, ed. by O. Madelung (Springer, Berlin, Heidelberg 1992) pp. 256–295
4.15 J.O. Hirschfelder, C.F. Curtiss, R.B. Bird: *Molecular Theory of Gases and Liquids* (Wiley, New York 1964)
4.16 A.N. Vereshchagin: *Polarizability of Molecules* (Nauka, Moscow 1980) (in Russian)
4.17 T.M. Miller: Atomic and Molecular Polarizabilities, in *CRC Handbook of Chemistry and Physics*, ed. by D.R. Lide, 76th edn. (CRC, Boca Raton, 1995, 1996) pp. 10-192–10-206

Chapter 5

5.1 G. Herzberg: *Molecular Spectra and Molecular Structure*, Vol. 1, *Spectra of Diatomic Molecules*, 2nd edn. (Van Nostrand-Reinhold, New York 1950)
5.2 G. Herzberg: *Molecular Spectra and Molecular Structure*, Vol. 2, *Infrared and Raman Spectra of Polyatomic Molecules* (Van Nostrand-Reinhold, New York 1945)
5.3 G. Herzberg: *Molecular Spectra and Molecular Structure*, Vol. 3, *Electronic Spectra and Electronic Structure of Polyatomic Molecules* (Van Nostrand-Reinhold, New York 1966)
5.4 R.F. Barrow, D.A. Long, D.J. Millen (eds): *Molecular Spectroscopy*, Vols. 1–7 (The Chemical Society, London 1973-1979)
5.5 W. Gordy, R.L. Cook: *Microwave Molecular Spectra*. 3rd edn. (Wiley, New York 1984)

5.6 L.A. Kuznetzova, N.E. Kuz'menko, Yu.Ya. Kuzyakov, Yu.A. Plastinin: *Probabilities of Optical Transitions in Diatomic Molecules* (Nauka, Moscow 1980) (in Russian)
5.7 N.E. Kuz'menko, L.A. Kuznetsova, Yu.Ya. Kuzyakov: *Franck-Condon Factors for Diatomic Molecules* (Moscow State Univ., Moscow 1984) (in Russian)
5.8 A.M. Pravilov: *Photoprocesses in Molecular Gases* (Energoatomizdat, Moscow 1992) (in Russian)
5.9 M. Roche, H.H. Jaffe: Chem. Soc. Rev. **5**, 165 (1976)
5.10 W.B. Person, G. Zerbi (eds.): *Vibrational Intensities* (Elsevier, Amsterdam 1980)
5.11 I. Kovács: *Rotational Structure in the Spectra of Diatomic Molecules* (Akademia Kiadó, Budapest 1969)
5.12 H.W. Kroto: *Molecular Rotation Spectra* (Wiley, London 1975)
5.13 M. Mizushima: *The Theory of Rotating Diatomic Molecules* (Wiley, New York 1975)
5.14 B.R. Judd: *Angular Momentum Theory for Diatomic Molecules* (Academic, New York 1975)
5.15 W.J. Orville-Thomas (ed.): *Internal Rotation in Molecules* (Wiley, London 1974)
5.16 E.F.H. Brittain, W.O. George, C.H.J. Wells: *Introduction to Molecular Spectroscopy* (Academic, London 1970)
5.17 L. Wallace: Astrophys. J. Suppl. **6**, 445 (1962)
5.18 L. Wallace: Astrophys. J. Suppl. **7**, 165 (1962)
5.19 A. Lofthus, P.H. Krupenie: J. Phys. Chem. Ref. Data. **6**, 113 (1977)
5.20 S.N. Suchard, J.E. Melzer: *Spectroscopic Data*, Vol, 2. Homonuclear Diatomic Molecules (IFI/Plenum, New York 1976)
5.21 Ya.B. Zeldovich, Yu.P. Reiser: *Physics of Impact Waves and High Temperature Hydrodynamic Events*, (Nauka, Moscow 1966) 2nd edn. (in Russian)
5.22 K. Yoshino, D.E. Freeman, J.R. Esmond, W.H. Parkinson: Planet. Space Sci. **31**, 339 (1983)
5.23 A.S.C. Cheung, K.Yoshino, W.H. Parkinson, D.E. Freeman: Cdn. J. Phys. **62**, 1752 (1984)
5.24 B.R. Lewis, L. Berzins, J.H. Carver: J. Quant. Spectrosc. Radiat. Transfer **36**, 209 (1986)
5.25 K. Yoshino, D.E. Freeman, J.R. Esmond, W.H. Parkinson: Planet. Space Sci. **35**, 1067 (1987)
5.26 R.D. Hudson, W.L. Carter: J. Opt. Soc. Am. **58**, 1621 (1968)
5.27 B.R. Lewis, L. Berzins, J.H. Carver, S.T. Gibson: J. Quant. Spectrosc. Radiat. Transfer **36**, 187 (1986)
5.28 V. Degen, R.W. Nicholls: J. Phys. B **2**, 1240 (1969)
5.29 V. Hasson, R.W. Nicholls, V. Degen: J. Phys. B **3**, 1192 (1970)
5.30 V. Hasson, R.W. Nicholls: J. Phys. B **4**, 1778 (1971)
5.31 P.H. Krupenie: J. Phys. Chem. Ref. Data. **1**, 423 (1972)
5.32 E.W. Schlag, S. Schneider, S.F. Fischer: Ann. Rev. Phys. Chem. **22**, 465 (1971)
5.33 R. Anderson: Atomic Data **3**, 227 (1971)
5.34 L.A. Kuznetsova; Spectrosc. Lett. **20**, 665 (1987)
5.35 T.A. Carlson: *Photoelectron and Auger Spectroscopy* (Plenum, New York 1975)
5.36 C.R. Brundle, A.D. Baker (eds.): *Electron Spectroscopy: Theory, Techniques and Applications* (Academic, New York 1977, 1981)
5.37 J. Heicklein: *Atmospheric Chemistry* (Academic, New York 1976)
5.38 H.S.W. Massey, D.R. Bates: *Atmospheric Physics and Chemistry* (Academic, New York 1982)
5.39 G.H.F. Dierckson, W. Huebner, P.W. Langhoff: *Molecular Astrophysics* (Reidel, Dordrecht 1985)
5.40 E.J. McCartney: *Absorption and Emission by Atmospheric Gases* (Wiley, New York 1983)
5.41 C.E. Brion, J.P. Thomson: J. Electron Spectrosc. **33**, 301 (1984)
5.42 I. Nenner, J.A. Beswick: *Handbook of Synchrotron Radiation*, ed. by. J.W. Marr (North-Holland, Amsterdam 1986) Vol. 2, p. 355
5.43 F.C. Brown: In *Synchrotron Radiation Research*, ed. by H.W. Winck, S. Doniach (Plenum, New York 1980) Chap. 4, p. 61
5.44 U. Fano, J.W. Cooper: Rev. Mod. Phys. **40**, 441 (1968)

5.45 H. Bethe: Ann. Phys. (Leipzig) **5**, 325 (1930)
5.46 C.E. Brion: Commun. At. Mol. Phys. **16**, 249 (1985)
5.47 J. Berkowitz: *Photoabsorption, Photoionization and Photoelectron Spectroscopy* (Academic, New York 1979)
5.48 K. Kirby, E.R. Constantinides, S. Babeu, M. Oppenheimer, J.A. Victor: At. Data Nucl. Data Tables. **23**, 63 (1979)
5.49 Y. Itikawa, M. Hayashi, A. Ichimura, K. Onda, K. Sakimoto, K. Takaynagi, M. Nakamura, H. Nishimura, and T. Takaynagi: J. Phys. chem. Ref. Data **15**, 985 (1986)
5.50 J.A.R. Samson, G.N. Haddad, J.L. Gardner: J. Phys. B **10**, 1749 (1977)
5.51 G.R. Wight, M.J. van der Wiel, C.E. Brion: J. Phys. B **9**, 675 (1976)
5.52 B.E. Cole, R.N. Dexter: J. Phys. B **11**, 1011 (1978)
5.53 A. Hamnett, W. Stoll, C.E. Brion: J. Electron Spectrosc. **8**, 367 (1976)
5.54 Y. Itikawa, A. Ichimura, K. Onda, K. Sakimoto, K. Takaynagi, I. Hatano, M. Hayashi, H. Nishimura, S. Tsurubuchi: J. Phys. Chem. Ref. Data. **18**, 23 (1989)
5.55 J.A.R. Samson, J.L. Gardner, G.N. Haddad: J. Electron Spectrosc. **12**, 281 (1977)
5.56 T. Gustafsson: Chem. Phys. Lett. **75**, 505 (1980)
5.57 C.E. Brion, K.H. Tan, M.J. van der Wiel, Ph.E. van der Leeuw: J. Electron Spectrosc. **17**, 101 (1979)
5.58 J.W. Gallagher, C.E. Brion, J.A.R. Samson, P.W. Langhoff: J. Phys. Chem. Ref. Data. **17**, 9 (1988)
5.59 G. Raseev, H. LeRouso: Phys. Rev. A **27**, 268 (1983)
5.60 S.V. Oneil, W.P. Reinhardt: J. Chem. Phys. **69**, 2126 (1978)
5.61 J.W. Davenport: Ind. J. Quantum Chem. Symp. **11**, 89 (1977)
5.62 S. Southworth, W.D. Brewer, S.M. Truesdale, P.H. Kobrin, D.W. Linde, D.A. Shirley: J. Electron Spectrosc. Relat. Phenom. **26**, 43 (1982)
5.63 G.V. Marr, R.M. Holmes, K. Codling: J. Phys. B **13**, 283 (1980)
5.64 C.M. Dutta, F.M. Chapman, Jr., E.F. Hayes: J. Chem. Phys. **67**, 1904 (1977)
5.65 Y. Itikawa, H. Takagi, H. Nakamura, H. Sato: Phys. Rev. A **27**, 1319 (1983)
5.66 L.A. Collins, B.I. Schneider: Phys. Rev. A **29**, 1695 (1984)
5.67 W. Thiel: Chem. Phys. **57**, 227 (1981)
5.68 N.T. Padial, G. Csanak, B.V. McCoy, P.W. Langhoff: J. Chem. Phys. **69**, 2992 (1978)
5.69 J.W. Davenport: Phys. Rev. Lett. **36**, 945 (1976)
5.70 A. Hamnett, W. Stoll, C.E. Brion: J. Electron Spectrosc. Relat. Phenom. **8**, 367 (1976)
5.71 J.A.R. Samson, J.L. Gardner: J. Electron Spectrosc. Relat. Phenom. **8**, 35 (1976)
5.72 E.W. Plummer, T. Gustafsson, W. Gudat, D.E. Eastman: Phys. Rev. A **15**, 2339 (1977)
5.73 S. Wallace, D. Dill, J.L. Dehmer: J. Phys. B **12**, L417 (1979)
5.74 G.V. Marr, J.M. Morton, R.M. Holmes, D.G. McCoy: J. Phys. B **12**, 43 (1979)
5.75 M.R. Hermann, C.W. Bauschlicher Jr., W.M. Huo, S.R. Langhoff, P.W. Langhoff: Chem. Phys. **109**, 1 (1986)
5.76 M.E. Smith, V. McCoy, R.R. Lucchese: J. Chem. Phys. **82**, 4149 (1985)
5.77 S. Wallace, D. Dill, J.L. Dehmer: J. Chem. Phys. **76**, 1217 (1982)
5.78 T. Gustafsson, H.J. Levinson: Chem. Phys. Lett. **78**, 28 (1981)
5.79 G.R. Wight, M.J. van der Wiel, C.E. Brion: J. Phys. B **10**, 1863 (1977)

Chapter 6

6.1 M. Hayashi: In *Swarm Studies and Inelastic Electron-Molecule Collisions*, ed. by L.C. Pitchford, B.V. McKoy, A. Chutjian, S. Trajmar (Springer, Berlin, Heidelberg 1987) p. 167
6.2 H. Tawara, Y. Itikawa, H. Nishimura, M. Yoshina: J. Phys. Chem. Ref. Data. **19**, 617 (1990)
6.3 G.F. Drukarev: *Collisions of Electrons with Atoms and Molecules* (Plenum, New York 1987)

6.4 N.F. Lane: Rev. Mod. Phys. **52**, 29 (1980)
6.5 K. Takayanagi, Y. Itikawa: Adv. Atom. Mol. Phys. **6**, 105 (1970)
6.6 Y. Itikawa: Phys. Rep. **143**, 69 (1986)
6.7 V.P. Zhigunov, B.N. Zakhar'ev: *Methods of Strong Coupling in Quantum Scattering Theory* (Atomizdat, Moscow 1974) (in Russian)
6.8 P.G. Burke, K.A. Berrington: *Atomic and Molecular Processes* (IOP, Bristol 1993)
6.9 Fr. Gianturco (ed.): *Atomic and Molecular Collision Theory* (Plenum, New York 1980)
6.10 R.R. Lusshese, K. Takatsuka, V. McKoy: Phys. Rep. **131**, 147 (1986)
6.11 L.G. Christophorou (ed.): *Electron–Molecule Interactions and Their Applications*, Vols. 1, 2 (Academic, Orlando 1984)
6.12 I. Shimamura, K. Takayanagi (eds.): *Electron–Molecule Collisions* (Plenum, New York 1984)
6.13 P.G. Burke, J.B. West (eds.): *Electron–Molecule Scattering and Photoionization* (Plenum, New York 1988)
6.14 T.D. Mark, G.H. Dunn (eds.): *Electron Impact Ionization* (Springer, Berlin, Heidelberg 1985)
6.15 S. Trajmar, D.F. Register, A. Chutjian: Phys. Rep. **97**, 219 (1983)
6.16 L.C. Pitchford, B.V. McKoy, A. Chutjian, S. Trajmar (eds.): *Swarm Studies and Inelastic Electron–Molecule Collisions* (Springer, Berlin, Heidelberg 1987)
6.17 I. Itikawa, M. Hayashi, A. Ichimura, K. Onda, K. Sakimoto, K. Takayanagi, M. Nakamura, H. Nishimura, T. Takayanagi: J. Phys. Chem. Ref. Data **15**, 985 (1986)
6.18 J. Ferch, W. Raith, K. Schroeder: J. Phys. B **13**, 1481 (1980)
6.19 J. Dalba, P. Fornasini, I. Lazzizzera, G. Ranieri, A. Zecca: J. Phys. **B 13**, 2839 (1980)
6.20 B. van Wingerden, R. Wagenaar, F.J. de Heer: J. Phys. B **13**, 3481 (1980)
6.21 R. Jones, R.A. Bonham: Unpublished (1981)
6.22 K.R. Hoffman, M.S. Dababneh, Y.F. Hsieh, W.E. Kauppila, V. Pol, J.H. Smart, T.S. Stein: Phys. Rev. A **25**, 1393 (1982)
6.23 M. Hayashi: Nagoya Univ. Report No IPPJ-AM-19 (1981)
6.24 Y. Itikawa, A. Ichimura, K. Onda, K. Sakimoto, K. Takayanagi, Y. Hatano, M. Hayashi, H. Nishimura, S. Tsurubuchi: J. Phys. Chem. Ref. Data **18**, 23 (1989)
6.25 C. Szmytkowsky, M. Zubek: Chem. Phys. Lett. **57**, 105 (1978)
6.26 J.B. Hasted, S. Kadifachi, T. Solovyeg: *Abstracts of the XIth Int'l Conf. on Physics of Electronic and Atomic Collisions*, Kyoto, (1979) p. 334
6.27 J. Ferch, C. Masche, W. Raith: J. Phys. B **14**, L97 (1981)
6.28 R.E. Kennerly, R.A. Bonham, M. McMillan: J. Chem. Phys. **70**, 2039 (1979)
6.29 I. Shimamura: Sci. Pap. Inst. Phys. Chem. Res. **82**, 1 (1989)
6.30 S.J. Buckman, A.V. Phelps: JILA Inf. Cent. Rep. No. 27 (Joint Inst. Laboratory Astrophys., Co, 1985)
6.31 J.T. England, L.P. Elford, R.W. Crompton: Austral. J. Phys. **41**, 573 (1988)
6.32 M.A. Khakoo, S. Trajmar: Phys. Rev. A **34**, 138 (1986)
6.33 H. Nishimura, A. Danjo, H. Sugahara: J. Phys. Soc. Jpn. **54**, 1757 (1985)
6.34 T.W. Shyn, W.E. Sharp: Phys. Rev. A **24**, 1734 (1981)
6.35 A.V. Phelps: JILA Inf. Cent. Rep. No. 28 (1985)
6.36 J.E. Land: J. Appl. Phys. **49**, 5716 (1978)
6.37 G.N. Haddad, H.B. Milloy: Austral. J. Phys. **36**, 473 (1983)
6.38 R.D. DuBois, M.E. Rudd: J. Phys. B **9**, 2657 (1976)
6.39 H. Tanaka, S.K. Srivastava, A. Chutjian: J. Chem. Phys. **69**, 5329 (1978)
6.40 J.C. Nikel, C. Mott, I. Kanik, D.C. McCollum: J. Phys. **21**, 1867 (1988)
6.41 S.J. Buckman, B. Lohmann: Phys. Rev. A **34**, 1561 (1986)
6.42 M. Hayashi: *Proc. IAEA Advisory Group Meeting on Atomic and Molecular Data for Radiotherapy*, Vienna (1988)
6.43 L.J. Kieffer: JILA Inf. Cent. Rep. No. 13 (Joint Inst. Laboratory Astrophys., Co (1973)
6.44 F. Linder, H. Schmidt: Z. Naturforsch. **26**a, 1603 (1971)

6.45 S.K. Scrivastava, A. Chutjian, S. Trajmar: J. Chem. Phys. **63**, 2659 (1975)
6.46 B. van Wingerden, E. Weigold, F.J. de Heer, K.J. Nygaard: J. Phys. B **10**, 1345 (1977)
6.47 T.W. Shyn, G.R. Carignan: Phys. Rev. A **22**, 923 (1980)
6.48 S.K. Srivastava, H. Chutjian, S. Trajmar: J. Chem. Phys. **64**, 1340 (1976)
6.49 D.F. Register, S. Trajmar, S.K. Srivastava: Phys. Rev. A **21**, 1134 (1980)
6.50 L.A. Collins, D.W. Norcross: Phys. Rev. A **18**, 467 (1978)
6.51 K. Jung, T. Antoni, R. Muller, K.-H. Kochem, H. Ehrdardt: J. Phys. B **15**, 3535 (1982)
6.52 T.W. Shyn, W.E. Sharp, J.R. Carignan: Phys. Rev. A **17**, 1855 (1978)
6.53 D.F. Register, H. Nishimura, S. Trajmar: J. Phys. B **13**, 1651 (1980)
6.54 K. Rohr: J. Phys. B **13**, 4897 (1980)
6.55 H. Tanaka, T. Okada, L. Boesten, T. Suzuki, T. Yamamoto, M. Kubo: unpublished (1982)
6.56 L. Vušković, S. Trajmar: J. Chem. Phys. **78**, 4947 (1983)
6.57 M. Fink, K. Jost, D. Hermann: J. Chem. Phys. **63**, 1985 (1975)
6.58 S.K. Srivastava, S. Trajmar, A. Chutjian, W. Williams: J. Chem. Phys. **64**, 2767 (1976)
6.59 I. Shimamura: In *Electron–Molecule Collisions* ed. by I. Shimamura, K. Takayanagi (Plenum, New York 1984) Chap. 2
6.60 R.W. Crompton, D.K. Gibson, A.I. McIntosh: Austral. J. Phys. **22**, 715 (1969)
6.61 D.K. Gibson: Austral. J. Phys. **23**, 683 (1979)
6.62 K. Onda: J. Phys. Soc. Jpn. **54**, 4544 (1985)
6.63 G.J. Schulz: In *Principles of Laser Plasmas*, ed. by G. Bekefi (Wiley, New York 1976) Chap. 2
6.64 M.A. Morrison: Adv. Atom. Mol. Phys. **24**, 51 (1988)
6.65 G.J. Schulz: Rev. Mod. Phys. **45**, 378 (1973)
6.66 G.J. Schulz: Rev. Mod. Phys. **45**, 423 (1973)
6.67 K. Onda and A. Temkin: Phys. Rev. A **28**, 621 (1983)
6.68 H. Tanaka, T. Yamamoto, T. Okada: J. Phys. B **14**, 2081 (1981)
6.69 D.G. Truhlar, M.A. Brandt, A. Chutjian, S.K. Srivastava, S. Trajmar: J. Chem. Phys. **65**, 2962 (1976)
6.70 D.G. Truhlar, M.A. Brandt, S.K. Srivastava, S. Trajmar, A. Chutjian: J. Chem. Phys. **66**, 655 (1977)
6.71 A.G. Engelhardt, A.V. Phelps, C.G. Risk: Phys. Rev. **135**, A1566 (1964)
6.72 Z. Pavlović, M.J.W. Boness, A. Herzenberg, G.J. Schulz: Phys. Rev. A **6**, 776 (1972)
6.73 A. Huetz, I. Cadez, F. Gresteau, R.I. Hall, D. Vichon, J. Mazeau: Phys. Rev. A **21**, 622 (1980)
6.74 M. Allan: J. Phys. B **18**, 4511 (1985)
6.75 S. Trajmar, D.C. Cartwright, W. Williams: Phys. Rev. A **4**, 1482 (1971)
6.76 S. Trajmar, W. Williams, A. Cuppermann: J. Chem. Phys. **56**, 3759 (1972)
6.77 F. Linder, H. Schmidt: Z. Naturforsch: A **26**, 1617 (1971)
6.78 M.J.W. Boness, G.J. Schulz: Phys. Rev. A **8**, 2883 (1973)
6.79 A. Chutjian, H. Tanaka: J. Phys. B **13**, 1901 (1980)
6.80 K. Rohr: In *Symp. Electron–Molecule Collisions* ed. by I. Shimamura, M. Matsuzawa (Univ. of Tokyo, 1979) p. 67
6.81 K.-H. Kochem, W. Sohn, N. Hebel, K. Jung, H. Ehrhardt: J. Phys. B **18**, 4455 (1985)
6.82 L. Andrić, I. Čadez, R.I. Hall, M. Zubek: J. Phys. B **16**, 1837 (1983)
6.83 N.F. Mott, H.S.W. Massey: *The Theory of Atomic Collisions* 3rd. edn. (Oxford Univ. Press, London 1965)
6.84 M. Inokuti: Rev. Mod. Phys. **43**, 297 (1971)
6.85 H. Bethe: Ann. Physik **5**, 325 (1930)
6.86 C.E. Brion, A. Hamnett: *The Excited State in Chemical Physics*, Part 2, Adv. Chem. Phys. **45**, 1 (1981)
6.87 J. Berkowitz: *Photoabsorption, Photoionization, and Photoelectron Spectroscopy* (Academic, New York 1979)
6.88 S. Trajmar, D.C. Cartwright, J.K. Rice, R.T. Brinkmann, A. Cuppermann: J. Chem. Phys. **49**, 5464 (1968)

6.89 D.C. Cartwright, S. Trajmar, A. Chutjian, W. Williams: Phys. Rev. A **16**, 1041 (1977)
6.90 D.C. Cartwright, A. Chutjian, S. Trajmar, W. Williams: Phys. Rev. A **16**, 1013 (1977)
6.91 A. Chutjian, D.C. Cartwright, S. Trajmar: Phys. Rev. A **16**, 1052 (1977)
6.92 I.Yu. Yurova, V.E. Ivanov: *Electron Scattering Cross Sections of Atmospheric Gases* (Nauka, Leningrad 1989) (in Russian)
6.93 M.J. Mumma, E.J. Stone, E.C. Zipf: J. Chem. Phys. **54**, 2627 (1971)
6.94 R. Berry, L.S. Stephen: Adv. Electron. Phys. **57**, 1 (1981)
6.95 H.F. Winters: J. Chem. Phys. **44**, 1472 (1966)
6.96 A. Neihaus: Z. Naturforscr. A **22**, 690 (1967)
6.97 D. Rapp, P. Englander-Golden, D.D. Briglia: J. Chem. Phys. **42**, 4081 (1965)
6.98 E.C. Zipf, R.W. McLaughlin: Planet. Space Sci. **26**, 449 (1978)
6.99 H.F. Winters: J. Chem. Phys. **63**, 3462 (1975)
6.100 H.F. Winters: Chem. Phys. **36**, 353 (1979)
6.101 S.J.B. Corrigan: J. Chem. Phys. **43**, 4381 (1965)
6.102 H. Tawara, Y. Itikawa, Y. Itoh, T. Kato, H. Nishimura, S. Ohtani, H. Takagi, K. Takaayanagi, M. Yoshino: IPPJ Rep. AM-46 (Institute of Plasma Physics, Nagoya Univ., Nagoya) (1986) p. 4–01
6.103 D.E. Shemansky, J.M. Ajello, D.T. Hall: Astrophys. J. **296**, 765 (1985)
6.104 H. Tawara: Rep. NIFS-Data-19 (Nat'l Inst. Fusion Sci., Nagoya 1992)
6.105 G.R. Mohlmann, S. Tsurubuchi, F.J. de Heer: Chem. Phys. **18**, 145 (1976)
6.106 H.D. Morgan, J.E. Mentall: J. Chem. Phys. **60**, 4734 (1974)
6.107 J.W. McGowan, J.F. Williams, D.A. Vroom: Chem. Phys. Lett. **3**, 614 (1969)
6.108 G.M. Lawrence: Phys. Rev. A **2**, 397 (1970)
6.109 C.I.M. Beenakker, F.J. de Heer, H.B. Kropp, G.R. Mohlmann: Chem. Phys. **6**, 445 (1974)
6.110 T.G. Finn, B.L. Carnahan, W.C. Wells, E.C. Zipf: J. Chem. Phys. **63**, 1596 (1975)
6.111 H. Tawara, T. Kato: At. Data Nucl. Data Tables **36**, 167 (1987)
6.112 A.K. Edwards, R.M. Wood, A.S. Beard, R.L. Ezell: Phys. Rev. A **37**, 3697 (1988)
6.113 D. Rapp, P. Englander-Golden: J. Chem. Phys. **43**, 1464 (1965)
6.114 D.E. Shemansky, A.L. Broadfoot: J. Quant. Specrosc. Radiat. Transfer **11**, 1401 (1971)
6.115 V.V. Skubenich, I.P. Zapesochnyy: Geomag. Aeron. **21**, 355 (1981)
6.116 O.J. Orient, S.K. Srivastava: J. Phys. B **20**, 3923 (1987)
6.117 E. Hille, T.D. Mark: J. Chem. Phys. **69**, 4600 (1978)
6.118 T.W. Shyn, W.E. Sharp: Phys. Rev. A **20**, 2332 (1979)
6.119 T.D. Mark, E. Hille: J. Chem. Phys. **69**, 2492 (1978)
6.120 A. Crowe, J.W. McConkey: J. Phys. B **7**, 349 (1974)
6.121 H. Chatham, D. Hils, R. Robertson, A. Gallagher: J. Chem. Phys. **81**, 1770 (1984)
6.122 B.L. Schram, M.J. van der Wiel, F.J. de Heer, H.R. Moustafa: J. Chem. Phys. **44**, 49 (1966)
6.123 B. Adamczyk, A.J.H. Boerboom, B.L. Schram, J. Kistemaker: J. Chem. Phys. **44**, 4640 (1966)
6.124 B. Adamczyk, K. Bederski, L. Wojcik: Biomed. Environ. Mass Spectrom. **16**, 415 (1988)
6.125 C.B. Opal, E.C. Beaty, W.K. Peterson: At. Data **4**, 209 (1972)
6.126 R.R. Coruganthu, W.G. Wilson, R.A. Bonham: Phys. Rev. A **35**, 540 (1987)
6.127 A.I. Zhukov, A.N. Zavilopulo, A. Snegursky, O.B. Shpenik: J. Phys. B **23**, 2373 (1990)
6.128 R.N. Compton, J.N. Bardsley: In *Electron–Molecule Collisions* ed. by I. Shimamura, K. Takayanagi (Plenum, New York 1984) Chap. 4
6.129 J.B.A. Mitchell, H. Hus: J. Phys. B **18**, 547 (1985)
6.130 P.M. Mul, J.W. McGowan, P. Defrance, J.B.A. Mitchell: J. Phys. B **16**, 3099 (1983)
6.131 R.A. Heppner, F.L. Walls, W.T. Armstrong, G.N. Dunn: Phys. Rev. A **13**, 1000 (1976)
6.132 J.M. Wadera, J.N. Bardsley: Phys. Rev. Lett. **41**, 1795 (1978)
6.133 L.G. Christophorou: Radiat. Phys. Chem. **12**, 19 (1978)
6.134 G.J. Schulz: Phys. Rev. **128**, 178 (1962)
6.135 B. Eliassen, V. Kogelschatz: J. Phys. B **19**, 1241 (1986)

Chapter 7

7.1 J.O. Hirschfelder, C.F. Curtiss, R.B. Bird: *Molecular Theory of Gases* (Wiley, New York 1954)

7.2 A. Dalgarno, W.D. Davison: Adv. Atom. Molec. Phys. **2**, 1 (1966)

7.3 A. Dalgarno: Adv. Chem. Phys. **12**, 143 (1967)

7.4 D. Langbein: *Theory of Van der Waals Interaction*, Springer Tracts Mod. Phys., Vol. 72 (Springer, Berlin, Heidelberg 1974)

7.5 G. Scoles: Ann. Rev. Phys. Chem. **31**, 81 (1980)

7.6 R. Duren: Adv. Atom. Molec. Phys. **16**, 55 (1980)

7.7 A.A. Radzig, B.M. Smirnov: *Reference Data on Atoms, Molecules and Ions*, Springer Ses. Chem. Phys., Vol. 31 (Springer, Berlin, Heidelberg 1986)

7.8 Yu.S. Barash: *Van der Waals Forces* (Nauka, Moscow 1988) (in Russian)

7.9 E.W. McDaniel, M.R. Flannery, E.W. Thomas, S.T. Manson: At. Data Nucl. Data Tables **33**, 1 (1985)

7.10 H. Tawara: NIFS-DATA-20 (Nat'l. Inst. Fusion Sci., Nagoya 1993)

7.11 H.B.G. Casimir, D. Polder: Phys. Rev. **73**, 360 (1948)

7.12 J.M. Standard, P.R. Certain: J. Chem. Phys. **83**, 3002 (1985)

7.13 D.R. Lide (ed.) *Handbook of Chemistry and Physics*, 73rd edn. (CRC, Boca Raton L, 1993)

7.14 A.M. Jones, M.P. Lord: *Macmillian's Chemical and Physical Data* (Macmillan, London 1992)

7.15 T.E. Sharp: At. Data **2**, 119 (1971)

7.16 G.H. Dieke: J. Molec. Spectrosc **2**, 494 (1958)

7.17 W. Kolos, L. Wolniewicz: J. Chem. Phys. **43**, 2429 (1965)

7.18 W. Kolos, L. Wolniewicz: J. Chem. Phys. **50**, 3228 (1969)

7.19 W. Kolos, L. Wolniewicz: J. Chem. Phys. **48**, 3672 (1968)

7.20 H.S. Taylor, F.E. Harris: J. Chem Phys. **39**, 1012 (1963)

7.21 I. Eliezer, H.S. Taylor, J.K. Williams Jr.: J. Chem. Phys. **42**, 2165 (1967)

7.22 D.R. Bates, R.H.G. Reid: Adv. Atom. Molec. Phys. **4**, 13 (1968)

7.23 H.H. Michels, R.H. Hobbs: Astrophys. J. **286**, L27 (1984)

7.24 J.B.A. Mitchell: *Electron-Molecule Processes in Fusion Edge Plasmas*, ed. by R.K. Janev (Plenum, New York 1995)

Subject Index

Springer and the environment

At Springer we firmly believe that an international science publisher has a special obligation to the environment, and our corporate policies consistently reflect this conviction.
We also expect our business partners – paper mills, printers, packaging manufacturers, etc. – to commit themselves to using materials and production processes that do not harm the environment. The paper in this book is made from low- or no-chlorine pulp and is acid free, in conformance with international standards for paper permanency.

Springer

MIX
Papier aus verantwortungsvollen Quellen
Paper from responsible sources
FSC® C105338

If you have any concerns about our products,
you can contact us on
ProductSafety@springernature.com

In case Publisher is established outside the EU,
the EU authorized representative is:
Springer Nature Customer Service Center GmbH
Europaplatz 3, 69115 Heidelberg, Germany

Printed by Libri Plureos GmbH
in Hamburg, Germany